Instructor's Resource Manual
with Video Outline and Lecture Demos

Robert C. Pfaff
St. Josephs's College

Sixth Edition
CHEMISTRY
The Central Science

Theodore L. Brown
University of Illinois

H. Eugene LeMay, Jr.
University of Nevada

Bruce E. Bursten
The Ohio State University

Prentice Hall, Englewood Cliffs, NJ 07632

© 1994 by PRENTICE-HALL, INC.
A Paramount Communications Company
Englewood Cliffs, N.J. 07632

All rights reserved.

10   9   8   7   6   5   4   3   2   1

ISBN 0-13-338567-1

Printed in the United States of America

TABLE OF CONTENTS

Sample Course Sequence   iv

Lecture Outlines and Sample Quiz Questions

Chapter 1	1
Chapter 2	8
Chapter 3	15
Chapter 4	22
Chapter 5	28
Chapter 6	36
Chapter 7	45
Chapter 8	52
Chapter 9	61
Chapter 10	71
Chapter 11	79
Chapter 12	88
Chapter 13	95
Chapter 14	104
Chapter 15	112
Chapter 16	119
Chapter 17	128
Chapter 18	136
Chapter 19	143
Chapter 20	149
Chapter 21	160
Chapter 22	171
Chapter 23	180
Chapter 24	188
Chapter 25	195
Chapter 26	203

Video Outline: Chemistry Concepts & Techniques   217

Lecture Demonstrations   219

Sample Course Sequences

To help in structuring your general chemistry courses, three possible course sequences are given here. Two of them are for two-semester courses. Of the two, one is intended for courses that are more physically oriented, such as those for engineering or "hard-science" students. The other is intended for courses that are more descriptively oriented, such as those for preprofessional students. Because both course outlines share a common basis in the fundamentals, only one sequence is included for the first half of the course; the differences arise in the second half of the course. The third sequence presented is for a two-quarter course that meets four times per week for 60 minutes.

In all cases, approximately five class meetings per semester (three class meetings per quarter) have been left unscheduled to allow for examinations and quizzes to be scheduled to suit individual desires and needs.

Chapter	Title	Physical[†]	Descriptive[‡]	Quarters[§]
1	Introduction: Some Basic Concepts	3	3	3
2	Atoms, Molecules, and Ions	3	3	3
3	Stoichiometry: Calculations with Chemical Formulas and Equations	4	4	4
4	Aqueous Reactions and Solution Stoichiometry	4	4	4
5	Energy Relationships in Chemistry: Thermochemistry	3	3	3
6	Electronic Structures of Atoms	3	3	3
7	Periodic Properties of the Elements	3	3	3
8	Basic Concepts of Chemical Bonding	4	4	4
9	Molecular Geometry and Bonding Theories	4	4	4
10	Gases	3	3	3
11	Intermolecular Forces, Liquids and Solids	3	4	3
12	Modern Materials	1	2	1
13	Properties of Solutions	2	3	1
14	Chemical Kinetics	4	3	4
15	Chemical Equilibrium	4	4	4
16	Acid–Base Equilibria	5	5	4
17	Additional Aspects of Aqueous Equilibria	4	3	4
18	Chemistry of the Environment	2	3	1
19	Chemical Thermodynamics	5	4	4
20	Electrochemistry	4	3	4
21	Nuclear Chemistry	5	4	4
22	Chemistry of Hydrogen, Oxygen, Nitrogen and Carbon	1	1	1
23	Chemistry of Other Nonmetallic Elements	1	1	1
24	Metals and Metallurgy	1	1	1
25	Chemistry of Coordination Compounds	1	2	1
26	Chemistry of Life: Organic and Biological Chemistry	2	3	2
	Class Periods	79	80	74

[†] Refers to the traditional two-semester sequence for engineering and physical science majors meeting three times per week for 50 minutes.

[‡] Refers to a two-semester sequence for preprofessional majors meeting three times per week for 50 minutes.

[§] Refers to a two-quarter sequence meeting four times per week for 60 minutes. Note that a three-quarter course meeting three times per week for 50 minutes follows the semester schedules given above.

Note to users of the Instructor's Resource Manual

The lecture outline portion of the manual is also available in electronic format --both Word 2.0 for Windows and Word Perfect 5.1-- and in DOS and in MAC. Disk size is 3.5 in each case.

For a copy, please phone or write: Supplements Editor, Chemistry, College Editorial, Prentice Hall, 113 Sylvan Avenue, Englewood Cliffs, N.J. 07632

Chapter 1 — Introduction: Some Basic Concepts

OVERVIEW: General chemistry represents the beginning of a long process of learning for most students. They often know what field they *think* they wish to pursue, but they know little about the preparation required of them. This chapter is an introduction to the *tools* of chemistry: categorization, elements and compounds, measurements, and algebra.

LECTURE OUTLINE

I. **Why Study Chemistry?**
   A. For many students, chemistry is required for their major, whether it is one of the physical sciences, engineering, premedicine, or an allied health field.
   B. Other students take chemistry because the subject interests them.
   C. Still other students take chemistry because they realize that the future will be more technical than the present and they want the skills to understand their world.

Discussion Question: Why is the study of chemistry useful to the students as they prepare for careers?

II. **Introduction to Matter**
   A. In a chemical context, *matter* can exist in one of three distinct forms, or states of matter.
      1. *Solids* are rigid, have fixed volume, and are noncompressible.
      2. *Liquids* can change shape but have reasonably fixed volume, and are noncompressible.
      3. *Gases* (Vapors) readily change in shape and volume, and are highly compressible.
   B. Substances have fixed compositions and distinct properties.
      1. *Pure substances* are substances that are free of impurities.
      2. *Impurities* are trace quantities of substances in a sample other than the principal component.
   C. Properties of substances can be divided into physical and chemical.
      1. *Physical properties* are those that describe a substance as it is. Physical properties include color, density, hardness, melting point, etc.
      2. *Chemical properties* are those that describe how a substance can be transformed. Chemical properties include flammability, solubility, corrosiveness, etc.
   D. Changes in substances can be chemical or physical, just as properties could be chemical or physical.
      1. *Physical changes* are those which do not change the identity of a substance. Physical changes include melting, boiling, etc.
      2. *Chemical changes* or chemical reactions do change the identity of a substance. Chemical changes include combustion, oxidation, metathesis, etc.
   E. Mixtures are combinations of more than one substance.
      1. *Homogeneous mixtures* or solutions are those mixtures in which all the species are in the same phase *and* are uniformly distributed throughout the sample.

2. *Heterogeneous mixtures* are those mixtures in which not all the species are in the same phase *or* they are not uniformly distributed.

Discussion Question: Think of products around your home. Are they generally pure or mixtures? Do they cause physical changes or chemical changes when they are used?

III. Elements and Compounds

A. *Elements* are substances that cannot be decomposed into simpler substances by chemical means. By contrast, compounds can be decomposed by chemical means.
  1. Only a few of the elements are abundant on Earth.
      a. Two elements comprise the majority of the Earth's crust.
         49.5% Oxygen
         25.7% Silicon
         7.5% Aluminum
         4.7% Iron
         3.4% Calcium
         9.2% Other elements
      b. The human body concentrates other elements.
         65% Oxygen
         18% Carbon
         10% Hydrogen
         7% Other elements
  2. Each element has an assigned one- or two-letter symbol.

B. *Compounds* consist of more than one element united chemically in definite mass proportions.
  1. The law of constant composition or law of definite proportions states that, for a given chemical compound, the mass proportions of the elements in it are always the same.
  2. When two pure samples have different mass proportions of the elements present, they must be different substances.

Discussion Question: If hydrogen and carbon are fairly rare on earth, why do they appear so prominently in the human body?

IV. Units of Measurement

A. The *metric system* of units was developed in France about 100 years ago. It is attractive since all related units differ by powers of ten. For example, there are 100 centimeters in a meter.

B. *SI* (Système International d'Unités) is the internationally recognized set of fundamental metric units for scientific measurement. There are seven SI quantities (mass, length, time, electrical current, temperature, light intensity, and amount of substance); all other quantities can be derived from these.

C. Length is based on the meter, which is about 39.37 inches. Subunits are used as appropriate, like kilometers for distances between cities, centimeters for objects we can hold, and nanometers and picometers for fundamental particles of matter.

D. Mass, not weight, describes the amount of matter in a sample. The SI standard is the kilogram (kg). Mass is measured by comparing a sample of known mass to the sample being measured. Weight is the interaction of a sample with gravity.

E. Temperature is the measure of the direction and magnitude of heat flow.
1. The Fahrenheit scale (°F) is the historical system used in Europe based on 32°F for the freezing point of water and 212°F for the boiling point of water.
2. The Celsius scale (°C) is the metric unit for temperature. It is based on 0°C for the freezing point of water and 100°C for the boiling point of water. $T°C = \frac{5}{9}(T°F - 32)$
3. The Kelvin scale (K) is the SI unit for temperature. As in Celsius, there are 100 degrees between the freezing point of water and the boiling point of water. The Kelvin scale simply uses a different zero-point. $T_K = T°C + 273.15$

F. Metric uses prefixes to indicate subunits. For example, 1 centimeter equals 1/100 meter, 1 millimeter equals 1/1,000 meter, and 1 micrometer equals 1/1,000,000 meter.

G. Volume is derived from length, most easily seen by considering the volume of a box. The volume is given by the product of the lengths of the three dimensions of the box. One Liter is the volume of a box that is 10 centimeters on each side; 1 L = 1000 cm^3 or 1 mL = 1 cm^3.

H. Density is the ratio of mass to volume of a sample. For a pure substance, density is constant.

I. "Chemistry at Work: Chemistry in the News" introduces chemistry as a current events topic.

J. *Intensive properties* are those which do not depend on the size of a sample. Examples of intensive properties include density and temperature. *Extensive properties* are those which do depend on the size of a sample. Examples of extensive properties include volume and mass.

V. **Uncertainty in Measurement**
A. *Exact numbers* are those that are infinitely precise. Most exact numbers are integers, such as 12 eggs in 1 dozen, and are not measurements. Some exact numbers are defined to be infinitely precise, such as 2.54 cm = 1 inch.

B. *Inexact numbers* result from measurements and so are subject to error.

C. Precision and accuracy represent the two types of error.
1. *Precision* is a measure of random error; the less the random error, the higher the precision. Precision then refers to the reproducibility of a measurement.
2. *Accuracy* is a measure of systematic error, the error that arises from the same mistake being made on each measurement; the less the systematic error, the higher the accuracy. Accuracy then refers to the agreement of a measurement with the "correct" value.

D. *Significant figures* reflect how precisely a measurement is made. Values are reported such that all of the well known digits are reported plus a last, estimated, digit. For example, a measurement with a metric ruler, graduated in mm, might be reported as "27.6 mm" showing what the scale actually has markings for (27 mm) and an estimate "between-the-lines" (.6 mm).

E. Significant figures in calculations try to preserve the precision of the measurements, neither overstating them nor understating them. In addition and subtraction, the result has the same number of decimal places as that of the term which had the fewest decimal places. For multiplication and division, the result has the same number of digits (not decimal places) as the measurement with the fewest significant figures.

Discussion Question: How accurate is an auto's odometer? How precise is it?

VI. Dimensional Analysis

A. *Conversion factors* are ratios relating the *same* quantity in two unit systems. For example, the statement 1 inch = 2.54 cm becomes $\frac{1 \text{ inch}}{2.54 \text{ cm}}$ or $\frac{2.54 \text{ cm}}{1 \text{ inch}}$.

B. *Dimensional analysis* is an approach to problem solving in which the units or dimensions of the quantities are examined to see what conversions are required.

C. As a summary of dimensional analysis, the text offers the following steps.
1. What data are we given in the problem?
2. What quantity do we wish to obtain in the problem?
3. What conversion factors do we have available to take us from the given quantity to the desire one?

VII. Strategies in Chemistry: The Importance of Practice

This section emphasizes the usefulness of practice and gives advice on how to obtain help in studying.

SAMPLE QUIZ QUESTIONS

1. Briefly describe the law of definite proportions.

2. Given that there are 1.094 yd in a meter and 1760 yd in a mile, calculate the speed in miles/hour of an automobile traveling at 70 km/hr.

3. Gold melts at 1063°C. What is this temperature on the Kelvin and Fahrenheit scales?

4. Gold has a density of 19.3 g/cm^3 while fool's gold (iron pyrite) has a density of about 5.0 g/cm^3. If a sample has a mass of 0.530 g and a volume of 1.06 x 10^2 mm^3, what is its density and possible identity?

5. A pure substance: (a) refers only to the elements; (b) has unique chemical properties; (c) may have two phases together; (d) has two or more components; (e.) has only one component.

6. Which of the following statements is true of an element but *not* of a compound: (a) It is a pure substance; (b) It has a constant composition from one sample to another; (c) It cannot be separated into simpler substances by chemical means; (d) It cannot be separated into simpler substances by physical means.

7. The freezing point of a substance is: (a) the place in a liquid where solid first forms upon freezing; (b) the temperature at which it changes from solid to liquid; (c) the same for all pure substances; (d) the temperature at which it changes from solid to gas; (e) the temperature at which it changes from liquid to gas.

8. Boiling is: (a) the temperature at which a liquid changes to a gas; (b) an exothermic process; (c) dependent on the size of the sample; (d) a physical change; (e) a chemical change.

9. A liquid: (a) takes on the shape and volume of its container; (b) is a fluid; (c) must be an element; (d) has the same properties as a gas; (e) maintains its own shape.

10. Which of the following is(are) homogeneous? (a) a chalkboard eraser; (b) a silverplate spoon; (c) bronze; (d) the instructor's desk; (e) a pencil.

11. Which of the following is(are) heterogeneous? (a) distilled water; (b) window glass; (c) a plastic trash bag; (d) brass; (e) the Mississippi River.

12. Which of the following are intensive properties? (a) mass; (b) color; (c) temperature; (d) density.

13. Which of the following is(are) example(s) of physical properties? (a) the ability of a substance to neutralize stomach acid; (b) corrosiveness; (c) flammability; (d) hardness; (e) toxicity.

14. Which of the following is(are) example(s) of chemical properties? (a) odor; (b) texture; (c) conductivity; (d) ability of a diamond to shatter; (e) ductility.

15. Which of the following is(are) example(s) of chemical change? (a) copper roofing turning green with age; (b) the disappearance of snow on a cold day; (c) the expansion of mercury in a thermometer when you have a fever; (d) the cracking of a glass jar by a hot liquid; (e) the condensation of water on a cold beverage container.

16. A physical change: (a) may alter the identity of a substance; (b) is always accompanied by a chemical change; (c) does not alter the identity of a substance; (d) cannot be reversed; (e) must involve a state change.

17. A chemical change: (a) generally can be reversed; (b) generally does not include physical change; (c) does not alter the identity of the substance; (d) generally cannot be reversed; (e) is not generally accompanied by an energy change.

18. The volume measure that is closest to the English unit the quart is the: (a) milliliter; (b) cubic centimeter; (c) liter; (d) kiloliter.

19. Precision refers to: (a) how close to the actual value a measurement is; (b) how skillful the experimenter is; (c) how careful the experimenter is; (d) how well successive measurements of a quantity agree; (e) how accurate successive measurements of a quantity are.

20. Accuracy refers to: (a) how close to the actual value a measurement is; (b) how skillful the experimenter is; (c) how many times the experiment is performed; (d) how well successive measurements of a quantity agree; (e) how precise successive measurements of a quantity are.

21. Give the result of the following calculation with the proper number of significant figures.

$$\frac{(56.1 - 48.3)}{13.2}$$

(a) 0.6; (b) 0.59; (c) 0.591; (d) 0.5909; (e) 0.59091

22. Give the result of the following calculation with the proper number of significant figures.

$$(6.321)(7.11 + 3.27 + 9.9)$$

(a) 130; (b) 128; (c) 128.3; (d) 128.32; (e) 128.316

23. Express the number 3.7020×10^{-2} as an ordinary number without an exponent: (a) 0.037020; (b) 370.20; (c) 3.7020; (d) 0.03702; (e) 370.2

24. Convert 33.1 mm to μm: (a) 0.0331 μm; (b) 3.31×10^7 μm; (c) 3.31×10^{-5} μm; (d) 3.31 μm; (e) 3.31×10^4 μm

25. Convert 2.9979×10^8 m to cm: (a) 2.9979×10^6 cm; (b) 3.00×10^{10} cm; (c) 2.9979×10^{10} cm; (d) 2.9979×10^{11} cm; (e) 3.00×10^6 cm

26. Convert 45.0 mL to qt. (a) 0.0450 qt; (b) 0.0426 qt; (c) 42.6 qt; (d) 0.0476 qt; (e) 47.6 qt

27. An object has a mass of 56.32 g and occupies a volume of 8.75 mL. Calculate the density of the object. (a) 6.44 g/mL; (b) 0.155 g/mL; (c) 493 g/mL; (d) 6.437 g/mL; (e) 0.1554 g/mL

28. An object has a mass of 192.77 g and a density of 7.953 g/mL. Calculate the volume that the object occupies.   (a) 0.04126 mL; (b) 1530 mL; (c) 24.24 mL; (d) 0.02424 mL; (e) 0.0413 mL

29. A liquid has a density of 0.8110 g/mL. Calculate the mass of a 156 L sample of the liquid. (a) 156 g; (b) 127,000 g; (c) 192 g; (d) 127 g; (e) 192,000 g

30. Copper melts at about 2300°C. What is that temperature in °F? (a) 1260°F; (b) 4172°F; (c) 2268°F; (d) 4200°F; (e) 4108°F

31. A wax melts at 375°F. What is that temperature in °C? (a) 707°C; (b) 343°C; (c) 191°C; (d) 643°C; (e) 226°C

Lecture Outline – Chapter 2
Brown, LeMay, & Bursten, *Chemistry: The Central Science*, 6th Edition

Chapter 2 — Atoms, Molecules, and Ions

OVERVIEW: Fortunately for students, chemistry is a pyramidal science, that is, all higher learning draws on the material already absorbed. The strategy of this text is to describe particles and events from small to large, starting with Dalton's atomic theory as a foundation. Atoms and elements are discussed within this context. To enhance students' understanding of this material, the chapter also covers subatomic particles, and the forces holding substances together.

After focusing on what differentiates the elements, the chapter introduces the periodic table, with emphasis on how the elements are categorized. Finally, introductory nomenclature is covered so that these substances can be discussed.

LECTURE OUTLINE

I. **The Atomic Theory of Matter**

   A. Fire, Earth, air, and water were the original "elements" as proposed by ancient Greek philosophers.

   B. *Dalton's atomic theory* was the first comprehensive attempt at applying the concept of atoms. We still believe that Dalton was fundamentally correct. His postulates were:
      1. Each *element* is composed of extremely small particles called *atoms*.
      2. All atoms of a given element are identical; the atoms of different elements are different and have different properties, including masses.
      3. Atoms of an element are not changed into different types of atoms by chemical reactions; atoms are neither created nor destroyed in chemical reactions.
      4. *Compounds* are formed when atoms of more than one element combine; a given compound always has the same relative number and kind of atoms.

   C. The *law of constant composition* came from Dalton's theory. It states that all samples of a given substance must have the same composition (#4).

   D. The *law of conservation of matter* or mass was an important part of Dalton's theory. It states that matter cannot be created nor destroyed, only rearranged (#3).

   E. The *law of multiple proportions* complements the law of constant composition. It states that, if two samples have different compositions, they must necessarily be samples of two different substances.

Discussion Question: Can students think of any situations or events that violate the law of conservation of matter?

II. **The Discovery of Atomic Structure**

   A. *Cathode rays* were discovered in the mid-1800s and were found to be beams of electrons streaming from the negatively charged cathode toward the positively charged anode.

   B. *Radioactivity* is the spontaneous emission of radiation from a substance. The radiation can be of any wavelength but emission of X rays, γ rays, α particles, and β particles are of most interest.
      1. X rays are high-energy light.

2. γ radiation is light of higher energy than X rays. γ radiation is the most damaging of all known forms for radiation.
3. α radiation consists of a stream of helium nuclei, $_2^4He^{2+}$.
4. β radiation consists of a stream of electrons that originate in the nucleus, not in the electron orbitals.

C. *Nuclear atom* is the term applied to models of the atom since the work of Rutherford. Rutherford deduced that most of the mass and all of the positive charge of the atom was concentrated in its *nucleus* while the negative charge and most of the atom's volume was accounted for by the electrons being dispersed outside the nucleus.

III. A Closer Look: Basic Forces

This section discusses the four basic forces in nature.
1. *Gravitational forces* draw objects together in proportion to their masses.
2. *Electromagnetic forces* act to attract or repel electrically charged or magnetic objects.
3. *Strong nuclear forces* hold the positively charged protons together in the nucleus.
4. *Weak nuclear forces* are, as yet, poorly understood but are apparent from observing radioactivity.

Discussion Question: Discuss the positive and negative perceptions of radioactivity. Is there a class consensus that the positives outweigh the negatives or vice versa?

IV. The Modern View of Atomic Structure

A. *Subatomic particles* are those fundamental particles which comprise an atom once assembled. In a chemical context, there are three important subatomic particles.
1. The *proton* is found in the nucleus, has a charge of +1, and has a mass of 1.0073 amu.
2. The *neutron* is also found in the nucleus, has no charge, and has a mass of 1.0087 amu.
3. The *electron* is found outside the nucleus, has a charge of -1, and has a mass of 5.486×10^{-4} amu.

B. *Isotopes* are atoms of a given element that have different numbers of neutrons.
1. An atom's *atomic number* is the number of protons in its nucleus. All atoms of the same element have the same atomic number.
2. An atom's *mass number* is the sum of the number of protons and neutrons in its nucleus. Therefore, the number of neutrons in the nucleus is given by the mass number minus the atomic number.
3. Since isotopes are atoms of the same element that have different numbers of neutrons, it follows that isotopes have the same atomic number but different mass numbers.
4. An atom of a specific isotope is called a *nuclide*.

Discussion Question: Are the fundamental subatomic particles, like mesons, neutrinos, or quarks, likely to be important in most chemical applications?

V. The Periodic Table

A. *Periods* are the rows in the periodic table.

B. *Families* or *groups* are the columns in the periodic table.

- C. The *metals* are the elements on the left and in the middle of the periodic table (excluding hydrogen). They are generally lustrous, ductile, malleable, and conduct heat and electricity.
- D. The *nonmetals* are on the right of the periodic table. Nonmetals generally lack the properties of metals, they are brittle as solids, do not conduct well, and are often dull in appearance. Many of the nonmetals are gases.
- E. *Semimetals* or *metalloids* occur at the boundary between metals and nonmetals. Their properties are intermediate between those of the metals and nonmetals. The boundary is often shown by a "stair-step" line on the periodic table, starting at boron and ending at astatine.

VI. Chemistry at Work: Radioactive Isotopes and the Chernobyl Accident

This section describes the nuclear plant accident at Chernobyl, Ukraine (then in the U.S.S.R.) in 1986.

VII. Molecules and Ions

- A. *Molecules* are tightly bound assemblies of two or more atoms. The assembled "package" physically behaves as a single unit.
    1. *Molecular formulas* represent the composition of a molecule using the atomic symbol of each element present and a subscript on each symbol telling how many atoms of that element are present in the molecule. Subscripts of "1" are usually omitted.
    2. *Empirical* or *simplest formulas* show the simplest *ratio* of the numbers of atoms of each element in a substance. When a substance does not exist in discrete molecules, only this ratio is meaningful. For example, C_6H_6 is a molecular formula showing the numbers of carbon atoms and hydrogen atoms in the molecule, CH is an empirical formula showing the simplest ratio of atoms, and $CaSO_4$ is an empirical formula showing the simplest ratio of atoms.
    3. *Structural formulas* can be used for describing molecules. They give all the information that a molecular formula does, plus it shows pictorially how the atoms are arranged in the molecule.
- B. *Ions* are atoms or molecules that have a net electrical charge, that is, the number of electrons is not equal to the number of protons.
    1. *Monatomic ions* are ions comprised of single atoms, while *polyatomic ions* are ions comprised of several atoms.
    2. *Cations* are those atoms or molecules that have a net positive charge. Cations are deficient in electrons to the extent of their charges. For example, Ca^{2+} is short two electrons and NH_4^+ is short one electron.
    3. *Anions* are those atoms or molecules that have a net negative charge. Anions have a surplus of electrons to the extent of their charges. For example, Cl^- has one extra electron and CO_3^{2-} has two extra electrons.
- C. The charge of an ion can be predicted from positions on the periodic table. With monatomic ions, atoms either gain or lose electrons until they have the same number of electrons as the nearest noble gas. Metals lose electrons, forming cations; nonmetals gain electrons, forming anions.
- D. Ionic compounds are those compounds formed from the combination of ions. In these compounds, the total cationic charge must be equal in magnitude to the total anionic charge; the overall charge for the compound must be zero.

VIII. Strategies in Chemistry: Pattern Recognition

This section emphasizes the value of noticing patterns of similarity in studying chemistry.

IX. Naming Inorganic Compounds

A. *Ionic compounds* are named by listing the name of the cation followed by the name of the anion. No attempt is made to tell how many of each is present; the charges of the ions set how many.

B. *Acids* are named based on the name of the anion. If the anion ends in *-ide*, the acid's name has the form *hydro____ic acid*, where the root of the anion's name goes in the blank. Similarly, if the anion's name ends in *-ite*, the acid's name has the form ____*ous acid*, and if the anion's name ends in *-ate*, the acid's name has the form ____*ic acid*.

C. *Molecular compounds* are generally those which are formed between nonmetals. Their names are given by listing the elements as they appear left-to-right and bottom-to-top in the periodic table, placing an *-ide* ending on the last element, and adding a prefix indicating number to each element's name. For example, PCl_5 is phosphorus pentachloride, SF_4 is sulfur tetrafluoride, and N_2O_5 is dinitrogen pentoxide.

Lecture Outline – Chapter 2
Brown, LeMay, & Bursten, *Chemistry: The Central Science*, 6th Edition

SAMPLE QUIZ QUESTIONS

1. A beta particle is: (a) an electron; (b) a helium nucleus; (c) a proton; (d) high energy radiation.

2. The nuclear nature of the atom is revealed by: (a) emission of beta particles in radioactivity; (b) the equality of protons and neutrons in a neutral atom; (c) the scattering of alpha particles by a gold foil; (d) spontaneous radioactivity as first observed by Becquerel.

3. Elements of the same family in the periodic table: (a) have equal numbers of electrons in their atoms; (b) have the same type of arrangement of electrons at the periphery of their atoms; (c) are either all metallic or all nonmetallic; (d) are arranged in short or long horizontal rows in the table.

4. Which of the following properties is characteristic of all metallic elements? (a) high melting point; (b) easily oxidized; (c) good electrical conductivity; (d) form low-melting oxides with oxygen

5. Which of the following elements is a semimetal? (a) Cr; (b) Ge; (c) Hg; (d) I_2

6. Based on its place in the periodic table, which element should have the most metallic character? (a) Mg; (b) Cd; (c) Pb; (d) Al

7. Based on its place in the periodic table, which element should have the strongest nonmetallic character? (a) Tl; (b) C; (c) As; (d) Br

8. Which of the following elements would you expect to show the greatest tendency to form negative ions? (a) Li; (b) F; (c) Ne

9. Which of the following elements is a halogen? (a) sodium; (b) magnesium; (c) chlorine; (d) sulfur

10. An element is: (a) a pure substance which cannot be separated into simpler substances by ordinary means; (b) electrically charged, either positive or negative; (c) a substance which has identical atoms; (d) a molecule; (e) part of a larger particle.

11. On the periodic table, a period: (a) is the term given to the elements on the right side of the table; (b) is the term given to the two bottom rows which are separated from the other elements; (c) is the same as a family; (d) is the term given to any set of elements with similar properties; (e) is the term given to the horizontal rows.

12. Ions: (a) are atoms with an excess of electrons; (b) are atoms or groups of atoms with a net electrical charge other than zero; (c) are formed by adding or removing neutrons to atoms; (d) are atoms with a deficiency of electrons; (e) are formed by adding or removing protons.

13. An atom's atomic number: (a) is the number of protons in the nucleus; (b) is the sum of the number of protons and the number of neutrons in the nucleus; (c) is the number of nucleons in the nucleus; (d) is the number of electrons circulating around the nucleus; (e) is the number of neutrons in the nucleus.

14. An atom's mass number: (a) is the number of protons in the nucleus; (b) is the sum of the number of protons and the number of neutrons in the nucleus; (c) is the number of nucleons circulating around the nucleus; (d) is the number of electrons circulating around the nucleus; (e) is the number of neutrons in the nucleus.

15. Isotopes are: (a) atoms with different numbers of protons and neutrons; (b) atoms with the same numbers of neutrons and electrons; (c) atoms with the same atomic number and different mass numbers; (d) atoms with the same numbers of protons and electrons; (e) atoms with different atomic numbers and the same mass number.

16. Isotopic mass: (a) is the mass number of an atom; (b) is the average of the masses of the isotopes of an element; (c) is the atomic number times the mass of each nucleon; (d) is expressed in the units of grams per amu; (e) is the mass of an isotope of an element.

17. The law of definite proportions: (a) applies only to molecules; (b) says that elements can combine in one ratio to form compounds; (c) is one of the parts of Dalton's theory which is no longer believed valid; (d) says that compounds with different ratios of the same elements are different compounds; (e) applies to mixtures as well as compounds.

18. A chemical formula: (a) is a representation of a compound which indicates the proportions in which atoms or ions are combined; (b) applies to mixtures; (c) is a representation of an element; (d) shows which elements are contained in the compound; (e) is an arbitrary group of symbols of elements.

19. Name the following compounds. (a) $FeCl_2$; (b) HNO_2; (c) $CrBr_2$; (d) $SiCl_4$; (e) $BaCO_3$; (f) $Al_2(SO_4)_3$

20. Give the chemical formulas for the following compounds: (a) ammonium sulfate; (b) magnesium bromide; (c) calcium phosphate

21. Name the compound AsI_3. (a) arsenic iodate; (b) arsenous iodide; (c) arsenic iodide; (d) arsenic triiodide; (e) arsenic(V) iodide

22. Name the compound P_4O_{10}. (a) phosphorus oxide; (b) decaoxygen tetraphosphide; (c) tetraphosphorus decaoxide; (d) phosphorus tetroxide; (e) potassium oxide

23. Which compound is dinitrogen pentoxide? (a) NO_5; (b) N_2O; (c) N_5O_2; (d) NO; (e) N_2O_5

24. Which compound is boron trihydride? (a) B_3H; (b) BH; (c) BH_3; (d) BeH_3; (e) BHe_3

25. Name the compound $Mg_3(PO_4)_2$. (a) manganese phosphate; (b) magnesium(III) phosphate; (c) magnesium phosphite; (d) magnesium phosphate; (e) manganese phosphide

26. Name the compound $FeSO_3$. (a) ferric sulfate; (b) ferrous sulfite; (c) ferric sulfite; (d) iron(II) sulfate; (e) iron(III) sulfide

27. Which of the following compounds is sodium perchlorate? (a) $NaClO_4$; (b) $SClO_3$; (c) $NaClO_2$; (d) $NaClO$; (e) SCl_2

28. Name the compound $CuCl_2$. (a) copper(I) chlorate; (b) cuprous chloride; (c) copper chloride; (d) cobalt(II) chlorite; (e) cupric chloride

29. Name the compound Al$_2$S$_3$. (a) aluminum(II) sulfide; (b) aluminum sulfite; (c) aluminum sulfate; (d) aluminum sulfide; (e) aluminum sulfurate

30. Which of the following compounds is calcium nitride? (a) Ca$_2$N$_3$; (b) CaN; (c) Ca$_3$N$_2$; (d) CaN$_2$; (e) Ca(NO$_3$)$_2$

31. Name the compound HIO$_4$. (a) iodic acid; (b) hydrogen iodite; (c) iodous acid; (d) hydrogen iodate; (e) periodic acid

32. Name the compound NaH$_2$PO$_4$. (a) sodium hydrogen phosphate; (b) sodium biphosphate; (c) sodium phosphite; (d) sodium dihydrogen phosphate; (e) sodium phosphoric acid

33. Name the compound H$_2$Se. (a) hydroselenic acid; (b) hydrogen selenate; (c) hydrogen hyposelenite; (d) hydrogen diselenide; (e) dihydrogen selenite

The following are included for those who have chosen to cover hydrates.

34. Name the compound CoCl$_2$·5H$_2$O. (a) copper(II) chlorate pentahydrate; (b) cobalt(II) chloride pentahydrate; (c) cuprous chloride hydrate; (d) cobalt(II) chloride heptahydrate; (e) cobaltic chloride pentahydrate

35. Which of the following compounds is scandium oxide octahydrate? (a) ScO$_2$·6H$_2$O; (b) SO$_3$·H$_2$O; (c) Sc$_3$O$_2$·8H$_2$O; (d) Sc$_2$O·9H$_2$O; (e) Sc$_2$O$_3$·8H$_2$O

Chapter 3 — Stoichiometry: Calculations with Chemical Formulas and Equations

OVERVIEW: This chapter introduces quantities of substances. As such, it begins with types of chemical reactions and how to balance chemical equations. There is a discussion on predisposed chemical behaviors so students can learn to anticipate what reactions are likely to occur for various substances.

Following the discussion on reactivity and chemical equations, masses of substances are introduced, first with isotope masses, then average atomic masses, and finally formula weights of compounds. For background, the use of the mass spectrometer to measure particle masses is also covered.

The masses of single atoms and molecules are too small to be of practical use, so the fundamental concept of the chemical mole is introduced. The emphasis here is on Avogadro's number and on the mole representing a sample for which the mass in grams is numerically the same as the mass in amu of a single formula unit of the substance.

The masses of substances, in grams, are combined with the law of definite proportions to introduce empirical formulas and molecular or true formulas.

With balanced chemical equations and formula weights finished, it is easy to extend the discussion to the mass relationships in chemical reactions, stoichiometry. This chapter deals only with mole and mass relationships of substances; solution reactions in which amounts of substances are described by volume and concentration are covered in Chapter 4. Stoichiometrically equivalent amounts are covered in detail and then the discussion is extended to limiting reactants and, finally, to percent yields of reactions.

LECTURE OUTLINE

I. Stoichiometry

A. *Stoichiometry* is the set of relationships among the amounts of substances in a chemical reaction. These amounts are usually, and most usefully, masses, but the term also applies to the relationships of moles in a reaction.

B. Stoichiometry makes use of the *law of conservation of mass*, as described by Dalton, which says that matter (and hence mass) cannot be created nor destroyed. Thus, all of the mass present at the beginning of a chemical reaction must be accounted for at the conclusion of the reaction.

Discussion Question: Ask students for examples of reactions at home or work that depend on the amounts of chemicals present.

II. Chemical Equations

A. *Reactants* are the substances present which combine to produce other substances. By convention, reactants are written on the left-hand side of a chemical equation, followed by an arrow, →. In some applications, most of which arise in the second half of the text, bi-directional arrows, ⇌, are used to show that the reaction can proceed in either direction.

B. *Products* are the substances which are formed in a chemical reaction. By convention, products are written on the right-hand side of a chemical equation, following the arrow.

Lecture Outline – Chapter 3
Brown, LeMay, & Bursten, *Chemistry: The Central Science*, 6th Edition

C. Miscellaneous information often is listed above or below the arrow, or at the end of an equation. The information commonly included at the arrow can be a required temperature, the presence of a catalyst, or a particular solvent. Thermodynamic information or equilibrium constants are commonly listed at the end of an equation. These latter concepts are covered in later chapters.

III. Patterns of Chemical Reactivity

A. The periodic table is an excellent source of information for predicting chemical reactions. Members of a group often react similarly, usually differing only in vigor. Changes in reactivity across a period is also usually predictable.

B. *Combustion* in air is a common type of reaction, where a compound containing carbon and hydrogen (and sometimes oxygen) reacts with O_2 to produce carbon dioxide and water: fuel + $O_2(g)$ → $CO_2(g)$ + $H_2O(l)$.

C. Other common reaction types which are important in learning the basics of reactivity are:
1. *Combination reactions* are those in which two reactants combine to form a single product.
2. *Decomposition reactions* are those in which a single reactant breaks apart to form two or more products.
3. *Single displacement reactions* are those in which an element replaces another in a compound.
4. *Double displacement (metathesis) reactions* are those in which two atoms or ions replace one another in compounds; they exchange partners.

Discussion Question: Perhaps students can think of reactions they've seen that can be categorized as combustion, combination, decomposition, single displacement, or double displacement.

IV. Atomic and Molecular Weights

A. The atomic mass scale is a *relative* scale. Originally, the basis was H = 1, and the masses of other elements were taken to be the ratios to hydrogen. Thus, He = 4, C = 12, O = 16, etc. The unit of measure devised for this scale was the atomic mass unit, amu.

B. *Average atomic masses* recognize that naturally occurring elements usually consist of a mixture of isotopes. The average mass of an element, then, is not the mass of one of the isotopes, but rather is a weighted average of the isotope masses, AW = Σ(fractional abundance)·(isotope mass).

C. *Formula weight* is a generic term for the mass of a formula, whatever it represents. The formula weight is simply the sum of the average atomic weights of the elements in the formula. *Molecular weight* is a formula weight for a molecule.

D. *The percentage composition*, by weight, of the elements in a formula can be gotten from the formula by determining the formula weight and the mass which is contributed by each element present. For example, in CH_4, the percentage composition is given by
$\frac{\text{mass C}}{\text{mass CH}_4} \cdot 100\%$ or $\frac{12.01 \text{ amu}}{16.05 \text{ amu}} \cdot 100\%$ for carbon and
$\frac{\text{mass H}}{\text{mass CH}_4} \cdot 100\%$ or $\frac{4.04 \text{ amu}}{16.05 \text{ amu}} \cdot 100\%$ for hydrogen.

V. A Closer Look: The Mass Spectrometer

The *mass spectrometer* is a device that breaks a molecule into smaller *fragments*, which are characteristic of the type of molecule. Each of these fragments has a characteristic mass and, using magnetic fields, the fragments are accelerated and focused to collide with a detector. By changing the magnetic field, fragments of different masses strike the detector. Therefore, you get a plot of mass vs. the number of fragments in the sample with that mass, called a mass spectrum. Mass spectra help identify compounds and to determine very accurate masses for isotopes.

VI. The Chemical Mole

A. The *chemical mole* is a number of particles, be they atoms, molecules, electrons, etc. One mole of particles is Avogadro's number, 6.022×10^{23}, particles. It is useful to think of the mole as a very large dozen. *Molar mass* is then a formula weight for a mole of particles. The basis of Avogadro's number is that the formula weight of a particle, in amu, is *numerically* equal to the molar mass of the particle, in grams. For example, the formula weight for O_2 is 16.00 amu; its molar mass is 16.0 g.

B. Mass, moles, and numbers of particles can be interconverted by

$$\text{grams} \leftrightarrow \text{moles} \leftrightarrow \text{particles}$$

where $\text{moles} = \frac{\text{mass}}{\text{FW}}$ and $\text{particles} = (\text{moles}) \cdot (6.022 \times 10^{23} \text{ particles/mol})$.

VII. Empirical Formulas from Analyses

A. *Empirical formulas* are derived from experimental information, most notably percent composition. The sample is chemically decomposed and the mass of each element present is determined and converted to a set of percentages.

B. The *molecular formula* of a compound can be found from the empirical formula and the molecular weight. The molecular formula is always a whole-number multiple of the empirical formula.

C. A common type of analysis for organic samples is *combustion analysis*. In this method, the sample, containing H, C, and possibly O, undergoes a complete combustion reaction. The gas stream runs through two tubes, one packed with a substance which absorbs CO_2 and one packed with a substance that absorbs H_2O. The resulting mass gains of each tube give the mass of CO_2 and H_2O produced. Stoichiometry is then used to determine the masses of each element in the sample.

VIII. Strategies in Chemistry: Solving Problems

This section gives useful advice on approaching problems:
A. Understand the problem.
1. Identify the unknown and the given data.
2. Set up the problem.
3. Solve the problem.
4. Check the solution.

IX. Quantitative Information from Balanced Equations

Stoichiometry generally refers to the mass relationships in chemical reactions. Using the formula weights and masses of substances, the matching or "stoichiometrically equivalent" amounts of other substances in the reaction can be determined by

$$\text{grams}_1 \rightarrow \text{moles}_1 \rightarrow \text{moles}_2 \rightarrow \text{grams}_2.$$

Lecture Outline – Chapter 3
Brown, LeMay, & Bursten, *Chemistry: The Central Science, 6th Edition*

X. Chemistry at Work: CO_2 and the Greenhouse Effect

This section examines the effect of increasing atmospheric CO_2 levels on the mean global temperature.

XI. Limiting Reactants

A. When two reactants are involved in a reaction, the experimenter can combine them in the stoichiometrically equivalent amounts or can combine them in some other ratio. If the reactants are present in *nonstoichiometric* amounts, the one which is fully consumed first is called the *limiting reactant* or *limiting reagent*.

B. When a limiting reagent is present, the other reactant or reactants are said to be *excess reactants* or *excess reagents*.

C. For several reasons, reactions often do not form as much product as the stoichiometric calculations predict.
  1. The amount of a product predicted by stoichiometric calculation is called the *theoretical yield* of the reaction. The theoretical yield is always as large or larger than the actual yield.
  2. The *percent yield* of a reaction is the actual yield expressed as a percentage of the theoretical yield, $\%\text{Yield} = \frac{\text{actual yield}}{\text{theoretical yield}} 100\%$.

SAMPLE QUIZ QUESTIONS

1. Balance the following equations:
   a. $KClO_3 \rightarrow KCl + O_2$
   b. $CO(NH_2)_2 + H_2O \rightarrow CO_2 + NH_3$
   c. $NH_3 + O_2 \rightarrow N_2O_4 + H_2O$
   d. $H_2SO_4(aq) + KOH(aq) \rightarrow K_2SO_4(aq) + H_2O(l)$
   e. $AgNO_3(aq) + Na_2CrO_4(aq) \rightarrow Ag_2CrO_4(s) + NaNO_3(aq)$

2. Convert the following descriptions into balanced chemical equations: (a) Gaseous sulfur dioxide reacts with water to form sulfurous acid; (b) Hydrogen gas and chlorine gas combine to form hydrogen chloride; (c) Solid sodium hydroxide reacts with carbon dioxide gas to form solid sodium carbonate and gaseous water.

3. Complete and balance the following equations:
   a. $Na + H_2O \rightarrow$
   b. $C_3H_8 + O_2 \rightarrow$
   c. $Ca(OH)_2 + H_3PO_4 \rightarrow$

4. Complete and balance the following reactions.
   a. $SO_2(g) + KOH(aq) \rightarrow$
   b. $P_4O_6(s) + H_2O(l) \rightarrow$
   c. $Mg(s) + Cl_2(g) \rightarrow$
   d. $Ba(s) + O_2(g) \rightarrow$
   e. $Ca(s) + HBr(aq) \rightarrow$

5. Antimony exists in nature as a mixture of 57.25% ^{121}Sb (mass = 120.904 amu) and 42.75% ^{123}Sb (mass = 122.904 amu). What is the average atomic weight of antimony?

6. If lithium nitrate is 20.3% nitrogen by weight, how many grams of nitrogen are in 750.0 g of lithium nitrate? (a) 140 g; (b) 55.8 g; (c) 7.38 g; (d) 34.0 g; (e) 55.82 g

7. Calculate the formula weight of $Mg(ClO_4)_2$. (a) 123.8 amu; (b) 159.3 amu; (c) 223.3 amu; (d) 75.8 amu; (e) 187.8 amu

8. Calculate the percentage composition of NH_4CN. (a) 44.4% C; 3.7% H; 51.9% N; (b) 27.3% C; 9.1% H; 63.6% N; (c) 29.3% C; 2.4% H; 68.3% N; (d) 40.0% C; 13.3% H; 46.7% N; (e) 20.7% C; 6.9% H; 72.4% N

9. When 1.5×10^{-2} g of an organic substance containing only C, H, and O is combusted in an excess of oxygen, 0.0267 g of CO_2 and 0.0176 g of H_2O are recovered. What is the empirical formula?

10. A sugar is found to be 40.2% carbon, 6.15% hydrogen, and 53.6% oxygen. What is the empirical formula of the sugar? (a) $C_3H_5O_3$; (b) C_7HO_9; (c) CH_2O; (d) CH_2O_2; (e) $C_6H_{11}O_6$

11. A compound is found to be 16.2% K, 39.8% Ir, and 44.1% Cl. What is the empirical formula of this compound? (a) KIr_2Cl_3; (b) $KIrCl_3$; (c) $KIrCl_6$; (d) K_2IrCl_6; (e) KIr_2Cl_2

Lecture Outline – Chapter 3
Brown, LeMay, & Bursten, *Chemistry: The Central Science*, 6th Edition

12. A compound is found to be 9.86% Mg, 13.0% S, 5.68% H, and 71.4% O. What is the empirical formula of this compound?   (a) $MgSH_{14}O_{11}$;   (b) $MgSH_6O_5$;   (c) $Mg_{10}S_{13}H_6O_{71}$;   (d) $Mg_2S_2HO_{12}$;   (e) $MgSH_{10}O_5$

13. A compound is found to be 47.1% carbon, 6.54% hydrogen, and 46.4% chlorine. The actual molar mass is 153.0 g/mol. What is the true formula of the compound?   (a) C_4H_6Cl;   (b) $C_6H_{10}Cl_2$;   (c) $C_4H_{10}Cl_3$;   (d) C_4H_7Cl;   (e) C_3H_5Cl

14. A compound is found to be 30.7% carbon, 1.28% hydrogen, and 68.1% bromine. The actual molar mass is 469.6 g/mol. What is the true formula of the compound?   (a) $C_6H_3Br_2$;   (b) $C_{12}H_5Br_5$;   (c) $C_{12}H_6Br_4$;   (d) $C_{18}H_6Br_3$;   (e) $C_6H_6Br_5$

15. Give the number of moles represented by the quantity 7.73×10^{15} particles.   (a) 7.79×10^7 mol;   (b) 4.65×10^{39} mol;   (c) 1.28×10^{-8} mol;   (d) 1.28×10^{15} mol;   (e) 2.15×10^{-40} mol

16. The molar mass of $K_3Fe(CN)_6$ is: (a) 120.9 g/mol;   (b) 199.1 g/mol;   (c) 259.1 g/mol;   (d) 329.1 g/mol;   (e) 250.9 g/mol.

17. How many moles of N and O are represented by 9.06×10^{21} molecules of N_2O_4?   (a) 3.00×10^{-2} mol N; 6.00×10^{-2} mol O;   (b) 1.50×10^{-2} mol N; 3.00×10^{-2} mol O;   (c) 3.00×10^{-2} mol N; 3.00×10^{-2} mol O;   (d) 1.50×10^{-2} mol N; 1.50×10^{-2} mol O;   (e) 6.00×10^{-2} mol N; 3.00×10^{-2} mol O

18. What is the mass of 9.85×10^{27} formula units of $KClO_4$?   (a) 2.28×10^6 g;   (b) 6.10×10^{-5} g;   (c) 1.64×10^4 g;   (d) 8.48×10^{-3} g;   (e) 139 g

19. What is the mass of 6.77 mol of Rb_3PO_4?   (a) 352 g;   (b) 52.0 g;   (c) 2380 g;   (d) 181 g;   (e) 1230 g

20. Calculate the mass of silver in 2.00 g of AgCl.

21. Gold can be extracted from ores using CN^-. The resulting complex ion formed by CN^- and gold can be treated with zinc metal to recover the gold: $Zn + 2Au(CN)_2^- \rightarrow Zn(CN)_4^{2-} + 2Au$. How much zinc is required to obtain 2.00 g of gold?

22. The decomposition of a sample of hydrogen peroxide, H_2O_2, produced 5.14 g of water and 4.57 g of oxygen. What was the mass of the hydrogen peroxide sample?

$$2 H_2O_2(l) \rightarrow 2 H_2O(l) + O_2(g)$$

(a) 9.71 g;   (b) 4.86 g;   (c) 0.57 g;   (d) 7.42 g;   (e) 14.85 g

23. According to the equation,

$$S_8(s) + 8 O_2(g) \rightarrow 8 SO_2(g)$$

how many moles of O_2 are required to react with 50.0 g of sulfur?   (a) 400 mol;   (b) 1.56 mol;   (c) 0.0243 mol;   (d) 0.195 mol;   (e) 49.8 mol

24. According to the equation for decomposing the mineral fluorapatite,

$$Ca_{10}F_2(PO_4)_6(s) + 7 H_2SO_4(l) \rightarrow 2 HF(g) + 3 Ca(H_2PO_4)_2(s) + 7 CaSO_4(s)$$

how many moles of H_2SO_4 are required to produce 125 g of $CaSO_4$? (a) 6.42 mol; (b) 0.131 mol; (c) 12.9 mol; (d) 0.918 mol; (e) 90.0 mol

25. According to the equation,

$$4\ KO_2(s) + 2\ H_2O(g) + 4\ CO_2(g) \rightarrow 4\ KHCO_3(s) + 3\ O_2(g)$$

what mass of oxygen can be made when 1.00 kg of KO_2 and 500.0 g of carbon dioxide are allowed to react with excess water? (a) 3.38×10^2 g; (b) 4.50×10^2 g; (c) 1.50×10^3 g; (d) 2.73×10^2 g; (e) 3.63×10^2 g

26. According to the equation,

$$4\ NH_3(g) + 6\ NO(g) \rightarrow 5\ N_2(g) + 6\ H_2O(l)$$

what mass of nitrogen can be made if 10.0 g of NH_3 and 30.0 g of NO are allowed to react? (a) 20.6 g; (b) 23.3 g; (c) 16.5 g; (d) 28.0 g; (e) 40.0 g

27. According to the equation,

$$2\ Al(s) + 3\ MgO(s) \rightarrow 3\ Mg(s) + Al_2O_3(s)$$

what mass of both products can be made if 75.0 g of aluminum and 150.0 g of MgO are allowed to react? (a) 101 g Mg; 142 g Al_2O_3; (b) 202 g Mg; 283 g Al_2O_3; (c) 90.4 g Mg; 127 g Al_2O_3; (d) 90.4 g Mg; 380 g Al_2O_3; (e) 33.8 g Mg; 142 g Al_2O_3

28. Consider the reaction for making ethyl alcohol from sugar,

$$C_6H_{12}O_6(aq) \rightarrow 2\ C_2H_5OH(aq) + 2\ CO_2(g)$$

The reaction typically has a percentage yield of 78.2%. What mass of sugar must be used in order to get an actual yield of 100.0 g of C_2H_5OH? (a) 5.01×10^2 g; (b) 1.96×10^2 g; (c) 1.28×10^2 g; (d) 2.50×10^2 g; (e) 3.91×10^2 g

29. Consider one of the reactions involved in the biological metabolism of sugar,

$$C_6H_{12}O_6(aq) + 6\ O_2(g) \rightarrow 6\ CO_2(g) + 6\ H_2O(l)$$

When 15.0 g of $C_6H_{12}O_6$ and 20.0 g of oxygen are allowed to react, 17.8 g of CO_2 are made. What is the percentage yield of the reaction? (a) 80.9%; (b) 93.8%; (c) 64.7%; (d) 75.0%; (e) 89.0%

Lecture Outline – Chapter 4
Brown, LeMay, & Bursten, *Chemistry: The Central Science*, 6th Edition

Chapter 4 — Aqueous Reactions and Solution Stoichiometry

OVERVIEW: Chapter 4 extends the topics of quantitative relationships introduced in Chapter 3 to aqueous systems. After a thorough treatment of molarity, including dilution, attention is turned to the description of dissolved solutes, focusing on water as the solvent.

As a particularly important type of aqueous solute, electrolytes are introduced. The differentiation between strong and weak electrolytes is emphasized. Acids, bases, and salts are discussed, including how to identify strong and weak acids and bases. As a natural sequence, acid-base neutralization reactions are then covered, as are complete ionic equations and net ionic equations.

Continuing solution chemistry, metathesis reactions are covered in depth, with emphasis on reactions which form insoluble salts, a gas, or a weak electrolyte. As part of this, the solubility rules for ionic salts are discussed.

In order to introduce oxidation-reduction chemistry, the single displacement reactions of metals (and hydrogen) are discussed. Coverage of these reactions leads to a treatment of the activity series.

The chapter to this point is primarily descriptive in nature, focusing on identifying processes in aqueous solution. The chapter concludes with the extension of stoichiometry to these solution reactions. As the most common experimental technique in solution stoichiometry, titrations are introduced in terms of the equivalence point. More complete treatment of titrations, including pH and titration curves is presented as part of Chapters 16 and 17.

LECTURE OUTLINE

I. **Aqueous Solutions**

   An *aqueous solution* is simply a liquid solution in which water is the solvent, that component of the mixture present in the highest amount.

II. **Solution Composition**

   A. In a *solution*, the *solvent* is the component present in greatest amount.

   B. In a solution the component or components that are not identified as the solvent are called the *solute* or *solutes*.

   C. For most purposes, the concentration of a solute in a solution is expressed in *molarity*, symbolized M.

   Molarity is the ratio of the moles of solute to the volume of the solution, expressed in liters, Molarity = $\dfrac{\text{moles solute}}{\text{volume of soln in liters}}$.

   D. When a portion of a solution is used as a component of yet another solution, the solute has been *diluted*. Dilution is commonly performed, for example, when 12 M HCl is combined with water to make 1 M HCl. In dilution, the resulting concentration is always less that the starting concentration.

   Arithmetically, dilution obeys the expression, $M_{\text{initial}} \cdot V_{\text{initial}} = M_{\text{final}} \cdot V_{\text{final}}$.

III. Electrolytes

A. Solutes which separate into two or more ions when dissolved in water are called *electrolytes*. Solutes which remain as uncharged molecules are called *nonelectrolytes*.

B. Electrolytes have varying abilities to conduct electricity through the solution. Those that conduct electricity very well are called *strong electrolytes* while those that conduct electricity only moderately well are called *weak electrolytes*. This conductivity is caused by the extent to which the molecules *ionize*. The strong electrolytes are those in which all the molecules ionize; the weak electrolytes are those in which only a portion of the molecules ionize.

When the ionization process proceeds at the same rate as the reverse, association, process, the concentrations of the ions and the molecules remain constant, a situation referred to as *chemical equilibrium*.

Discussion Question: Many students are familiar with a blood test for electrolytes. What substances are included in a blood electrolyte screening?

IV. Acids, Bases, and Salts

A. *Acids* are solutes that can ionize to produce aqueous hydrogen ions, $H^+(aq)$, or cause the hydrogen ion concentration to increase when dissolved in water.

B. *Monoprotic acids* are those acids that can release a single hydrogen ion when dissolved in water.

C. *Diprotic acids* are those acids that can release two hydrogen ions when dissolved in water. Similarly, *polyprotic* acids are those acids that can release *more than one* hydrogen ion when dissolved in water.

D. *Bases* are solutes that accept hydrogen ions. For example, $H^+(aq) + OH^-(aq) \rightarrow H_2O(l)$ or $H^+(aq) + NH_3(aq) \rightarrow NH_4^+(aq)$, where the OH^- and NH_3 are the bases. OH^- is the most common base in aqueous solution.

E. Acids and bases are electrolytes. Therefore, if an acid is a strong electrolyte, it can be called a *strong acid*; if an acid is a weak electrolyte, it can be called a *weak acid*. Similarly for bases, if a base is a strong electrolyte, it can be called a *strong base*; if a base is a weak electrolyte, it can be called a *weak base*.

F. *Salts* are ionic compounds formed from acids by the replacement of one or more H^+ ions with some other cation. For example, the acid H_3PO_4 can form the salts H_2NaPO_4, HNa_2PO_4, and Na_3PO_4 by replacing one, two, or three H^+ ions with sodium ions.

G. Identifying strong and weak electrolytes is a matter of following a few simple rules.
1. Most salts are strong electrolytes.
2. Most acids are weak electrolytes. However, HCl, HBr, HI, HNO_3, H_2SO_4, $HClO_3$, and $HClO_4$ are strong acids.
3. The common strong bases are the hydroxides of the alkali metals and the heavy alkaline earths. Ammonia, NH_3, is a weak electrolyte.
4. Most other substances are nonelectrolytes.

H. *Neutralization reactions* are chemical reactions between an acid and a base to produce water and a salt of the remaining ions.

Discussion Question: Can students think of acid-base reactions common in their everyday world?

Lecture Outline – Chapter 4
Brown, LeMay, & Bursten, *Chemistry: The Central Science, 6th Edition*

V. Ionic Equations

A. Varying amounts of detail can be included in chemical equations, depending on what information is relevant to the problem at hand.

B. In *molecular equations*, all the species are written as associated molecules, without regard for the form in which they are actually present.

C. In *complete ionic equations*, strong electrolytes are written as the dissociated ions; molecules, solids, and weak electrolytes are written as the associated species.

D. *Spectator ions* are those ions present during a reaction that have no role in the reaction. For example, in the reaction of NaOH*(aq)* with HCl*(aq)*, the H^+ and OH^- combine to form water but the Na^+ and Cl^- are not involved; Na^+ and Cl^- simply remain freely floating in the solution and so are spectator ions.

E. A *net ionic equation* is one in which the spectator ions have been omitted because they aren't involved. Therefore, the net ionic equation for the reaction between NaOH*(aq)* and HCl*(aq)* is simply: $H^+(aq) + OH^-(aq) \rightarrow H_2O(l)$.

Discussion Question: Discuss the issue of when it is appropriate to use net ionic equations and when it is necessary to use complete ionic equations or molecular equations.

VI. Metathesis Reactions

A. *Metathesis reactions* are those in which two ionic reactants exchange ion partners.

B. *Precipitation reactions* are those metathesis reactions in which an insoluble solid product, or precipitate, is formed.
   1. *Precipitates* are simply insoluble solid products of metathesis reactions.
   2. *Solubility* is the amount of substance that can be dissolved in a given amount of solvent, often expressed in g/100 mL or in mol/L. A substance is considered *insoluble*, in this text, if its solubility is less than 0.01 mol/L.

C. Through observations, we know what combinations of ions result in insoluble compounds. These observations are summarized in the *solubility rules*.

D. Some metathesis reactions form H_2O or another *weak electrolyte* or *nonelectrolyte*. Because weak electrolytes exist predominantly as the associated species, they must be treated as such. Therefore, weak electrolytes (and nonelectrolytes) are written as the associated species in complete ionic equations and in net ionic equations.

E. In some metathesis reactions, a product is formed which is a *gas*. These are usually binary compounds of nonmetals. Common examples include H_2S, CO_2, and HCN.

VII. Chemistry at Work: Antacids

This section describes antacids from the perspective of being relevant and easily identified with by students.

VIII. Reactions of Metals

A. *Corrosion* of a metal is the conversion of the metal to a metal compound by reaction with some substance in its environment, usually O_2, H_2O, acids, or salts.

B. *Oxidation* is the loss of electrons. In the current context, a metal being corroded is being oxidized. *Reduction* is the gain of electrons from the substance being oxidized.

- C. Many metals are oxidized by acids to form a salt and H_2, such as $Mg + 2HClO_4 \rightarrow H_2 + Mg(ClO_4)_2$.
- D. *Single displacement reactions* are those in which one element replaces another in a compound. The oxidation of metals by acids is an example, as is the oxidation of metals by other metal cations.
- E. The *activity series* is an experimental ranking of elements, mostly metals, such that an element will replace the ion of an element below it on the series in a compound.

IX. A Closer Look: The Aura of Gold

This section discusses the important consequences of gold's resistance to oxidation, including money and jewelry.

X. Solution Stoichiometry

- A. When reactants are in solution, the number of moles of the substance is most conveniently related to the volume of the solution and its concentration, # mol = $M \cdot V$.
- B. *Titrations* are solution reactions set up such that a fixed amount of one reactant is present in a flask and the other reactant is slowly added in measured volumes. In this manner, the addition can be stopped when the stoichiometrically equivalent amount of the *titrant* is added.
  1. *Standard* solutions are solutions in which the concentration of the solute is well known. Stoichiometrically, the amounts of other reactants can be found by comparison to the standard.
  2. In a titration, the *equivalence point* is that volume of titrant at which the stoichiometrically equivalent amount has been added.
  3. *Indicators* are used in acid-base neutralization reactions to estimate when the equivalence point has been reached. Indicators are dyes that change color depending on how acidic or how basic the solution is.
  4. The *end point* of a titration is the volume of titrant at which the indicator changes color. The end point is an estimate of the equivalence point.

XI. Strategies in Chemistry: Analyzing Chemical Reactions

The emphasis of this section is in gaining intuition for chemical reactions by doing experiments, making observations, and organizing information. The key is in organizing information.

Lecture Outline – Chapter 4
Brown, LeMay, & Bursten, *Chemistry: The Central Science, 6th Edition*

SAMPLE QUIZ QUESTIONS

1. If 1.0 g of NaCl is dissolved in a sufficient amount of water to form 100 mL of solution, what is the molarity of the solution?

2. If 250 mL of 0.200 M NaCl is diluted in enough water to make 600 mL of solution, what is the new concentration of the solution?

3. Molarity is defined as: (a) grams solute per liter of solution; (b) moles solute per liter of solvent; (c) moles solute per liter of solution; (d) moles solute per kilogram of solvent.

4. The molarity of a solution prepared by adding 7.1 g Na_2SO_4 (MW = 142 amu) to enough water to make 100 mL of solution is: (a) 2.0 M; (b) 0.50 M; (c) 0.050 M; (d) 0.020 M.

5. Calculate the molarity of a solution containing 45.0 g of C_2H_5OH and 55.0 g of H_2O.

6. What is meant when nitric acid is described as a strong acid and nitrous acid as a weak acid?

7. Briefly describe the difference between a nonelectrolyte and an electrolyte.

8. An electrolyte: (a) is composed of molecules; (b) conducts electricity when molten or dissolved; (c) is generally a solid which must be molten to be useful; (d) conducts heat; (e) is generally a nonpolar liquid.

9. Electrolytes: (a) which dissociate completely are called weak; (b) are uncommon substances; (c) are nonconductors; (d) which dissociate slightly are called weak; (e) generally do not dissolve in water.

10. Write the net ionic equation for each of the following reactions: (a) Sodium chloride and silver nitrate solutions are mixed, forming silver chloride as a precipitate; (b) Sulfuric acid is added to sodium carbonate; (c) Aqueous HBr is added to a solution of sodium acetate.

11. Provide a balanced net ionic or complete ionic equation to describe each of the following syntheses (use whatever other reagents are necessary, in addition to those specified). (a) Cadmium sulfate beginning with cadmium chloride; (b) Calcium iodide, beginning with calcium carbonate

12. Identify the driving force in each of the following chemical reactions, and write the corresponding net ionic equation:
    a. $NaNO_2(aq) + HBr(aq) \rightarrow NaBr(aq) + HNO_2(aq)$
    b. $Ba(NO_3)_2(aq) + Na_2SO_3(aq) \rightarrow BaSO_3(s) + 2NaNO_3(aq)$
    c. $Na_2SO_3(aq) + 2HBr(aq) \rightarrow 2NaBr(aq) + H_2O(l) + SO_2(g)$
    d. $2KMnO_4(aq) + 5H_2SO_3(aq) \rightarrow 2MnSO_4(aq) + K_2SO_4(aq) + 2H_2SO_4(aq) + 3H_2O(l)$

13. Identify the water-soluble salt from among the following: (a) AgCl; (b) $CoSO_4$; (c) $PbSO_4$; (d) $CdCO_3$

14. Which of the following cations has fewest common insoluble salts? (a) Ca^{2+}; (b) Zn^{2+}; (c) K^+; (d) Ni^{2+}

15. Which of the following reactions does not represent an oxidation process?
    a. $BaCl_2(aq) + 2NaOH(aq) \rightarrow Ba(OH)_2(s) + 2NaCl(aq)$
    b. $Co(s) + H_2SO_4(aq) \rightarrow CoSO_4(aq) + H_2(g)$
    c. $Zn(s) + CuSO_4(aq) \rightarrow ZnSO_4(aq) + Cu(s)$
    d. $Li(s) + 2H_2O(l) \rightarrow 2LiOH(aq) + H_2(g)$

16. Which of the following displacement reactions does not proceed?
    a. $Zn(s) + 2AgNO_3(aq) \rightarrow Zn(NO_3)_2(aq) + H_2(g)$
    b. $Fe(s) + H_2SO_4(aq) \rightarrow FeSO_4(aq) + H_2(g)$
    c. $Pb(s) + Mg(NO_3)_2(aq) \rightarrow Pb(NO_3)_2(aq) + Mg(s)$
    d. $Mn(s) + 2HCl(aq) \rightarrow MnCl_2(aq) + H_2(g)$

17. An active metal reacts with water forming H_2. Reaction of 2.5 g of metal results in formation of 0.062 mol of H_2. Which of the following elements might it be? (a) Ca; (b) Al; (c) K; (d) Cd

18. How many mL of 0.10 M $Ba(OH)_2$ will be neutralized by 60 mL of 0.20 M HCl? (a) 200 mL; (b) 120 mL; (c) 60 mL; (d) 30 mL

19. The number of moles of NaOH neutralized by 25.0 mL of 0.150 M H_2SO_4 solution is: (a) 0.00375; (b) 0.00188; (c) 0.0075; (d) 3.75

Lecture Outline – Chapter 5
Brown, LeMay, & Bursten, *Chemistry: The Central Science*, 6th Edition

Chapter 5 — Energy Relationships in Chemistry: Thermochemistry

OVERVIEW: Chapter 5 introduces energy changes in chemical reactions. Because energy changes, particularly enthalpy changes are so important in describing the driving forces in chemical reactions, they are introduced here, while the more detailed study of thermodynamics is delayed until Chapter 19. The chapter begins with a general discussion of energy, including force, heat, work, potential energy, and kinetic energy. The relationship between a system and its surroundings is also covered in detail.

In a context more relevant for chemistry, the first law of thermodynamics is covered in order to describe enthalpy change and total energy change, ΔH and ΔE. The terms *exothermic* and *endothermic* are also introduced, as are state functions. The discussion of *P-V* work leads directly to enthalpy.

Enthalpies of reaction occupy the remainder of the chapter, beginning with describing ΔH as being the difference in the absolute enthalpies of the products and reactants. Hess's law is used to illustrate the summation of enthalpies for step-wise reactions. Heats of formation are covered as the basis by which all enthalpies can be compared.

Calorimetry is discussed as the experimental approach to finding enthalpies. Specific heat and heat capacity are covered, as are constant-pressure calorimetry and constant-volume (bomb) calorimetry.

The chapter ends with discussions of practical use: the fuel values of foods and commercial fuels, especially their abundance, production, and future outlook.

LECTURE OUTLINE

I. **Thermodynamics**
   A. The term *thermodynamics* is from Greek and means, literally, "heat power." Thermodynamics is the study of energy and its transformations.
   B. The relationship between energy changes and chemical reactions is a specific part of thermodynamics called *thermochemistry*.

II. **The Nature of Energy**
   A. *Force* is a "push" or "pull" exerted on an object.
   B. *Work* is the energy required to overcome a force, work = force·distance.
   C. *Heat* is the energy transferred from one object to another because of a difference in temperature.
   D. Kinetic and Potential Energy
      1. *Kinetic energy* is the energy of an object due to its motion. As such, kinetic energy represents the *performance* of work, $KE = \frac{1}{2}mv^2$.
      2. *Potential energy* is the energy stored by an object due to its position relative to other objects. As such, potential energy represents the *capacity* to do work.
   E. Energy can be expressed in a wide variety of units.
      1. The *joule* (J) is the metric unit of energy and is defined in terms of SI fundamental quantities and is the energy possessed by a 2 kg object moving at a velocity of 1 m/s. $1\ J = 1\ kg\text{-}m^2/s^2$

2. The *calorie* (cal) is an archaic metric unit of energy and is the energy required to raise the temperature of 1 g of water from 14.5°C to 15.5°C. 1 cal = 4.184 J exactly
3. The *Calorie* (Cal or kcal) is a unit generally used only for the fuel value of foods. 1 kcal = 1000 cal = 4.184 kJ

F. In thermodynamics, the environment of a chemical reaction is separated into two parts, the system and the surroundings.
1. The *system* is the part of the environment we single out for study. The system generally is comprised only of the participants (reactants, products, solvents, etc.) in the reaction.
2. The *surroundings* are all other parts of the universe, including the vessel.

G. Systems, including chemical systems, tend to attain as low an energy as possible. Due to this tendency, processes that lower a system's internal energy are *spontaneous*. Processes that increase a system's internal energy are *nonspontaneous*.

Discussion Question: In practical terms, can an object's energy be totally potential or totally kinetic?

III. The First Law of Thermodynamics

A. The *law of conservation of energy* states that energy is neither created nor destroyed; it can only be transformed, such as from potential energy to kinetic energy, or exchanged between the system and the surroundings.

B. The law of conservation of energy is also known as the *first law of thermodynamics*, due to its fundamental importance.

C. *Internal energy* is the total energy within a system, including potential energy of the electrons and the nuclei and the kinetic energy of all motions.
1. Although internal energy cannot be measured, changes in internal energy can be measured.
2. The *change in internal energy*, ΔE, is defined as the difference in the internal energy at the end of a process and that at the beginning of the process, $\Delta E = E_{final} - E_{initial}$.
3. Positive values of ΔE signify energy being gained by the system from the surroundings, while negative values of ΔE signify energy being lost by the system to the surroundings.

D. The change in internal energy is related to heat and work.
The change in internal energy is equal to the sum of the *heat added to the system* and the *work done to the system*, $\Delta E = q + w$. Therefore, heat added to the system has a positive value while heat withdrawn from the system has a negative value. Similarly, work done to the system has a positive value while work done by the system has a negative value.

E. *State functions* are those energy properties which depend only on the current *state* of the system and are independent on the history of the system or the pathway leading to the current state. The change in internal energy, the flow of heat, and the work performed are all state functions.

Discussion Question: Discuss how the energy derived from a battery can be predominantly heat or predominantly work, depending on how the battery is discharged.

IV. Heat and Enthalpy Changes

Lecture Outline – Chapter 5
Brown, LeMay, & Bursten, *Chemistry: The Central Science*, 6th Edition

- A. *Enthalpy* is the amount of heat energy possessed by substances. The enthalpy of a system cannot be measured but changes in enthalpy can be measured. The enthalpy change in a process then corresponds to the heat change of the system when the process occurs at constant pressure:
  $\Delta H = q_p$ and $\Delta H = H_{final} - H_{initial}$
- B. *Endothermic reactions* are those in which heat is added, or flows, into the system, that is, the heat change of the process has a positive value.
- C. *Exothermic reactions* are those in which heat is withdrawn, or flows, from the system, that is, the heat change of the process has a negative value.
- D. Just like the internal energy change, the enthalpy change of a process is a *state function*.

V. A Closer Look: Relating Energy and Enthalpy

Work performed *by* a gas sample on its surroundings is described by $w = -P\Delta V$, where P is the applied pressure and ΔV is the change in volume that accompanies the work. If *volume* is *constant*, $P\Delta V = 0$, and $\Delta E = q_v$, where the subscript, v, indicates that the heat is taken at constant volume. If *pressure* is *constant*, $P\Delta V \neq 0$, and $\Delta E = q_p - P\Delta V$. Therefore, $q_p = \Delta E + P\Delta V$. The quantity, q_p, is defined as the enthalpy change of the process, ΔH.

Discussion Question: In a practical sense, does it make any real difference if a process is performed under constant-pressure or constant-volume conditions?

VI. Enthalpies of Reaction

- A. The enthalpy change of a process is the sum of the absolute enthalpies of the products minus the absolute enthalpies of the reactants,
  $\Delta H = H(\text{products}) - H(\text{reactants})$.
- B. Processes with enthalpy changes that are *negative* have heat flowing *from* the system and are called *exothermic* processes.
- C. Processes with enthalpy changes that are *positive* have heat flowing *into* the system and are called *endothermic* processes.
- D. Chemical equations that also show the enthalpy change of the reactions are called *thermochemical equations*.
- E. *Enthalpy diagrams* are often used to organize thermochemical data and to show their interrelationships.
- F. Enthalpy is an *extensive property*.
- G. The enthalpy change for a reaction is equal in magnitude but *opposite* in sign to ΔH for the reverse reaction.
- H. The enthalpy change for a reaction depends on the states of the reactants and the products.

VII. Chemistry at Work: Using Exothermic Reactions to Warm Food

Exothermic chemical reactions have been used to warm food, most notably in military rations called MREs (Meal, Ready to Eat). The Japanese have experimented with using chemical food warmers in vending machine foods.

VIII. Calorimetry

Lecture Outline – Chapter 5
Brown, LeMay, & Bursten, *Chemistry: The Central Science,* 6*th* Edition

- A. *Calorimetry* is the measurement of heat flow. These measurements are made using an apparatus called a *calorimeter*.
- B. The *heat capacity* of an object, C, is the amount of heat energy required to raise its temperature by 1°C, $q = C \cdot \Delta T$.
- C. The *molar heat capacity* of a substance is the amount of heat required to raise the temperature of 1 mole of the substance by 1°C.
- D. The *specific heat* of a substance, S, is the amount of heat required to raise the temperature of a 1 g sample of the substance by 1°C, $q = S \cdot m \cdot \Delta T$.
- E. In *constant-pressure calorimetry*, the pressure remains essentially constant because the apparatus is open to the atmosphere. At constant pressure, the heat change of the reaction is the *enthalpy change*, ΔH.
- F. In *bomb calorimetry*, the apparatus is sealed and the experiment is a *constant-volume* process. The heat change of the reaction is the *internal energy change*, $\Delta E = q_{evolved} = -C_{calorimeter} \cdot \Delta T$.

Discussion Question: How do the concepts of heat being an extensive property and specific heat fit together?

IX. Hess's Law

Hess's law states that if a reaction is carried out in a series of steps, ΔH for the reaction will be equal to the sum of the enthalpy changes for each step.

X. Enthalpies of Formation

- A. The enthalpy of the reaction that *forms* a substance from its constituent elements is called the *enthalpy of formation*, ΔH_f, or the *heat of formation*.
- B. The *standard* state of a substance is the state (i.e., solid, liquid, or gas) that is most stable for the substance at the temperature of interest, usually 25°C (298 K), and at standard atmospheric pressure, 1 atm.
- C. The *standard enthalpy of formation*, $\Delta H°_f$, of a substance is the enthalpy of the reaction that forms it from its constituent elements in their standard states.
- D. Enthalpies of reaction can be calculated by applying Hess's law to the enthalpies of formation of the participants,

$$\Delta H°_{rxn} = \Sigma\, n\Delta H°_f(\text{products}) - \Sigma\, m\Delta H°_f(\text{reactants}).$$

XI. Foods and Fuels

- A. *Foods* and *fuels* both produce energy when used. Generally, fuels are utilized in combustion reactions while foods are processed metabolically.
- B. The *fuel value* of a material is the energy *released* when a 1 g sample of it is combusted. Values are commonly given as absolute values but it is understood that all combustions are exothermic.
- C. *Foods* are used by the body to produce energy. The energy-producing reactions generally are stoichiometrically the same as combustions, reacting O_2 and the food to produce H_2O and CO_2, but are controlled by the body.
- D. *Fuels* differ by source, composition, and fuel value. The combusted fuels contain carbon, hydrogen, and possibly small amounts of other elements. The *fossil fuels* include oil, natural gas, and coal.

Discussion Question: Are foods really fuels?

E. *Syngas* or *synthesis gas* is the product mixture obtained by passing superheated steam over pulverized coal. It consists of CO, H_2, and CH_4, although other substances may be obtained depending on the conditions the reaction is performed under. Syngas is then subsequently reacted to produce a wide range of substances, like CH_3OH and acetic anhydride. It is believed that syngas can help store energy from solar power as well as decrease the need to burn sulfur-containing fossil fuels.

Discussion Question: What does the class see as the future of fuels?

XII. Chemistry at Work: Hydrogen as a Fuel

This section discusses the current feasibility of using H_2 as a commercial fuel.

SAMPLE QUIZ QUESTIONS

1. Explain how a piece of wood can have potential energy even when stationary on the ground.

2. What is a state function? Explain briefly using an example.

3. Calculate the change in internal energy of a system when 450 J of heat are supplied to an expanding gas if the gas does 300 J of work on its surroundings. (a) 750 J; (b) 150 J; (c) -150 J; (d) -750 J

4. What is meant by the term enthalpy?

5. If ΔH = -94 kJ for a particular process, (a) it must be slow; (b) it is endothermic; (c) it is exothermic; (d) it does not occur without heating.

6. The enthalpy change for the reaction

    $$Mg(s) + 2HCl(aq) \rightarrow MgCl_2(aq) + H_2(g)$$

    conducted at 1 atm pressure, is -466.9 kJ. Which of the following is true? (a) The internal energy change is a more negative number than the enthalpy; (b) The surroundings does work on the system; (c) The enthalpy change would not change with a change in external pressure; (d) The internal energy change and the enthalpy change are identical in this system.

7. Consider the diagram below. Which of the following statements concerning this diagram is(are) correct?

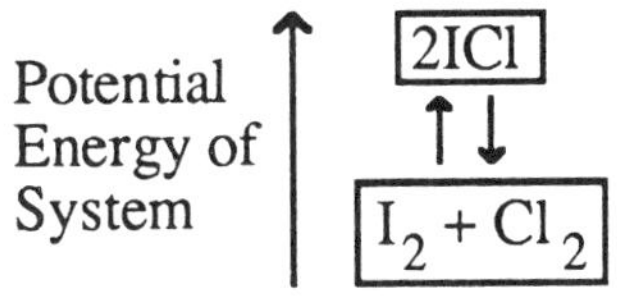

    Potential Energy of System ↑ : 2ICl ⇅ I$_2$ + Cl$_2$

    (a) The energy of the system increases when ICl is converted to I$_2$ + Cl$_2$; (b) The energy of the surroundings increases when ICl is converted I$_2$ + Cl$_2$; (c) Conversion of I$_2$ + Cl$_2$ into ICl is an exothermic process; (d) Conversion of ICl into I$_2$ + Cl$_2$ is an exothermic process.

8. Given the following data

	$\Delta H°$ (in kJ)
1/2 Br$_2$(l) → Br(g)	111.75
1/2 Cl$_2$(g) → Cl(g)	121.38
1/2 Br$_2$(l) + 1/2 Cl$_2$(g) → BrCl(g)	14.7

    Determine $\Delta H°$rxn for

    $$Br(g) + Cl(g) \rightarrow BrCl(g)$$

Lecture Outline – Chapter 5
Brown, LeMay, & Bursten, *Chemistry: The Central Science*, 6th Edition

9. Given the following data:

$$HCl(g) + KOH(s) \rightarrow KCl(s) + H_2O(l) \qquad \Delta H° = -203.6 \text{ kJ}$$
$$H_2SO_4(l) + 2KOH(s) \rightarrow K_2SO_4(s) + 2H_2O(l) \qquad \Delta H° = -342.4 \text{ kJ}$$

   calculate the heat of reaction for:

$$2\ KCl(s) + H_2SO_4(l) \rightarrow 2\ HCl(g) + K_2SO_4(s)$$

10. Copper metal can be obtained by roasting copper oxide, CuO, with carbon monoxide. What is the value of $\Delta H°$ in kJ for this reaction, which goes according to the following equation.

$$CuO(s) + CO(g) \rightarrow Cu(s) + CO_2(g)$$

   The following thermochemical data are known.

$$2CO(g) + O_2(g) \rightarrow 2CO_2(g) \qquad \Delta H° = -566.1 \text{ kJ}$$
$$2Cu(s) + O_2(g) \rightarrow 2CuO(s) \qquad \Delta H° = -310.5 \text{ kJ}$$

11. Consider the following reactions:

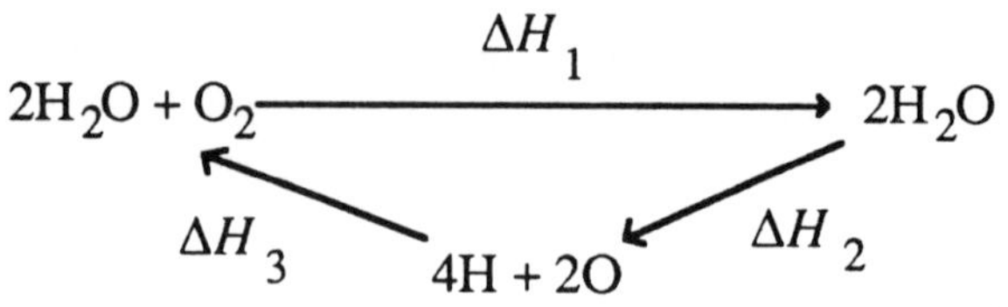

   $2H_2O + O_2 \xrightarrow{\Delta H_1} 2H_2O$

   $\Delta H_3 \nwarrow \qquad \nearrow \Delta H_2$

   $\quad 4H + 2O$

   Which of the following relationships is correct? (a) $\Delta H_1 = \Delta H_2 + \Delta H_3$; (b) $\Delta H_1 + \Delta H_2 + \Delta H_3 = 0$; (c) $\Delta H_1 = 0$; (d) $\Delta H_3 = -\Delta H_2$

12. Given the following ΔH values

$$H_2O_2(l) \rightarrow H_2O(l) + \tfrac{1}{2}O_2(g) \qquad \Delta H° = -98.3 \text{ kJ}$$
$$H_2(g) + \tfrac{1}{2}O_2(g) \rightarrow H_2O(l) \qquad \Delta H° = -285.8 \text{ kJ}$$

   Calculate $\Delta H°_f$ for $H_2O_2(l)$.

13. Calculate the enthalpy change for the following reaction:

$$Ca(s) + SO_3(g) + 2H_2O(l) \rightarrow CaSO_4 \cdot 2H_2O(s)$$

   $\Delta H°_f(SO_3) = -395.0$ kJ/mol; $\Delta H°_f(H_2O) = -285.8$ kJ/mol; $\Delta H°_f(CaSO_4 \cdot 2H_2O) = -1762.3$ kJ/mol.

14. The standard heat of combustion of CH_4 is $\Delta H°$ for which of the following reactions? (a) $C + 2H_2 \rightarrow CH_4$; (b) $CH_4 + 2O_2 \rightarrow CO_2 + 2H_2O(g)$; (c) $CH_4 + 2O_2 \rightarrow CO_2 + 2H_2O(l)$; (d) $CH_4 + \tfrac{3}{2}O_2 \rightarrow CO + 2H_2O(l)$

15. What is the heat evolved on a mole basis if 1.00 g of toluene, C_7H_8, combusted in a bomb calorimeter with a heat capacity of 16.03 kJ/°C, causes a temperature increase of 2.67°C.

Lecture Outline – Chapter 5
Brown, LeMay, & Bursten, *Chemistry: The Central Science*, 6*th* Edition

16. The heat of combustion of octane, C_8H_{18}, is -5460 kJ/mol. How much heat is generated when 5.0 g of C_8H_{18} is combusted?

17. A 7.5 gram sample of hexane, C_6H_{14}, was burned in a bomb calorimeter. The heat capacity of the calorimeter apparatus was 17.3942 kJ/°C. The temperature rose from 22.363°C to 25.663°C during the reaction. Calculate the heat of combustion for hexane, expressed both in kJ/g and in kJ/mol.

18. Sulfur (2.56 grams) was burned in a calorimeter with excess oxygen by the reaction

$$S(s) + O_2(g) \rightarrow SO_2(g)$$

The temperature increased from 21.35°C to 26.72°C. The bomb had a heat capacity of 923 J/°C and the calorimeter contained 815 g of water. Calculate the heat of combustion (per mole) of sulfur.

19. Why are fats a better choice than starches for energy storage in animals?

20. At present in the U.S., the largest single source of energy consumed is: (a) coal; (b) wood; (c) hydroelectric; (d) petroleum; (e) nuclear

Lecture Outline – Chapter 6
Brown, LeMay, & Bursten, *Chemistry: The Central Science*, 6th Edition

Chapter 6 — Electronic Structure of Atoms

OVERVIEW: Chapter 6 introduces the topic of electronic structure. The following chapters extend these ideas into molecules and their bonding. The authors have chosen the historical approach in the discussion, beginning with the first models of the atom and building from them as discoveries were made. Therefore, Rutherford's nuclear atom forms the basis of the atomic models and is extended by the work of Planck (the quantum theory), Einstein (the photoelectric effect), Bohr (quantized electron energy), de Broglie (matter waves), and Schrödinger (the wave equation).

The first topic in the chapter is the electromagnetic spectrum. It is with this basic coverage of wavelength, frequency, and energy that developments in the atomic models can be explained. The nature of electromagnetic radiation is used to explain the work of Planck and Einstein and its importance. Planck first proposed the idea of quantized energies. Einstein applied these ideas to the photoelectric effect. Spectra are discussed as the experimental basis for much of the work on atomic theory.

As the first detailed model of the atom, Bohr's description of hydrogen is covered completely. It is noted that Bohr's model has serious limitations, but that the properties ascribed to electrons are crucial to the modern view of atoms.

A discussion of de Broglie's matter waves and the Heisenberg uncertainty principle completes the preparation for the quantum mechanical model of Schrödinger. Although the math is omitted, a good understanding of the nature of the quantum mechanical atom is given. The meanings of the quantum numbers as variables to the wave equation are also covered in detail.

The quantum mechanical model is then used to cover electron configurations, which are arguably the most important information to be gotten from this material. Electron configurations are discussed in terms of the quantum numbers and the energy progression of the allowed combinations. Finally, the relationship of electronic structure to an element's position on the periodic table is covered.

LECTURE OUTLINE

I. **Electronic Structure**

   The *electronic structure* of an atom is the detailed description of the arrangement of the electrons in the atom.

II. **The Wave Nature of Light**

   A. *Electromagnetic radiation* refers to the electrical and magnetic waves that travel at 2.9979×10^8 m/s (the speed of light). The form to which we are most accustomed is visible light. More generally, radio waves, microwaves, infrared (heat), visible, ultraviolet, X ray, and γ ray radiation are included.

   B. *Wavelength*, λ, is the distance between two successive peaks or troughs in a wave. The longer the wavelength, the lower the energy of the radiation.

   C. *Frequency*, ν, is the number of complete waveforms that pass through a point in one second. Thus, frequency has the units of s^{-1}, /s, or hertz (Hz).

   D. The higher the frequency, the higher the energy of the radiation. Frequency and wavelength of electromagnetic radiation are related by $c = \lambda \nu$.

Discussion Question: We have all heard of the health complications of γ, X ray, and ultraviolet radiation. What can be implied about the health dangers of infrared, microwave, and radiowave radiation?

III. Quantum Effects and Photons

A. *Max Planck* proposed that radiation is not continuous, but rather consists of small "chunks."
   1. *Blackbody radiation* is radiation (light) given off by objects simply because of being hot.
   2. The *quantum* was the term Planck gave to the small "chunks" of energy given off in blackbody radiation. Further, he found that the frequencies, ν, of these quanta were whole-number multiples of a fundamental frequency. The energies, ΔE, are given by $\Delta E = h\nu, 2h\nu, 3h\nu,...$, where h is now known as *Planck's constant*, 6.626×10^{-34} J-s.

B. The *photoelectric effect* is the observation that many metallic surfaces produce electricity (electrons are ejected) when exposed to light. It was discovered that, for each metal, there is a minimum frequency below which no electricity is produced. It was also found that, above the minimum frequency, the number of electrons ejected depended only on the light intensity and the energy of the ejected electrons depended only on the frequency of the light.

The packet of energy sufficient to eject an electron is called a *photon*. The more intense the light, the greater number of photons, and the greater number of ejected electrons. The kinetic energy of the electrons is given by the frequency of the light times Planck's constant less the energy holding the electron in place in the metal (the *binding energy*), $E_{photon} = h\nu - E_b$.

IV. Bohr's Model of the Hydrogen Atom

A. A *spectrum* is produced when radiation from a source is separated into its component wavelengths.
   1. *Monochromatic radiation* or light is radiation of a single wavelength or frequency.
   2. *Continuous spectra* are those in which all the colors of a particular region of radiation are present and overlapping.
   3. *Line spectra* are those in which only a few discrete wavelengths occur.

B. Johann Balmer discovered that some of the lines in the line spectrum of hydrogen could be described by $\nu = (3.29 \times 10^{15} \text{ s}^{-1})(\frac{1}{4} - \frac{1}{n^2})$, $n = 3, 4, 5, 6$.

C. Niels Bohr first used Planck's quantum theory to interpret the line spectrum of hydrogen. *Bohr's model* of the hydrogen atom described the nucleus as being surrounded by a number of orbits of fixed (quantized) radius, numbered $n = 1, 2, 3,..., \infty$.

D. Bohr concluded that the *energy* of the electron in an orbit of hydrogen is *quantized*, that is, it can only have a few discrete values. If the energy of the electron is quantized, then the energy difference between two orbits must also be quantized. The frequency of a line in the spectrum corresponds to the energy difference between two orbits, $\Delta E = h\nu$.

E. The *energy* of a Bohr *orbit* (and an electron in it) is given by $E_n = -R_H(\frac{1}{n^2})$, where R_H is the *Rydberg constant*, 2.179×10^{-18} J.

Lecture Outline – Chapter 6
Brown, LeMay, & Bursten, *Chemistry: The Central Science*, 6*th* Edition

- F. If the electron is in the lowest energy orbit, $n = 1$, it is said to be in the *ground state*.
- G. If the electron is in *any* orbit other than $n = 1$, it is said to be in an *excited state*.
- H. The energy of a line in the emission spectrum is the difference in the energies of the two orbits involved, $\Delta E = E_{final} - E_{initial}$, or $\Delta E = h\nu = R_H(\frac{1}{n_i^2} - \frac{1}{n_f^2})$.
- I. The *radius* of a Bohr orbit is given by radius = $n^2(5.30 \times 10^{-11}$ m).
- J. The *ionization energy* of hydrogen is the energy required to remove the electron from the atom, that is, the energy of the $n = 1$ to $n = \infty$ transition.
- K. The Bohr model only worked precisely for hydrogen. As it was applied to other elements, the agreement grew worse with increasing atomic number.

V. A Closer Look: Discovery of Helium

This section discusses how spectroscopy was used to discover the element helium.

Discussion Question: How does the identity of the gas in a light bulb influence the color of the light? What colors of lights have students seen, in addition to those cited in the text?

VI. The Dual Nature of the Electron

- A. De Broglie proposed that particles behave under some circumstances as if they were waves. This is similar to the idea that light behaves under some circumstances as if it was a particle. *Matter wave* is the term given by de Broglie to matter behaving as a wave with wavelength, $\lambda = \frac{h}{mv}$.
- B. The *momentum* of any moving particle is given by mv, where m is the mass and v is the velocity.
- C. Werner *Heisenberg* proposed the *uncertainty principle*, which states that it is impossible for us to know simultaneously *both* the exact momentum of an electron and its exact location in space.

VII. A Closer Look: Measurement and the Uncertainty Principle

This section discusses how the uncertainty principle applies to measurement. There is a minimum uncertainty in any measurement that cannot be reduced below a certain level.

Discussion Question: If de Broglie was correct, and we do believe he was, what is the wavelength of a typical student? Does this help explain why classical physics did not account for the wave nature of large objects?

VIII. Quantum Mechanics and Atomic Orbitals

- A. Erwin *Schrödinger* proposed a *mathematical model* of the atom. By using measured energies and known forces, he used math to solve equations to yield the structure of the atom, rather than start with a preconceived "picture" of the atom's structure. This approach is called *quantum mechanics* or *wave mechanics*.
- B. The *wave equation* is Schrödinger's basic expression, incorporating the known interactions in an atom.

C. The solutions to the wave equation are called *wave functions*, symbolized ψ. From Heisenberg, the wave functions cannot describe the exact position of an electron but only the *probability* of finding it in a given location.

D. The probability of finding the electron in a given location is the *electron density* and is given by the square of the wave function for that location, ψ^2.

E. The solutions to the wave equation are called *orbitals* and each has a characteristic energy.
   1. An orbital is a region of high probability of finding the electron; it is not a path or trajectory. Thus, an orbital in the quantum mechanical model is *not* the same as a Bohr orbit.
   2. The variables in the wave equation are called *quantum numbers*. The Bohr model used only one variable or quantum number, n. The quantum mechanical model uses three quantum numbers to describe each orbital.
      a. The *principal quantum number* (n) has values of 1, 2, 3,..., ∞. It describes the relative *size* of the orbital.
      b. The *azmuthal quantum number* (l) has values of 0, 1,..., $n-1$. It describes the *shape* of the orbital. The value of l is often referred to by a letter equivalent, $0 = s$, $1 = p$, $2 = d$, $3 = f$, etc.
      c. The *magnetic quantum number* (m_l) has values of $-l$,... -1, 0, 1,...,l. It describes the *orientation* of the orbital in space.
      d. A collection of orbitals with the same value of n is called an *electron shell*.
      e. A collection of orbitals with the same values of n and l is called a *subshell*. A subshell can then be referred to by using the value of n and the letter equivalent of l, such as 1s, 2p, 3d, and so on.

IX. Representations of Orbitals

A. The s orbitals are those for which $l = 0$. All s orbitals are spherical. There is one s orbital in each s subshell.

B. *Nodal surfaces* or *nodes* are those regions of space around the nucleus, other than at very large distances (the outside edge), where the function ψ^2 goes to zero. Nodes represent places where the probability of finding the electron is zero.

C. The p orbitals are those for which $l = 1$. All p orbitals are "dumbbell" or "figure-eight" shaped. There are three p orbitals in each p subshell.

D. The d and f orbitals are those for which $l = 2$ and $l = 3$, respectively. There are five d orbitals in each d subshell and seven f orbitals in each f subshell. For the d orbitals of a subshell, four are "four-leaf clovers"; the fifth looks like a p orbital with the addition of a ring around the center. The shapes of the f orbitals are beyond this discussion but are more complex, yet.

X. Orbitals in Many-Electron Atoms

A. The hydrogen atom is special in that it has only one proton and one electron. All other atoms have more protons and electrons, and also have neutrons in the nucleus. In general, then, the nuclear attraction on an electron will be greater than in hydrogen. The *nuclear charge* affects the energy of an electron as well as the orbitals of an atom.
   1. Bohr's equation can be extended to many-electron atoms by noting that the energy of an electron is proportional to the square of the nuclear charge, Z: $E_n = -R_H(\frac{Z^2}{n^2})$. However, the relationship worsens as the atomic number increases.

Lecture Outline – Chapter 6
Brown, LeMay, & Bursten, *Chemistry: The Central Science, 6th Edition*

    2.   Electrons moving around the nucleus do not experience the same nuclear attraction; those electrons closer to the nucleus experience a greater force than those that are farther away. The nuclear charge actually "felt" by an electron is called the *effective nuclear charge*, Z_{eff}. Z_{eff} for a given electron is given by the true nuclear charge, Z, less the amount by which electrons closer to the nucleus *screen* it, S, $Z_{eff} = Z - S$.

  B. Due to screening and the resulting effective nuclear charge, the *energies* of the orbital *subshells* within a given principle energy shell (value of *n*) vary. Within a given value of *n*, Z_{eff} decreases as *l* increases and so the energies of the subshells increase as *l* increases, $ns < np < nd < nf$. Orbitals that have the same energy are said to be *degenerate*.

Discussion Question: Should the varying geometries of electron orbitals affect the shapes of molecules?

  C. Electron Spin and the Pauli Exclusion Principle
    1.   It was discovered that spectral lines were actually composed of two very closely spaced lines. This meant that there were twice as many energy states as previously thought. This observation was explained by proposing that electrons *spin*, like tops, and can spin clockwise or counterclockwise.
    2.   In order to quantitatively account for electron spin, a fourth quantum number was devised. This fourth quantum number is the *electron spin quantum number* (m_s) and has allowed values of $\pm \frac{1}{2}$.

  D. The *Pauli exclusion principle* states that no two electrons can have the same set of four quantum numbers (n, l, m_l, m_s). In practical terms, this means that only two electrons can go into the same orbital (sharing the same values of n, l, and m_l) and can only do so if they have opposite spins, one $+\frac{1}{2}$ and one $-\frac{1}{2}$.

XI. A Closer Look: Experimental Evidence for Electron Spin

This section looks in greater detail at the discoveries that required the introduction of the spin quantum number.

XII. Electron Configurations

  A. The most stable electron arrangements or *configurations* of the elements are those that minimize energy. Therefore, each element has a *ground state* electron configuration in which all the electrons are in the lowest-energy orbitals available. Any other electron configuration is an *excited state*.

  B. The process of writing an electron configuration is fairly straightforward.
    1.   The *Pauli exclusion principle* must be obeyed.
    2.   *Hund's rule* states that, if electrons are being assigned to a set of degenerate orbitals, they will go into separate orbitals if that option is available. For example, two electrons in the 2p subshell will exist in separate orbitals (of the three available) rather than both going into one of the orbitals.
    3.   *Valence electrons* or *outer-shell electrons* are those electrons present beyond the preceding noble gas. Aluminum, for example, has the ground state electron configuration $1s^2 2s^2 2p^6 3s^2 3p^1$, and has three valence electrons, $3s^2 3p^1$. Note that aluminum is in group 3A.
    4.   The *core* (or *kernel*) *electrons* are those that are not valence electrons, that is, they are the electrons which comprise the preceding noble gas electron configuration. Electron configurations are often written to gather the core

electrons together and to emphasize the valence electrons, such as [Ne]$3s^2 3p^1$ for aluminum.

5. The *transition elements* or *transition metals* are those elements in the groups 3B through 2B (the scandium group through the zinc group). All these elements are metals and include all the structural metals except aluminum and all the coinage metals. As one proceeds across the transition metals, electrons are being added to a *d* subshell.

6. The *rare-earth* or *lanthanide elements* are those which follow lanthanum in the periodic table, atomic numbers 58 (cerium) through 71 (lutetium). These elements are commonly written as a row separated from the body of the periodic table. As one proceeds across the lanthanides, electrons are being added to the 4*f* subshell.

7. The *actinide elements* are those which follow actinium in the periodic table, atomic numbers 90 (thorium) through 103 (lawrencium). These elements are commonly written as a row separated from the body of the periodic table. As one proceeds across the actinides, electrons are being added to the 5*f* subshell.

XIII. Chemistry at Work: Nuclear Spin and Magnetic Resonance Imaging

This section discusses practical applications of nuclear spin, which is analogous to electron spin.

1. Nuclear magnetic resonance (NMR) is a standard instrumental technique in which the strength of the magnetic and radio frequency fields required to force a nucleus to change its spin are determined.

2. Magnetic resonance imaging (MRI) is the medical application of NMR to the diagnosis of diseases. The word "nuclear" was dropped from the name for primarily public relations reasons.

XIV. Electron Configurations and the Periodic Table

A. The left-most two groups in the periodic table, excluding hydrogen, are the *active metals*. All members of group 1A have electron configurations that end in ns^1. All members of group 2A have electron configurations that end in ns^2.

B. The left-most two groups in the periodic table comprise the *s* block while the right-most six groups in the periodic table comprise the *p* block, based on the subshell to which electrons are being added. Together, the *s* block and the *p* block are referred to as the *representative elements*.

C. The transition metals, groups 3B through 2B, comprise the *d* block.

D. The *f*-block metals are the lanthanides and the actinides.

Discussion Question: What types of properties of the elements seem to relate to an element's position on the periodic table? What properties do the *s* block elements seem to have in common? The *p* block? The *d* block?

Lecture Outline – Chapter 6
Brown, LeMay, & Bursten, *Chemistry: The Central Science*, 6th Edition

SAMPLE QUIZ QUESTIONS

1. Chlorophyll, the green plant pigment, has its maximum absorption of visible light at 600 nm. What is the frequency and energy of a photon of light at this wavelength?

2. Which of the following has the shortest wavelength? (a) X rays; (b) visible light; (c) UV radiation; (d) radio waves

3. What is the smallest increment of energy that can be delivered by light with a wavelength of 300 nm? ($h = 6.63 \times 10^{-34}$ J-sec)

4. What happens to the wavelength of a beam of electrons if its speed is doubled? (a) It doubles; (b) It increases by a factor of four; (c) It decreases by a factor of two; (d) It decreases by a factor of four.

5. The wavelength (λ) of light: (a) is directly proportional to energy; (b) is shorter in the infrared region of the spectrum than in the visible; (c) applies to white light; (d) determines its speed; (e) is inversely proportional to energy.

6. Which of the following relationships is correct? (a) $E = \lambda c/h$; (b) $E = \lambda h/c$; (c) $E = hc/\lambda$; (d) $E = \lambda/hc$; (e) $E = h/\lambda c$

7. An atomic emission spectrum: (a) cannot be used to identify an element; (b) contains only a few, discrete wavelengths; (c) contains a continuous series of wavelengths; (d) contains fewer lines for heavier elements; (e) usually consists of white light.

8. What do we mean when we say that the energy of an electron bound in an atom is quantized?

9. Radiation that has a wavelength greater than 650 nm will not free electrons from the surface of cesium metal, no matter how intense the radiation. Explain this observation.

10. What is meant by the "emission spectrum" of a substance? Why does the emitted light contain only specific wavelengths?

11. The idea that there is a fundamental limitation in our ability to precisely know both the location and the momentum of a particle is known as: (a) Pauli's exclusion principle; (b) Heisenberg's uncertainty principle; (c) Hund's rule; (d) De Broglie's hypothesis.

12. In the Bohr model of the atom, (a) electrons move around the nucleus in spherical orbits; (b) electrons can occupy positions at any distance from the nucleus; (c) an electron's energy depends on the speed at which it travels; (d) an electron cannot move to another orbit; (e) electrons move around the nucleus in elliptical orbits like planets around the sun.

13. In the Bohr model of the atom, when an electron moves from a lower shell to a higher shell: (a) the atom enters the ground state; (b) one quantum of energy is emitted; (c) the atomic line spectrum is produced; (d) one quantum of energy is absorbed; (e) another electron moves from a higher shell to a lower shell.

14. In the Bohr model of the atom, (a) the excited state is stable; (b) energy absorbed is $\Delta E = E_{\text{higher shell}} - E_{\text{lower shell}}$; (c) the lowest energy shell is called "L"; (d) the electron may

fall into the nucleus under the right circumstances; (e) energy absorbed is $\Delta E = E_{\text{lower shell}} - E_{\text{higher shell}}$.

15. Using the Bohr model of the atom, determine the energy of the transition of an electron from $n=3$ to $n=7$ and calculate the wavelength of the light associated with the transition. State whether the light is given off or absorbed.

16. The Bohr model of the atom: (a) worked well for all but the heaviest elements; (b) accounted for interactions between atoms; (c) worked well for hydrogen but did not satisfactorily work for other elements; (d) remains the most useful model of the atom; (e) addressed the repulsion between electrons in an atom.

17. De Broglie's hypothesis: (a) stated that particles should exhibit some of the properties of waves; (b) stated that waves should exhibit some of the properties of particles; (c) defined the quantum; (d) disproved the Bohr model of the atom; (e) was considered interesting but did little to change atomic theory.

18. Write a correct set of four quantum numbers for an electron in a 5f orbital.

19. Heisenberg: (a) suggested that electrons have some of the properties of waves; (b) showed how to determine the location and the energy of an electron; (c) proved the existence of defined electron orbits in atoms; (d) was able to determine the location of an electron in an atom given its energy; (e) stated that it is impossible to know both the location and the energy of an electron.

20. Principal energy levels have electron capacities given by: (a) $2n^2$; (b) $2n^3$; (c) $4n^2$; (d) $2n+1$; (e) n^2

21. If an electron exhibits a wavelength of 0.300 nm, what is its velocity? (m = 9.11 x 10^{-31} kg; h = 6.63 x 10^{-34} J-sec)

22. Which of the following configurations for hydrogen represents the highest energy state? (a) 2s; (b) 2p; (c) 4s; (d) 3d

23. For hydrogen, the quantum number l indicates: (a) the general shape of the orbital; (b) the energy of the orbital; (c) the orientation of the orbital; (d) the size of the orbital.

24. If $n = 3$, which of the following values of l are not possible? (a) 0; (b) 1; (c) 2; (d) 3

25. In the quantum mechanical model of the atom, the azmuthal quantum number, l, (a) describes the size of the orbital; (b) describes the spatial orientation of the orbital; (c) describes the spin of the electron in the orbital; (d) describes the shape of the orbital; (e) is allowed the values 0, 1,..., n.

26. The allowed values of the magnetic quantum number, m_l, are: (a) 1, 2, 3,...; (b) 0, 1,..., +(l-1); (c) 0, 1, 2,..., n-1; (d) +n, +(n-1),..., 0,..., -(n-1), -n; (e) +l, +(l-1),..., 0,..., -(l-1), -l.

27. The 2p orbital is an orbital for which $l =$ (a) 0; (b) 1; (c) 2; (d) 3

28. The number of 3d orbitals is: (a) 3; (b) 4; (c) 5; (d) 6

29. Show, by means of sketches, the difference between an s and a p orbital.

30. Show, by means of sketches, how the following orbitals differ from one another: (a) $2s$ and $3s$; (b) $2p_x$ and $2p_y$

31. The $2s$ and $3s$ orbitals of hydrogen differ in: (a) the number of nodal surfaces; (b) energy; (c) extent in space; (d) all of the above.

32. An atomic orbital: (a) illustrates well defined paths for electrons in an atom; (b) is a statistically determined volume in which an electron should be found; (c) is of little use in describing an atom; (d) is no longer considered a valid concept; (e) is the same as an orbit in the Bohr model of the atom.

33. Write an acceptable electron configuration for each of the following. (a) Cr; (b) Cu; (c) As; (d) Ir; (e) I

34. In modern atomic theory, (a) each electron shell has a capacity of the same number of electrons; (b) zero is an allowed value of n; (c) the first principal energy level, $n = 1$, has a capacity of eight electrons; (d) the concept of electron shells is a central idea; (e) heavy atoms still are difficult to describe.

35. The orbitals which exist at $n = 3$ are

    (a) $2s2p_x2p_y2p_z3s3p_x3p_y3p_z3d_{xy}3d_{yz}3d_{xz}3d_{z^2}3d_{x^2-y^2}$;
    (b) $1s2s2p_x2p_y2p_z3d_{xy}3d_{yz}3d_{xz}3d_{z^2}3d_{x^2-y^2}$;
    (c) $1s2s2p_x2p_y2p_z3s3p_x3p_y3p_z3d_{xy}3d_{yz}3d_{xz}3d_{z^2}3d_{x^2-y^2}$;
    (d) $3s3p_x3p_y3p_z3d_{xy}3d_{yz}3d_{xz}3d_{z^2}3d_{x^2-y^2}$;
    (e) $1s2s2p_x2p_y2p_z3s3p_x3p_y3p_z$.

Lecture Outline – Chapter 7
Brown, LeMay, & Bursten, *Chemistry: The Central Science, 6th Edition*

Chapter 7 — Periodic Properties of the Elements

OVERVIEW: Chapter 7 builds the periodic table both from the historical perspective of Meyer and Mendeleev, but also from the perspective of the quantum mechanical model of the atom and electron configurations.

The early work of Lothar Meyer and Dmitri Mendeleev in constructing the periodic table is covered first, including Mendeleev's predictions of elements in places where no known element would match the expected properties. G.N. Lewis's idea of shells is also introduced as a corollary to the rows of the periodic table.

Important periodic trends are covered next, beginning with effective nuclear charge and atomic radius. Having developed the importance of effective nuclear charge, the trends in ionization energy and electron affinity follow nicely.

The concept of metallic character is introduced and used as the context for the division of the elements into the metals, the nonmetals and the semimetals.

Finally, a survey of the properties of the most important main-group families, 1A, 2A, 6A, 7A, and 8A, is given. This survey focuses on the properties introduced earlier in the chapter, such as ionization energy and metallic character.

LECTURE OUTLINE

I. Trends in Atomic Behavior

  A. Columns on the periodic table are called *groups* or *families*. Members of the same group tend to have similar chemical properties and have the same *valence shell* electron configurations.

  B. Rows on the periodic table are called *periods*. Members of the same period tend to have chemical properties which change gradually as you proceed across the period.

II. Development of the Periodic Table

  A. Lothar Meyer and Dmitri Mendeleev both discovered meaningful patterns of properties among the approximately 63 elements known in 1865. Both listed the elements in the order of increasing atomic weight and saw that the properties repeat, a phenomenon called *periodicity*. Mendeleev offered some bold, but correct, proposals about places in the scheme that seemed inconsistent and so is generally given credit for the development of the *periodic table*.

  B. *Mendeleev's periodic table* left holes where a known element would not properly fit. The classic example is germanium, which was unknown. There was no element that fit the properties expected of the element below silicon in the same group and to the left of arsenic. Mendeleev left that position empty and proposed that the element that belonged there, which he called eka-silicon, was simply yet to be discovered. Within a few years, it was found and its properties matched Mendeleev's predictions almost perfectly.

  C. At the time of Mendeleev, scientists did not know about the structure of the atom and about subatomic particles and so they did not know about atomic numbers. We now know that the atomic number is the number of protons in the nucleus and therefore it is the charge of the nucleus. The periodic table is actually arranged in

Lecture Outline – Chapter 7
Brown, LeMay, & Bursten, *Chemistry: The Central Science, 6th Edition*

order of increasing *atomic number*, not increasing atomic weight, but there are only a few places where the order is affected.

Discussion Question: What other ways of arranging the elements to yield repeating patterns of behavior can students find?

III. Electron Shells in Atoms

A. From Chapter 6, sets of electrons that have the same value of the principle quantum number, n, are called *shells*. Shells form the basis of the models of atoms, ions, and molecules introduced by Gilbert N. Lewis.

B. Sets of electrons that have the same values of both the principle quantum number, n, and the azmuthal quantum number, l, are called *subshells*. Although Lewis's ideas are useful, his description of shells as being spherical is in error. Most subshells do not contain spherical orbitals.

IV. Sizes of Atoms

A. From Chapter 6, *effective nuclear charge* increases from left to right in a period (row) and from bottom to top in a group (column) on the periodic table.

B. The greater the effective nuclear charge, the greater the attractive force between the nucleus and its electrons.

C. As the attractive force between a nucleus and its electrons increases, the average distance between the nucleus and its electrons decreases. The average distance between the nucleus and its outermost electron is expressed as the *atomic radius* of the atom. It can then be said that atomic radius *decreases* from left to right in a period and from bottom to top in a group on the periodic table.

V. Ionization Energy

A. The ionization energy, I, is the energy required to remove the outermost electron from a gaseous atom or ion. The *first ionization energy*, I_1, is the energy for the removal of an electron from a neutral, gaseous atom: $M(g) \rightarrow M(g)^+ + e^-$. The *second ionization energy*, I_2, is the energy of the removal of a *second* electron from a gaseous ion with a 1+ charge: $M(g)^+ \rightarrow M(g)^{2+} + e^-$. Metallic atoms tend to lose enough electrons to gain the electron configuration of the preceding noble gas.

B. There are *periodic trends* in the ionization energies, also tied to the effective nuclear charge. As the effective nuclear charge increases, it requires more energy to remove the outermost electron from an atom. Consequently, ionization energy is also related to the atomic radius, with ionization energy increasing as atomic radius decreases. Therefore, the first ionization energy increases from left to right in a period and from bottom to top in a group of the periodic table.

Discussion Question: Does there seem to be a relationship between ionization energy and an element's ability to be corroded or to cause corrosion?

VI. Electron Affinities

A. *Electron affinity*, E, is the energy change of the reaction of adding an electron to a gaseous atom or ion: $M(g) + e^- \rightarrow M(g)^-$. These reactions tend to be exothermic and so the values of E are generally negative.

B. In general, electron affinity tends to decrease (become more negative) from left to right in a period on the periodic table. Going down a group, there is little change in the electron affinities.

VII. Metals, Nonmetals, and Metalloids

A. The elements can be conveniently divided as being metals, nonmetals, or metalloids. From Chapter 2, metals conduct heat and electricity, are lustrous, malleable, and ductile. By contrast, nonmetals are poor conductors, are often dull in appearance, and shatter if forged. The metalloids or semimetals tend to be intermediate in their properties.

B. The *metals* occupy approximately the left-hand two-thirds of the periodic table.
   1. Metals tend to lose electrons to other substances in reactions to become positively charged ions, or *cations*.
   2. Metals also form *basic oxides*, that is, metal oxides react with water to produce a basic or alkaline solution.

C. The *nonmetals* occupy approximately the right-hand one-third of the periodic table.
   1. Nonmetals tend to gain electrons from other substances in reactions to become negatively charged ions, or *anions*.
   2. Nonmetals form *acidic oxides*, that is, nonmetal oxides react with water to form acidic solutions.

D. The *metalloids* occur at the boundary between the metals and the nonmetals, usually indicated by a stair-step line from boron to bismuth. Metalloids can either gain electrons or lose electrons in chemical reactions.

E. *Metallic character* is the extent to which an element behaves as a metal. Elements are said to have greater metallic character as their conductivity, malleability, etc., increase. Metallic character increases from right to left in a period and from top to bottom in a group of the periodic table. *Nonmetallic character* could also be discussed, but it is the reverse of metallic character and so is seldom used.

Discussion Question: If the "best" metal is cesium (or francium), why are we so dependent on other metals? What factors, other than metallic character, are important in choosing an electrical conductor or a structural metal?

VIII. Group Trends: The Active Metals

A. Several *group trends*, or orderly variations in chemical behavior with position in a group, can be seen, particularly in the representative or main-group elements.

B. The *alkali metals*, group 1A, are soft, gray metals. The term *alkali* means "ashes" since these elements were first isolated from wood ashes.
   1. The alkali metals have low ionization energies and so lose an electron easily to form 1+ ions. *Electrolysis* is a technique in which electricity is applied to force an electron back onto the cation and so produce the neutral metal. The lower in the group an element is, the greater the electrical energy required to electrolyze its cation.
   2. When the alkali metals react with hydrogen (the only nonmetallic element in the group), hydrogen is present as the *hydride ion*, H^-.
   3. Reactivity tends to increase from top to bottom in the group.

C. The *alkaline earth metals*, group 2A, are slightly harder, more dense, and less reactive than the alkali metals. The first ionization energies are low, but not as low as those of the alkali metals. They form 2+ ions readily. Reactivity tends to increase from top to bottom in the group.

D. Several groups in the periodic table have the same number, being differentiated by the letter designation, A or B. In comparison of A and B groups, similarities in properties are seen, particularly in the charges of the ions. Metals of group 1 form 1+ ions; group 2 metals form 2+ ions, and so on. Obviously, this similarity isn't followed when the B group is metallic and the A group is nonmetallic.

IX. Chemistry and Life: Osteoporosis and Calcium Intake

This section describes osteoporosis, a calcium disorder, in the context of calcium's behavior as a group 2A metal.

X. Group Trends: Selected Nonmetals

A. *Hydrogen* is unique. Its $1s^2$ electron configuration usually places it in group 1A although it is not a metal. It reacts with nonmetals to make molecular compounds, many of which are gases or liquids. The majority of hydrogen's chemistry is of its 1+ ion, H^+.

B. The *oxygen family*, group 6A, is nonmetallic at the top (oxygen) and metallic at the bottom (polonium). Polonium is rare and is radioactive; the top three members (oxygen, sulfur, and selenium) are typical of nonmetals and are comparatively abundant. Oxygen and sulfur both occur as *allotropes*, molecules of the elements with different structures (O_2 and O_3, S_2, S_4, S_6, and S_8). Oxygen has the lowest electron affinity (greatest tendency to gain electrons) and the highest ionization energy.

C. The *halogens*, group 7A, are all nonmetals. Astatine is radioactive and is seldom included in discussions of the halogens. Halogens are more nonmetallic than the oxygen family, having lower (more negative) electron affinities and higher ionization energies. All are diatomic in the gas phase, although bromine is normally a liquid and iodine is normally a solid.

D. The *noble gases*, group 8A, are fairly nonreactive. Some compounds of xenon are known, as are just a few compounds of krypton. They are all *monatomic* gases. They have the highest ionization energies of any family

XI. A Closer Look: Discovery of the Noble Gases

This section puts the discovery of the noble gases in the context of their inertness.

Lecture Outline – Chapter 7
Brown, LeMay, & Bursten, *Chemistry: The Central Science, 6th Edition*

SAMPLE QUIZ QUESTIONS

1. Why is the first ionization potential of lithium so much smaller than the second ionization potential?

2. Explain why sulfur and selenium are chemically similar to one another.

3. A period of elements: (a) begins with a noble gas; (b) contains 18 elements; (c) refers to the columns on the periodic table; (d) refers to the rows on the periodic table; (e) contains five or six elements.

4. A group of elements: (a) begins with a noble gas; (b) contains 18 elements; (c) refers to the columns on the periodic table; (d) refers to the rows on the periodic table; (e) contains five or six elements.

5. The valence shell: (a) for 4th period elements consists of $4s$, $3d$, and $4p$ electrons; (b) is the outermost occupied electron shell in an atom; (c) consists of only s and p electrons; (d) contains eight electrons; (e) is the lowest energy shell in an atom.

6. The 1st period: (a) has eight elements in it; (b) contains a metal and a nonmetal; (c) fills the $1s$ and $1p$ subshells; (d) contains the most reactive elements; (e) has two elements in it.

7. The 3rd period: (a) is a short period; (b) is a long period; (c) contains all nonmetals; (d) fills the $2s$ and $2p$ subshells; (e) is unique in that its members all occur as diatomic molecules.

8. The lanthanides: (a) fill the $5f$ subshell; (b) are members of the 6th period; (c) fill the $5d$ subshell; (d) are semimetals; (e) are sometimes referred to as outer transition elements.

9. The 7th period: (a) includes the elements with atomic numbers 87 through 118; (b) includes the lanthanides; (c) contains metals and nonmetals; (d) has a capacity of 32 electrons; (e) fills the $7s$, $6f$, $5d$, and $7p$ subshells.

10. In which atom would the outermost electron experience a greater nuclear charge? (a) C; (b) N; (c) O; (d) F

11. Would you expect the radius of the sulfur atom to be larger or smaller than that of the sulfide ion, S^{2-}? Explain.

12. Atomic size tends to: (a) increase across a period from the left due to an increase in the number of protons; (b) decrease across a period from the left due to an increase in the number of neutrons; (c) remain almost constant across a period from the left since the principal energy level stays constant; (d) decrease across a period from the left due to an increase in the number of protons; (e) decrease down a group due to the increase in the number of protons.

13. Which of the following is larger than gallium? (a) germanium; (b) calcium; (c) silicon; (d) selenium; (e) aluminum.

14. Ionization energy tends to: (a) increase across a period from the right; (b) decrease going up a group; (c) increase going up a group; (d) decrease from the lower left corner of the periodic table to the upper right corner; (e) remain fairly constant across a period.

Lecture Outline – Chapter 7
Brown, LeMay, & Bursten, *Chemistry: The Central Science, 6th Edition*

15. The 2nd ionization energy of an element: (a) is greater than the 1st ionization energy of the element; (b) is less than the 1st ionization energy of the element; (c) is the energy required to remove two electrons from an atom of the element; (d) is the energy released when a second electron is removed from an atom of the element; (e) is lower for sodium than for magnesium.

16. Which of the following has a 1st ionization energy less than Rh? (a) Zn; (b) Pd; (c) I; (d) Co; (e) Ba.

17. Which of the following has the lowest electron affinity? (a) P; (b) N; (c) O; (d) F

18. Electron affinity tends to: (a) increase across a period from the right; (b) decrease going down a group; (c) increase going down a group; (d) decrease from the lower left corner of the periodic table to the upper right corner; (e) remain fairly constant across a period.

19. Which of the following has an electron affinity greater than In? (a) Ge; (b) Hg; (c) Cd; (d) Au; (e) Sr.

20. The trends in ionization energy and electron affinity tend to: (a) run opposite each other on the periodic table; (b) show that the greatest metallic character is in the upper right corner of the periodic table; (c) contradict the trend in atomic size; (d) parallel each other on the periodic table; (e) show little change across a period.

21. The representative elements: (a) are labeled as the "B" groups on the periodic table; (b) are the nonmetals; (c) are those which are members of the *s* and *p* blocks; (d) are generally nonreactive; (e) are those which best illustrate the properties of the elements.

22. The alkali metals: (a) include hydrogen; (b) have a common valence shell electron configuration of ns^2; (c) are dense, hard metals; (d) are highly reactive; (e) are relatively inert.

23. The elements of group 8A in the periodic table: (a) are fairly reactive; (b) tend to form anions like all nonmetals; (c) are also known as the inert gases; (d) are not known to make compounds; (e) all make compounds.

24. The metals: (a) tend to be brittle; (b) are all solids; (c) tend to be malleable; (d) all make cations with 1+ charges; (e) make cations by losing electrons from shells below the valence shell.

25. The semimetals: (a) are members of the *p* block; (b) have primarily metallic properties; (c) tend to conduct electricity well; (d) include the element radon; (e) are relatively inert elements.

26. Metallic character tends to: (a) include brittleness; (b) increase to the upper left corner of the periodic table; (c) parallel nonmetallic character; (d) require diatomic molecules; (e) increase to the lower left corner of the periodic table.

27. Which of the following statements is true? (a) The melting points of the alkali metals increase with increasing atomic number; (b) The alkaline earth metals are harder and have higher melting points than the alkali metal of the same period; (c) Reaction of an alkali metal with water leads to formation of the corresponding metal hydride; (d) The chemical reactivity of the alkaline earth elements decreases with increasing period.

28. Indicate the type of outer electrons being filled in each region of the periodic tale.

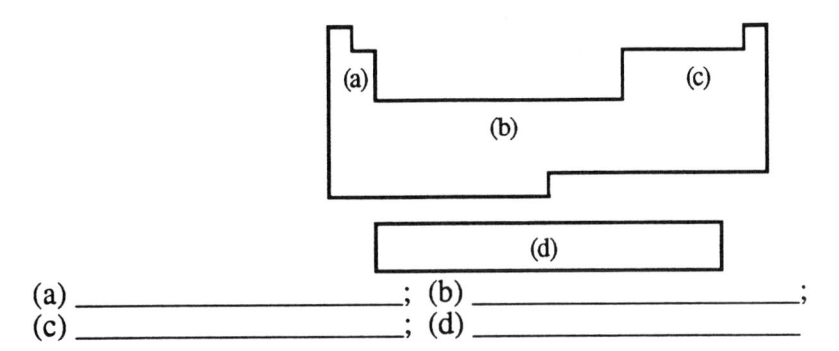

(a) _____ ; (b) _____ ;
(c) _____ ; (d) _____

29. If an element has an s^2p^5 electron configuration, it must be a: (a) halogen; (b) noble gas; (c) transition metal; (d) member of the oxygen family.

30. All members of the alkaline earth family, group 2A of the periodic table, possess: (a) two outer-shell s electrons; (b) two outer-shell p electrons; (c) two outer-shell s and two outer-shell p electrons; (d) none of these are correct

31. Write balanced chemical equations for the reactions that occur when: (a) Na_2O is dissolved in water; (b) Na metal is added to water.

Lecture Outline – Chapter 8
Brown, LeMay, & Bursten, *Chemistry: The Central Science*, 6*th* Edition

Chapter 8 — Basic Concepts of Chemical Bonding

OVERVIEW: Chapter 8 introduces bonding in chemical substances. It separates bonding into ionic bonding, covalent bonding, and metallic bonding, although it emphasizes the electrostatic nature of both. A detailed treatment of metallic bonding is deferred until Chapter 24. The discussion of ionic and covalent bonding is based on the octet rule. First, the common ions of the representative elements are described in terms of achieving the electron configuration of the "nearest" noble gas. The comparative sizes of the ions are also covered.

The rest of the chapter focuses on the covalent bond. It is first described as the sharing of electron pairs. To facilitate and strengthen that understanding, a full discussion of Lewis structures is given. The authors include not just single-bonded octet species, but also cover multiple-bonding and exceptions to the octet rule. Resonance is discussed as an extension to multiple bonding. Formal charge is introduced for those wishing to cover it.

Bond polarity is treated, although molecular polarity is held until Chapter 9, when molecular geometries are covered. Electronegativity is discussed within the framework of bond polarity.

The concept of bond dissociation energy is also covered. The relative strengths of the covalent bond are tied to the number of bonding pairs and the sizes of the atoms. The estimation of enthalpy changes for reactions based on bond energies is also discussed.

Finally, oxidation numbers are covered as a way to describe the effects of bond polarity on the character of the atoms in a compound.

LECTURE OUTLINE

I.   **The Ionic Bond, The Covalent Bond, and The Metallic Bond**

   A.   An *ionic bond* is the term given to the electrostatic (charge-based) attractive forces which hold oppositely charged ions together.

   B.   A *covalent bond* is the sharing of electrons between two atoms that acts to hold the atoms together.

   C.   A *metallic bond* is found in metals where the atoms of the metal are bound to several of their neighbors, holding the atoms together but allowing the electrons to move fairly freely.

II. **Lewis Symbols and the Octet Rule**

   A.   The valence or valency of an element is a measure of its ability to form bonds. Valence was originally defined to be the number of bonds the element could form with hydrogen, although it has been extended to include other atoms besides hydrogen.

   B.   *Valence electrons* are those that take part in chemical bonding. These electrons are the ones in the outer-most shell or the valence shell.

   C.   Gilbert N. Lewis devised symbols for the elements, called *Lewis symbols* or *electron-dot symbols*. Ignoring the transition elements, Lewis symbols place one dot for each valence electron around the symbol of the element. The dots are historically placed in one of four regions, keeping the electrons separated unless it is necessary to pair them (Hund's rule).

   D.   Lewis proposed the *octet rule*, which says that atoms *tend* to gain, lose, or share enough electrons to become surrounded by eight valence electrons (an octet).

Using the electron-dot symbols, when each of the four regions has two electrons in it (the Pauli exclusion principle), the total valence occupancy is eight. Note that the word "rule" is perhaps too strong; there is a *tendency* for atoms to gain an octet.

Discussion Question: Discuss why the octet rule seems to be so widely obeyed by chemical substances.

III. Ionic Bonding

A. The *ionic bond* is formed when ions of opposite charge, anions and cations, are attracted and held to one another by electrostatic attractions.

B. In order to maximize the attractions among ions, ionic solids exist in *lattices*, which are regularly repeating three-dimensional arrays of ions.

C. In order for an ionic bond (or any bond) to form the energetics must be favorable. The atoms must reduce their total energies when entering into the bond.

The *lattice energy* of an ionic substance is the energy required to separate the crystalline solid into the constituent gaseous ions. It is a measure of the stability of the crystalline state. Lattice energies tend to increase as the charges of the ions increase and as the sizes of the ions decrease.

D. The *representative elements* generally form ions that have noble-gas electron configurations. Thus, the metals lose one or more electrons in order to achieve the electron configuration of the immediately preceding noble gas, while nonmetals gain one or more electrons in order to achieve the electron configuration of the immediately following noble gas.

E. The *transition-metals* do not always form ions with the electron configurations of noble gases. In forming cations, the transition metals lose their outer-shell *s* electrons first, followed by *d* electrons, if necessary. Many stable ions are formed by emptying the *s* orbital or by leaving the *d* subshell full (d^{10}), empty (d^0), or half full (d^5).

F. *Polyatomic ions* are formed when molecules gain or lose electrons. It is often unclear which atom in the molecule has gained or lost electrons; however, the molecule as a whole has more or fewer electrons than are present in the neutral molecule.

Discussion Question: How do lattice energies relate to the observed melting points of ionic compounds?

IV. A Closer Look: The Born-Haber Cycle

Born-Haber cycles are graphical representations of the energy changes that accompany a chemical process. The cycles show the energy increases of some steps and the energy decreases of other steps. As such, the graph, using energy as the vertical axis, illustrates Hess's law.

V. Sizes of Ions

A. Recall that *atoms* increase in size going from right-to-left on a period and top-to-bottom in a family in the periodic table.

B. *Cations* are smaller than their parent atoms because the effective nuclear charge on the outer-most electrons is greater in the cation. The number of protons remains the same but the number of screening electrons decreases.

Lecture Outline – Chapter 8
Brown, LeMay, & Bursten, *Chemistry: The Central Science*, 6th Edition

- C. *Anions* are larger than their parent atoms because the effective nuclear charge on the outer-most electrons is smaller in the anion. The number of protons remains the same but the number of screening electrons increases.
- D. For ions of the *same charge*, size increases going down a family in the periodic table.
- E. *Isoelectronic series* are groups of atoms and ions which have the *same* electron configuration. Within isoelectronic series, the more positive the charge, the smaller the species and the more negative the charge, the larger the species.

Discussion Question: Given the coverage on the sizes of ions, what relative size should hydride, H-, have?

VI. Covalent Bonding

- A. *Electron-pair sharing* between two atoms, usually with one of the electrons from each atom, arises from the electrons being *simultaneously* attracted to both nuclei.
- B. A bond formed between two atoms by the sharing of electrons is called a *covalent bond*.
- C. Lewis proposed a way to depict covalent bonding in molecules using his electron-dot symbols of the elements. These structural formulas are called *Lewis structures*.
- D. *Multiple bonds* are formed when more than one pair of electrons is shared between the same two atoms.
    1. *Single bonds* are covalent bonds in which one pair of electrons is shared by the two atoms.
    2. *Double bonds* are covalent bonds in which two pairs of electrons are shared by the two atoms.
    3. *Triple bonds* are covalent bonds in which three pairs of electrons are shared by the two atoms.

Discussion Question: Under what conditions could a *quadruple bond* (four pairs of electrons) form? What type of element is most likely to participate in quadruple bonding?

VII. Bond Polarity and Electronegativity

- A. Ideally in a covalent bond, the bonding electrons are *equally shared* between the two atoms. The ionic bond can be viewed as the opposite extreme; in the ionic bond, the bonding electrons are *not at all shared* but rather are separated between the two ions. When sharing of the bonding electrons exists but is not an equal sharing, the bond is called a *polar covalent bond* or simply a *polar bond*. The case where the sharing is equal is then sometimes referred to as a *nonpolar* bond.
- B. *Electronegativity* is a calculated quantity that describes an element's ability to compete for electrons in a covalent bond; the higher an element's electronegativity, the better it competes for electrons.
- C. High electronegativity is associated with ease of adding an electron to the valence shell (a very negative electron affinity) and resistance to having an electron removed from the valence shell (a very positive ionization energy). Consequently, the elements of the upper-right in the periodic table have the highest electronegativities while those in the lower-left have the lowest electronegativities.

D. The electronegativity *difference* between the two atoms of a bond is related to the polarity of the bond formed. The greater the electronegativity difference, the more polar the bond.

Discussion Question: Some sources suggest that, if the difference in electronegativities is greater than 2.0, the bond should be considered ionic, while if the electronegativity difference is less than 0.5, the bond should be considered nonpolar. Electronegativity differences between those two limits should then describe polar bonds. Comment on how appropriate such a scheme is.

VIII. Drawing Lewis Structures

Lewis structures for covalent molecules or for polyatomic ions (which are held together covalently) are constructed using a fairly simple set of rules.
1. Sum the *valence electrons* from all atoms in the species.
2. Write the atomic symbols for the atoms involved so as to show which atoms are connected to which.
   Draw a *single* bond between each pair of bonded atoms.
3. Complete the octets of the atoms bonded *to* the central atom (the peripheral atoms).
4. Place any leftover electrons on the central atom, even if it results in the central atom having more than an octet.
5. If there are *not enough electrons* to give the central atom an octet, form multiple bonds by pulling terminal electrons from a peripheral atom and placing them into the bond with the central atom.

IX. A Closer Look: Formal Charge and Lewis Structures

A. *Formal charges* are a way of assigning all the valence electrons in a molecule to "parent" atoms. Formal charges can be simply assigned.
   1. All *bonding* electrons are divided equally between the atoms that form bonds.
   2. All *nonbonding* electrons are assigned to the atom on which they reside.
B. The formal charge is then the number of valence electrons in the isolated atom (usually the group number in the periodic table) minus the number of electrons assigned by the rules above.
C. When several different Lewis structures seem plausible, the one in which the *formal charges are minimized* is generally the preferred one.

X. Resonance Structures

A. There are times when more than one Lewis structure involving multiple bonds seems equally stable, such as O=C=O vs. O≡C–O vs. O–C≡O. These structures differ only in the placement of the electrons. Satisfactory Lewis structures for the same substance that differ *only* in the placement of electrons are called *resonance forms*.
B. Resonance forms rapidly interconvert so that it is seldom possible to measure the individual structures; the structure appears to be a *blend* of all the forms.

Discussion Question: Why should SO_2 show two equivalent S–O bonds at room temperature, but show two different S–O bonds at low temperatures?

XI. Exceptions to the Octet Rule

A. Although most of the second-period elements, most notably carbon, nitrogen, and oxygen, are always observed with octets, other elements do not reliably achieve octets, while still others rarely achieve octets.

B. There are instances of molecules that contain an *odd number of electrons*, although these are uncommon and tend to be reactive.

C. Light elements (H, Li, Be, B) tend to be surrounded by *less than an octet of electrons*. Hydrogen only surrounds itself with two electrons (a duet) in molecules, due to the valence-shell capacity of $n = 1$ being two electrons. Lithium and the other active metals rarely form covalent molecules, making the octet rule rather meaningless. Beryllium often is surrounded by four electrons and boron is often surrounded by six electrons, due mostly to their small size. Formal charge assignments generally support these deviations from the octet rule.

D. A larger group of compounds are those in which the central atom is surrounded by *more than an octet of electrons*. Elements of the third period or lower in the periodic table are capable of *expanding* their octets, due to the availability of d orbitals for bonding.

XII. Strengths of Covalent Bonds

A. The enthalpy required to break a covalent bond is called the *bond-dissociation energy* or the *bond energy*. Because the same bond, such as C–H, occurs in a wide range of molecules, the accepted value of the bond energy is an average of the enthalpies observed in each of the different molecules. Bond energies are, therefore, *average* enthalpies.

B. Inasmuch as most reactions can be described as a series of bond breakings and bond formations, *bond energies relate well with the enthalpies of reactions*. To a good approximation, then, $\Delta H_{rxn} = \Sigma$(bond energies of bonds broken) - Σ(bond energies of bonds formed).

C. Between atoms of comparable size, the *greater the bond strength*, the *shorter the bond length*.

XIII. Chemistry at Work: Explosives

A. *Explosives* are liquid or solid substances that fit the following criteria.
   1. They decompose very *exothermically*.
   2. The *products of the decomposition must be gases*, so that a tremendous pressure is generated by the reaction.
   3. The decomposition must occur very *rapidly*.

B. *Explosive balance* is the situation in which a substance satisfies the three criteria above, requires little energy to detonate, requires no other substances for the reaction, and produces only N_2, CO_2, and H_2O. These products have very strong bonds and so the enthalpy is extremely negative. Nitroglycerin, for example is nearly in a perfect explosive balance; it deviates only in that a small amount of O_2 is formed.

XIV. Oxidation Numbers

A. Oxidation numbers or oxidation states arise from an arbitrary assignment of the bonding electrons to the more-electronegative elements.
   1. The oxidation number of an element *in its elemental form* is zero.
   2. The oxidation number of a *monatomic ion* is the same as its charge.
   3. In binary compounds, the element with the *greater electronegativity* is assigned a negative oxidation number equal to its common anion charge.

4. The *sum* of the oxidation numbers equals zero for an electrically neutral compound and equals the net charge for a polyatomic ion.

Given these rules, the oxidation numbers of the remaining elements in a compound are deduced from the negative assignments and the net charge of the species.

B. Oxidation numbers form the basis of a more extensive set of nomenclature rules. Rather than using numerical prefixes, this system uses the oxidation number of the less-electronegative element in the name. For example, PCl_5 is commonly called phosphorus pentachloride. In this new scheme, phosphorus is in a +5 oxidation state, to balance the five Cl^-, and the name becomes phosphorus(V) chloride. This system is very similar to the Stock notation used for transition metal compounds, *e.g.*, $Fe(CN)_2$ is named iron(II) cyanide.

XV. A Closer Look: Binary Oxides

A. *Binary oxides* are compounds of oxygen that contain one other element.

B. *Basic oxides* are those which produce basic solutions when dissolved in water. The basic oxides tend to be ionic solids, consisting of a metal with oxygen, and have high melting points.

C. *Acidic oxides* are those which produce acidic solutions when dissolved in water. The acidic oxides tend to be molecular compounds, consisting of a nonmetal with oxygen, and have low melting points. They may indeed be liquids or gases under normal conditions. When two nonmetals have the same oxidation state, the acidity of the oxides increase with increasing electronegativity of the element. For a given nonmetallic element, the acidity of its oxides increases with increasing oxidation state of the element.

D. *Amphoteric oxides* are those that behave as acids or bases, depending on their environment. Amphoteric oxides are usually insoluble in water, but are soluble in acids or bases. The element paired with oxygen tends to be a semimetal.

SAMPLE QUIZ QUESTIONS

1. Write the electron configuration for: (a) Mn^{2+}; (b) Cr^{3+}; (c) S^{2-}.

2. Predict the formula of the compound formed between lithium and phosphorus.

3. Explain briefly why MgO has a greater lattice energy than KCl.

4. Ionic compounds: (a) are held together by the attraction of electrons by both nuclei; (b) tend to have fairly low melting points; (c) are composed of isolated molecules; (d) are composed of orderly arrays of ions; (e) are generally composed of nonmetals.

5. A crystal lattice: (a) is a three-dimensional pattern of particles in solids; (b) is irregular in its geometry; (c) makes a solid compound easy to boil; (d) ensures that ionic compounds have net charges other than zero; (e) exists only for ionic compounds.

6. When sodium reacts with chlorine, what is oxidized and what is reduced?

7. When magnesium undergoes chemical reactions, it generally: (a) gains two electrons; (b) loses two electrons; (c) shares two electrons; (d) becomes a different element.

8. Which of the following ions is smallest? (a) F^-; (b) Na^+; (c) Mg^{2+}; (d) Al^{3+}

9. Ionic compounds: (a) may have a net charge other than zero; (b) are held together by the attraction of oppositely charged ions in a random arrangement; (c) often are composed of molecules; (d) are held together by the attraction of electrons to two nuclei; (e) are held together by the attraction of oppositely charged ions arranged in a regular pattern.

10. Na^+: (a) is smaller than Na; (b) is larger than Na; (c) has the same electron configuration as Ar; (d) is an anion; (e) behaves like a noble gas.

11. In LiF: (a) Li^+ is larger than F^-; (b) both atoms achieve an octet by converting to ions; (c) Li becomes an anion while F becomes a cation; (d) the bonding is covalent; (e) LiF exists as isolated molecules.

12. Molecular compounds: (a) may have a net charge other than zero; (b) are held together by the attraction of oppositely charged ions in a random arrangement; (c) often are composed of molecules; (d) are held together by the attraction of electrons to two nuclei; (e) are held together by the attraction of oppositely charged ions arranged in a regular pattern.

13. Write the Lewis structures for: (a) H_2S; (b) NaBr.

14. Which of the following electron dot formulas is not correct?

(a) :Ï-Cl:

(b) H-C-Cl with :Cl: above and :Cl: below

(c) :F-S-F:

(d) :O=O:

(e) :H-O-H:

15. Which of the following is a correct electron dot formula for SiBr$_4$?

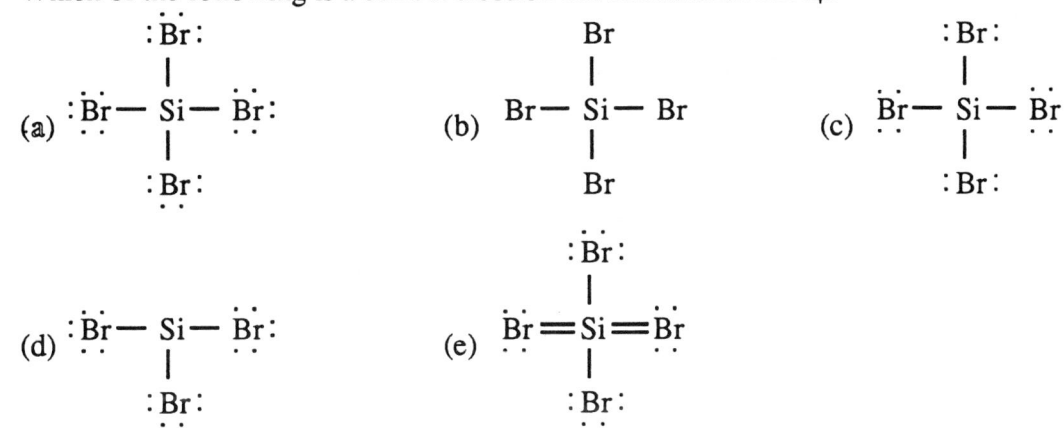

(a) Br—Si—Br with Br above and Br below (all Br with 3 lone pairs)

(b) Br—Si—Br with Br above and Br below (no lone pairs shown)

(c) Br—Si—Br with Br above and Br below (mixed lone pairs)

(d) Br—Si—Br with Br below only (three Br shown)

(e) Br=Si=Br with Br above and Br below (double bonds)

16. Which of the following electron dot formulas is correct?

(a) $[:N\equiv O:]^-$ (b) $[:O=N=O:]^-$ (c) $[:N\equiv O:]^+$

(d) $[:O=O:]^{2-}$ (e) $[:O=O:]^{4-}$

17. Which of the following is a correct electron dot formula for PO$_2{}^{3-}$?

(a) $[:\ddot{O}-\ddot{P}=\ddot{O}]^{3-}$ (b) $[:\ddot{O}-P-\ddot{O}:]^{3-}$ (c) $[:O=P=O:]^{3-}$

(d) $[:\ddot{O}-\ddot{P}-\ddot{O}:]^{3-}$ (e) $[:O=\ddot{P}-\ddot{O}:]^{3-}$

18. Draw the Lewis resonance structures for the nitrite ion, NO$_2{}^-$.

19. Covalent bonds: (a) are not stable; (b) always involve two electrons; (c) must contain one electron from each atom; (d) are formed by sharing electrons between two nuclei; (e) are usually found in bonds between two metals.

20. Using tabulated bond energies, estimate ΔH for the following reaction:

$$H-\underset{\underset{H}{|}}{\overset{\overset{H}{|}}{C}}-O-H + O=O \rightarrow H-\overset{\overset{O}{\|}}{C}-O-H + H-O-H$$

·21. Which of the following compounds is least ionic in nature? (a) K$_2$S; (b) Sb$_2$S$_3$; (c) P$_2$S$_3$; (d) SnS

22. Which of the following bonds is the most polar? (a) Be–B; (b) P–N; (c) Te–I; (d) Y–O; (e) H–C

23. Which set of oxidation numbers is correct?

    (a) Co(CN)₃   (b) KMnO₄   (c) MgS₂O₃   (d) PbCr₂O₇   (e) Cs₃PO₄
       +3 +2 -3   +1 +7 -3   +1 +3 -2   +4 +7 -2   +1 +3 -2

24. What is the oxidation state of nitrogen in each of the following? (a) N_2; (b) NO; (c) NH_4^+

25. The oxidation state of chlorine in $NaClO_3$ is: (a) -1; (b) +3; (c) +5; (d) +7

26. Name the following compounds using oxidation numbers: (a) SF_6; (b) TeO_3

Lecture Outline – Chapter 9
Brown, LeMay, & Bursten, *Chemistry: The Central Science, 6th Edition*

Chapter 9 — Molecular Geometry and Bonding Theories

OVERVIEW: This chapter covers the geometries of molecules, and the polarity that geometry can impart on a molecule, as well as the more theoretically rigorous views of covalent bonding, valence bond theory and molecular orbital theory.

The valence-shell electron-pair repulsion model is covered first in order to give students a feeling for the three-dimensional nature of molecules. The model is first built using the regular tetrahedron of CCl_4 to illustrate the features of shape. All electron-pair geometries from two regions of electrons (linear) through six regions of electrons (octahedral) are discussed, as are the relative repulsions of single bonds, unbonded pairs, and multiple bonds.

With this appreciation for molecular geometry, bond polarity is revisited. In this treatment, the arithmetic dipole moment is introduced, along with percent ionic character. The focus, however, is the influence of geometry on overall polarity.

The remainder of the chapter deals with covalent bonding theories, beginning with the valence bond theory. The concepts of overlap, electron probability, and σ and π symmetry are fully developed. The most important construct of valence bond theory is hybridization, which is explained in detail. Orbital energy-level diagrams are used to support the discussions on promotion and hybridization. The agreement in the predicted geometries between VSEPR and valence bond is emphasized. The use of *d* orbitals and the hybridized orbital sets they contribute to are also discussed. Delocalized bonding in aromatic and in conjugated π systems completes the coverage of valence bond theory.

Molecular orbital theory is the most rigorous bonding model usually given in a general chemistry text. MO is introduced as a corollary of the atomic orbitals of wave mechanics. The general properties of molecular orbitals are discussed using H_2 as a model. Energy-level diagrams are used throughout. The nature of π bonding is covered extensively, including the inversion of the σ_{2p} and π_{2p} energies in the second-row diatomic molecules. The molecular orbital energy-level diagram for O_2 gives an excellent opportunity to discuss paramagnetism and diamagnetism, and to explain why molecular orbital theory is necessary to explain some observations. The chapter closes with a section on organic dyes that emphasizes the relationship between conjugated π bonds and the magnitude of the HOMO-LUMO energy gap.

LECTURE OUTLINE

I. **Molecular Geometries**
   A. The *geometry* of a molecule, along with its size, determines in large part its chemical behavior.
      1. One of the most common geometries, especially in organic chemistry, is the *tetrahedron*. Tetrahedra have four corners and four faces, each of which is an equilateral triangle. The central atom is bonded to four peripheral atoms, each of the four at a corner of the tetrahedron.
      2. A *bond angle* is the geometric angle formed by the bonds of two peripheral atoms with the central atom. In a tetrahedron, the bond angles are 109.5°.
      3. The geometry of molecules, especially nonmetallic molecules, can be predicted using *VSEPR* (*valence-shell electron-pair repulsion model*).
   B. The *valence-shell electron pair-repulsion* (*VSEPR*) model is an approach to predicting geometries that considers how many electron pairs need to exist around the central atom.

Lecture Outline – Chapter 9
Brown, LeMay, & Bursten, *Chemistry: The Central Science*, 6th Edition

       1. The best arrangement of a given number of electron pairs is the one that *maximizes the separation* among them.
       2. In maximizing the separation, the *repulsions among the electron pairs are minimized*.
       3. The electron pairs are differentiated as being *bonding electron pairs* or *nonbonding electron pairs* (*lone pairs*).
       4. For the purposes of VSEPR, a *multiple bond counts as a single bonding pair* or as a single region of electrons.
       5. For any molecule, the *electron-pair geometry* is the geometric figure described *by the regions of electrons* around the central atom. Single bonds, nonbonding electron pairs, and multiple bonds each count as a region.
       6. For any molecule, the *molecular geometry* is the geometric figure described *by the central atom and the peripheral atoms to it*.

    C. When there are *four or fewer valence-shell electron pairs* around a central atom, the resultant electron-pair geometries are: linear for two regions of electrons (180°), trigonal planar for three regions of electrons (120°), and tetrahedral for four regions of electrons (109.5°). Several molecular geometries can occur, depending on whether any of the electron pairs are nonbonding.

    D. Nonbonding electrons and multiple bonds affect observed bond angles.
       1. *Nonbonding electron pairs* repel other electron pairs more than do single-bonding electron pairs.
       2. *Electrons in multiple bonds* repel other electron pairs more than do single-bonding electron pairs.
       3. The effective order of the magnitude of repulsion by electron pairs is *single bonds < unbonded electron pairs ≈ multiple bonds*.

    E. Molecules with *expanded valence shells* have geometries that are also predicted by VSEPR.
       1. With five regions of electrons around the central atom, the electron-pair geometry is a *trigonal bipyramid*. The trigonal bipyramid has a three-member *equatorial plane* in which the electron pairs are separated by 120°. The remaining two electron pairs, the *axial positions*, lie above and below the equatorial plane such that the bond angle formed from an axial position to the central atom to any of the equatorial positions is 90°. Molecular geometries derived from the trigonal bipyramid include the *trigonal bipyramid*, the *see-saw* (or asymmetric tetrahedron), and the *T-shape*.
       2. With six regions of electrons around the central atom, the electron-pair geometry is an *octahedron*. The octahedron has a four-member *equatorial plane* in which the electron pairs are separated by 90°. The remaining two electron pairs, the *axial positions*, lie above and below the equatorial plane such that the bond angle formed from an axial position to the central atom to any of the equatorial positions is 90°. As such, all six peripheral positions are equivalent. Molecular geometries derived from the octahedron include the *octahedron*, the *square pyramid*, and the *square plane*.

    F. When molecules have no single central atom, a geometry is determined for *each* central atom, rather than trying to describe the overall geometry of the molecule.

Discussion Question: There are a number of tasks and devices that we can use with one hand but not the other. Have students suggest some of these things and discuss them.

II. Polarity of Molecules

A. A *polar molecule* is one in which the centers of positive and negative charge do not coincide. Viewed as a whole, then, a polar molecule has one end with a slight negative charge and one end with a slight positive charge.
   1. Any diatomic molecule with a polar bond must be polar.
   2. A molecule's *dipole moment* is a measure of the polarity. The dipole moment, μ, is given by the charge at either end of the dipole, Q, and the distance that separates the charges, r. ($\mu = Qr$)
   3. The common unit of dipole moment is the *debye*, D. (= 3.33×10^{-30} C-m)

B. The *polarity of polyatomic molecules* is a function of a molecule's geometry and the polarity of its bonds. If all the individual bond polarities cancel, the molecule as a whole is nonpolar. If the individual bond polarities do not cancel, the molecule as a whole is polar.
   1. Determining whether the individual bond polarities cancel is a *vector addition* problem. Vectors are quantities that have both a *magnitude component* and a *direction component*. Each bond polarity has a magnitude, given by the dipole moment of the bond, and a direction, given by the geometry of the molecule. In order for the molecule to be nonpolar, the sum of the individual dipole moments must be zero.
   2. For *binary compounds*, if the molecule is highly symmetrical, as in CH_4, PCl_5, or SF_6, the molecule will be nonpolar. If, on the other hand, the molecule lacks high symmetry, particularly if it has asymmetrical unbonded electron pairs, it will be polar. Note that the square plane (six regions of electrons, two nonbonding regions) and the linear (five regions of electrons, three nonbonding regions) are nonpolar because the unbonded pairs are symmetrical.

Discussion Question: Even though it has not yet been formally covered, ask students how polarity might affect the ability of a solvent to dissolve a solute and the ability of a substance or mixture to conduct electricity.

III. Covalent Bonding and Orbital Overlap

A. *Valence bond theory* is a description of covalent bonding that combines Lewis's formulation and the atomic orbital idea of wave mechanics.

B. A covalent bond forms when an atomic orbital of one atom merges, or coexists in the same space, with an atomic orbital of another atom. This merging of atomic orbitals is called *orbital overlap* and the resulting covalent bonding orbital is called a *valence bond orbital*.

C. As with atomic orbitals, the valence bond orbital between two atoms is a region of *high probability* of finding the electron. There is also an optimum distance of separation between the two nuclei, called the *bond length*. As the nuclei approach one another, the attractive forces between the electrons and both nuclei increases while the repulsive forces between the nuclei increases. The bond length is the distance of separation at which the *total energy* is minimized.

IV. Hybrid Orbitals

A. Orbital overlap does not address the issues of geometry. Atomic orbitals do not lie in the proper directions to form most observed geometries. *Hybridization* is the process of mathematically mixing two or more atomic orbitals, *on a single atom*, giving rise to a set of blended orbitals called *hybrid orbitals*. The number of hybrid orbitals formed is *always* the same as the number of atomic orbitals used.

B. *sp* hybrid orbitals are formed from the mixing of an *s* orbital and a *p* orbital, generally with the same value of the principle quantum number, *n*. The arrangement of the two *sp* hybrid orbitals is linear, with a 180° angle between them.

C. sp^2 hybrid orbitals are formed from the mixing of an *s* orbital and two *p* orbitals. The arrangement of the three sp^2 hybrid orbitals is trigonal planar, with a 120° angle between them.

D. sp^3 hybrid orbitals are formed from the mixing of an *s* orbital and three *p* orbitals. The arrangement of the four sp^3 hybrid orbitals is tetrahedral, with a 109.5° angle between them.

E. Hybridization can also involve *d* orbitals. The five dsp^3 (or sp^3d) hybrid orbitals are arranged in a trigonal bipyramid, with bond angles of 120° among the three equatorial positions and 90° from either axial position to the equatorial plane. The six d^2sp^3 (or sp^3d^2) hybrid orbitals are arranged in an octahedron, with bond angles of 90° between any two positions.

F. Hybridization explains the directions in which the bonds point but does not address the equivalence of bonds. For example, carbon makes four bonds but has only two unpaired electrons in the free atom, suggesting that it can make only two bonds. The solution to the dilemma is *promotion*, a process by which an electron pair is separated into two unpaired electrons. Because two unpaired electrons cannot exist in the same orbital, one of them is *promoted* into a higher-energy orbital.

G. The valence bond theory cannot predict geometries; it can only explain the geometry that is observed. Consequently, the VSEPR geometry correlates perfectly with the hybridization. To summarize the steps in determining the hybridization in a molecule:
   1. Draw the Lewis structure for the molecule or polyatomic ion.
   2. Determine the electron-pair geometry using the VSEPR model.
   3. Specify the hybrid orbitals needed to accommodate the electron pairs based on their geometric arrangement.

V. **Multiple Bonds**

A. The imaginary line that passes through both nuclei is called the *internuclear axis*.

B. In the language of valence bond theory, *σ bonds* are those in which the electron density is circularly symmetrical to the internuclear axis.

C. In the language of valence bond theory, *π bonds* are those in which the electron density is above *and* below the internuclear axis. The internuclear axis is a region of zero electron density.

D. Any region around a molecule, except the outer-most edges, where there is zero electron density is called a *node*. In the case of a π bond, the node is actually a plane, or *nodal plane*, which coincides with the internuclear axis and is perpendicular to the two lobes of the π bond.

E. The *extent of overlap* tends to be greater in σ bonds than in π bonds. Consequently, σ bonds tend to be *stronger* than π bonds.

F. In general, single bonds are σ bonds. Double bonds consist of a σ bond and a π bond while triple bonds consist of a σ bond and *two* π bonds. Multiple bonding is more common with small atoms, especially C, N, and O.

G. In many hybridizations, there are left-over unhybridized orbitals. In the *sp* hybrid, two *p* orbitals are left unhybridized. In the sp^2 hybrid, one *p* orbital is left unhybridized. In the *sp³d* (or *dsp³*) hybrid, four *d* orbitals are left unhybridized. In the *sp³d²* (or *d²sp³*) hybrid, three *d* orbitals are left unhybridized. These left-over unhybridized orbitals are the orbitals that are available for π bonding.

- H. π bonds are generally formed between unhybridized *p* orbitals on the two atoms involved, with two regions of electron density on opposite sides of the internuclear axis. Because the ability to overlap requires that the two *p* orbitals lie parallel to each other, π bond formation freezes the geometry and rotation about the bond is not possible. Double bonds consist of one σ bond and one π bond; triple bonds consist of one σ bond and two π bonds.

- I. The π bonds discussed to this point have been *localized*. That is to say that the π electron density is entirely associated with two atoms and their bond.
    1. *Localized π bonding* is the normal case, such as in ethylene (C_2H_4), in which the π electrons must exist as part of a given bond.
    2. When double bonds exist in a molecule such that the double bonds can occur at several locations, *delocalized π bonding* can occur. In delocalized bonding, the electrons can migrate among the π regions available to them. The effect is that the adjacent single and double bonds appear to be blended into an intermediate form.
    3. A peculiar condition exists when the alternating double and single bonds are in a cyclic arrangement with an even number of vertices. When six π electrons are involved on a six-member carbon ring, it is called *aromaticity*. Aromatic compounds, based on benzene, C_6H_6, show unusual stability, failing to react or doing so very slowly in reactions that other double bonded species react in very rapidly.

- J. In general, valence bond theory describes the bonding in molecules very well.
    1. Every pair of bonded atoms shares one or more pairs of electrons.
    2. The electrons in σ bonds are localized in the region between two bonded atoms.
    3. When atoms share more than one pair of electrons, the additional pairs are in π bonds.
    4. Electrons in π bonds that extend over more than two atoms are said to be delocalized.

VI. Chemistry and Life: The Chemistry of Vision

- A. The human eye interprets light and color by a series of chemical transformations. The eye contains *photoreceptors* known as *rods* and *cones*.
    1. The rods are sensitive to dim light and are used in night vision.
    2. The Cones are sensitive to colors.

- B. The receptors contain a complex molecule called *rhodopsin*. It consists of a protein (opsin) and a pigment (retinal). Light is registered by changes in the bonding in retinal and conveyed to the brain electrically.

- C. Our current knowledge of the physiology of sight is still very limited.

VII. Molecular Orbitals

- A. Molecular orbital theory explains *why* covalent bonds form in terms of *energy*. *Molecular orbitals* result from performing wave mechanical calculations on molecules. The most straightforward approach to molecular orbitals is to consider the combination of atomic orbitals from each of the atoms of the bond. Molecular orbitals have many of the same properties as atomic orbitals. They can contain up to two electrons with paired spins. They are best visualized using density contour diagrams. Whenever two atomic orbitals interact, two molecular orbitals are formed.

Lecture Outline – Chapter 9
Brown, LeMay, & Bursten, *Chemistry: The Central Science, 6th Edition*

- B. The general features of molecular orbital theory can be illustrated by the hydrogen molecule.
    1. The first molecular orbital is formed by combining the 1s atomic orbitals so that the electron density is concentrated between the nuclei. An electron here is strongly attracted to both nuclei. This molecular orbital is called a *bonding molecular orbital*.
    2. Another molecular orbital is formed by combining the 1s atomic orbitals so that the electron density is distributed away from the space between the nuclei. An electron here does *not* promote bonding because a *node* exists between the two nuclei. This molecular orbital is called an *antibonding molecular orbital*.
    3. The molecular orbitals formed from the 1s atomic orbitals are symmetrical with respect to the internuclear axis and so are *sigma (σ) molecular orbitals*. A molecular orbital is commonly labeled with the atomic orbital that formed it. For H_2, the bonding σ molecular orbital is σ_{1s} while the antibonding σ molecular orbital is σ^*_{1s}.
    4. The relationship among the two atomic orbitals and the two molecular orbitals is often shown with an *energy-level diagram* or *molecular orbital diagram*. On these diagrams, the atomic orbitals are shown on the right and the left and the molecular orbitals are shown in the middle. The vertical direction is an energy scale, so the higher on the diagram an orbital is drawn, the higher its energy is.
    5. Like atomic orbitals, molecular orbitals can accommodate two electrons of *opposite spin*. The bonding molecular orbital is lower in energy than either of the contributing atomic orbitals. Electrons in the bonding molecular orbital are called *bonding electrons*.
    6. The antibonding molecular orbital is higher in energy than either of the contributing atomic orbitals. Electrons in the antibonding molecular orbital are called *antibonding electrons*.
- C. Molecular orbital theory treats the molecule and does not identify particular bonds. Consequently, it is not proper to speak of single bonds, double bonds, etc. Instead, molecular orbital theory defines the *net bonding* present in the molecule, called the *bond order*. Usually, bond order agrees with the Lewis structure.

$$\text{Bond Order} = \frac{1}{2}(\text{number of bonding e}^- - \text{number of antibonding e}^-)$$

VIII. Second-Period Diatomic Molecules

- A. Molecular orbital theory can be applied to very large molecules with several elements present. In this coverage, only the simple *homonuclear diatomic* molecules of the second period are discussed. These molecules have two atoms (diatomic) of the same element (homonuclear). We must consider some rules for constructing molecular orbital diagrams:
    1. The number of molecular orbitals formed equals the number of atomic orbitals combined.
    2. Atomic orbitals combine most effectively with other atomic orbitals of similar energy.
    3. The effectiveness with which two atomic orbitals can combine is proportional to their overlap with one another.
    4. Each molecular orbital can accommodate as many as two electrons, with paired spins.
    5. When molecular orbitals have the same energy, one electron enters each orbital with parallel spin before pairing occurs (Hund's rule).

B. Because lower-level orbitals are invariably full, molecular orbital diagrams generally show only the valence shell atomic orbitals and electrons. Dilithium and diberyllium need only utilize the 2s atomic orbitals and so, qualitatively, the molecular orbital diagram for these looks like that of H_2. Li_2 has two electrons in the σ_{2s} orbital and thus has a bond order of one. Be_2 has two electrons in the σ_{2s} orbital and two electrons in the σ^*_{2s} orbital and thus has a bond order of zero.

C. The other members of the series utilize the 2p atomic orbitals. As in the other bonding pictures examined, there are two orientations in which p orbitals can interact. One orientation is for the p_z orbitals to overlap end-to-end, which gives rise to a σ molecular orbital. The other orientation is for the p orbitals to lie parallel to each other, the overlap being between the individual lobes of the p orbitals on each atom, and giving rise to a π molecular orbital. There are two π interactions, p_x and p_y, perpendicular to each other and to the internuclear axis. As before, each of these interactions forms a bonding molecular orbital *and* an antibonding molecular orbital.

D. The molecular orbital diagrams for B_2 through Ne_2 utilize the 2p orbitals. The energy order, based on the extent of overlap, is $\sigma_{2s} < \sigma^*_{2s} < \sigma_{2p_z} < (\pi_{2p_x}, \pi_{2p_y}) < (\pi^*_{2p_x}, \pi^*_{2p_y}) < \sigma^*_{2p_z}$. This turns out to be true for O_2, F_2, and Ne_2. For B_2, C_2, and N_2, the π_{2p} orbitals are lower in energy than the σ_{2p} orbital because of an interaction, called *mixing*, between the σ_{2s} and the σ_{2p}. The bond order increases from one at B_2 to three at N_2 and decreases again to zero at Ne_2.

E. The electron configurations, or the placements of the electrons in the molecular orbitals relates with molecular properties of the species, particularly the *magnetic* properties.
   1. *Paramagnetism* is the interaction of a substance with a magnetic field due to *unpaired electrons* in the substance. Oxygen is paramagnetic and the liquid actually will form a bridge between the poles of a magnet. This result is unexpected from the Lewis structure and from valence bond theory. Molecular orbital theory, however, correctly predicts the existence of two unpaired electrons in the π^*_{2p} orbitals.
   2. *Diamagnetism* is the interaction of a substance with a magnetic field due to *paired electrons* in the substance. Because all substances, except hydrogen ion, contain paired electrons, all substances, except hydrogen ion, show a slight interaction with a magnetic field due to diamagnetism.

IX. Chemistry at Work: Organic Dyes

Light excites electrons in molecules. Within the framework of molecular orbital theory, an electron in a lower-energy molecular orbital can be excited into a higher-energy empty molecular orbital. Organic dyes, or coloring agents, are designed so that this energy gap corresponds to the energy of the color that is complimentary to the color desired.
   1. The molecular orbital of highest energy which has an electron in it is called the highest occupied molecular orbital or HOMO.
   2. The molecular orbital of lowest energy which is empty is called the lowest unoccupied molecular orbital or LUMO.
   3. When there are several double bonds in a molecule, formally occurring every second bond, the situation is called conjugated π bonding. The more double bonds that are conjugated, or the greater the conjugation, the closer together the energy-spacing of the molecular orbital. Consequently, as the number of conjugated double bonds increases, the energy of light required to cause a HOMO-to-LUMO transition decreases. When that energy corresponds to visible light, the compound appears colored to our eyes.

Discussion Question: What is the relationship between the color of light absorbed and the color of light transmitted? Students with art experience can get this discussion going.

SAMPLE QUIZ QUESTIONS

1. What is meant by a delocalized orbital?

2. Why is the C–C bond in acetylene, C_2H_2, so much shorter than that in ethane, C_2H_6?

3. Rationalize, using VSEPR theory, the observation that the bond angle in NH_3 is 107° while that in H_2O is smaller, 104°.

4. Predict the geometries of the following molecules: (a) SeO_3; (b) XeF_2.

5. Show, using drawing, how *p* orbitals can interact to form a pi-bond.

6. What is the hybridization used by the central atom in the following molecules? (a) OF_2; (b) PCl_5

7. Which exerts a greater electrostatic repulsion on neighboring electron pairs, an unshared electron pair or a bonding electron pair? Explain.

8. What hybrid orbital set is appropriate to a central atom that has three bonding electron pairs and two unshared electron pairs in its valence shell? (a) $sp + p^2$; (b) sp^3; (c) dsp^3; (d) sp^2

9. Which of the following molecules is nonpolar? (a) SO_2; (b) SO_3; (c) NO; (d) H_2O

10. Using MO theory, explain why the peroxide ion, O_2^{2-}, has a longer O–O distance than does the O_2 molecule. Is the peroxide ion paramagnetic or diamagnetic?

11. Which of the following elements exists in an allotropic form having four atoms bound in a tetrahedral-shaped molecule? (a) S; (b) C; (c) P; (d) Cl

12. The 109° bond angle is characteristic of which of the following hybridizations? (a) *sp*; (b) sp^2; (c) sp^3; (d) d^2sp^3

13. Which of the following bond angles is characteristic of sp^2 hybridization? (a) 180°; (b) 120°; (c) 109°; (d) 90°

14. Which of the following molecules possesses two π bonds and a σ bond? (a) N_2; (b) O_2; (c) Cl_2; (d) C_2

15. What distinguishes a σ molecular orbital from a π molecular orbital?

16. The bond order of NO, as predicted by MO theory, is: (a) 1; (b) 1.5; (c) 2; (d) 2.5.

17. When atomic orbitals are combined to form molecular orbitals, the number of molecular orbitals: (a) is equal to the number of atomic orbitals; (b) is twice the number of atomic orbitals; (c) is half the number of atomic orbitals.

18. Consider the following list of elements: He, H, C, N, P, S, O. From this list, identify the element that
    (a) consists of diatomic molecules with triple bonds. _____
    (b) forms highly reactive four-membered tetrahedra. _____
    (c) is found in nature as isolated atoms. _____

(d) forms molecules consisting of eight-membered rings. _____
(e) has an allotrope that consists of sheets of atoms, each possessing sp^2 hybridization. _____

19. The shape of PCl_3 is: (a) trigonal; (b) a tetrahedron; (c) bent; (d) linear; (e) a pyramid.

20. The shape of ClO_4^- is: (a) bent; (b) linear; (c) a tetrahedron; (d) trigonal; (e) a pyramid.

21. The shape of H_2Se is: (a) bent; (b) linear; (c) a tetrahedron; (d) trigonal; (e) a pyramid.

23. Which of the following bonds is the least polar: (a) S–F; (b) In–P; (c) B–C; (d) Na–Cl; (e) Cs–Br?

24. Which of the following compounds is polar: (a) BF_3; (b) H_2Te; (c) CBr_4; (d) CO_2; (e) N_2?

25. Which of the following compounds is not polar: (a) O_3; (b) SiO_2; (c) NCl_3; (d) OF_2; (e) $KClO$?

Chapter 10 — Gases

OVERVIEW: Chapter 10 deals with gases and their physical properties. Beginning with a discussion of how gases differ from solids and liquids, pressure is introduced as an important condition for a sample of a gas.

The gas laws, including Boyle's law, Charles's law, and Avogadro's law, are all detailed from experiment. The gas laws themselves are developed from the graphs.

The ideal-gas equation is introduced as a combination of the gas laws. It is then used to determine molar mass and gas density. Partial pressures and Dalton's law of partial pressures are also approached from the ideal-gas equation.

As reinforcement of stoichiometry, the ideal gas law is used to determine amounts of gases involved in chemical reactions.

In order to put a theoretical basis in the gas laws, the kinetic-molecular theory is fully developed. The kinetic-molecular theory is then used to discuss effusion, diffusion, and deviations from ideality.

LECTURE OUTLINE

I. **Characteristics of Gases**

   A. A *Gas* is a substance that expands to fill its container and attain the container's shape and is highly compressible. The common gases are all nonmetallic, have simple molecular formulas, and have low molar masses. Only substances that are gaseous under normal conditions of temperature and pressure are called gases.

   B. When a substance that is liquid or solid under normal conditions exists as a gas, the gas is called a *vapor*. Water, for example, is a liquid under normal conditions; the gaseous form of water is a vapor.

   C. Gases form *homogeneous mixtures* with one another, regardless of the amounts and characteristics of the components.

II. **Pressure**

   A. *Pressure* is the force a gas exerts on the walls of its vessel per unit area: $P = \dfrac{F}{A}$.

   In more general terms, pressure is exerted by a gas on any surface with which it makes contact.
   1. *Newtons* are the SI units of force. (1 N = 1 kg-m/s^2)
   2. *Pascals* are the SI units of pressure. (1 Pa = 1 N/m^2)

   B. The most commonly utilized gas is the mixture we call air and so *atmospheric pressure* is the gas pressure most commonly measured. Atmospheric pressure is measured with a *barometer* and so is also called *barometric pressure*.
   1. A scientific barometer consists of a column of mercury in a tube, the top of which is sealed and evacuated and the bottom of which is open and submerged in a bowl of mercury. As the surrounding air presses down on the surface of the bowl, mercury is forced up the tube. Consequently, air pressure is proportional to the height of the mercury in the tube. Traditional pressure units take the form (but not the significance) of length units. For example, meteorologists report air pressure as inches of mercury, inches Hg.

Lecture Outline – Chapter 10
Brown, LeMay, & Bursten, *Chemistry: The Central Science*, 6th Edition

    2.   Due to fluctuations in air pressure, a *standard atmospheric pressure* has been established, equal to the average atmospheric pressure at sea level at the equator.
        a.   The standard pressure, 1 *atmosphere* (atm), is the pressure required to support a mercury column to a height of 760 mm.
        b.   Expressed in pascals, the standard pressure is 1.01325×10^5 Pa or 101.325 kPa.
        c.   Gas pressure is often expressed in mm Hg, a unit called the *torr* in honor of the inventor of the barometer, Evangelista Torricelli.

  C.   The pressure of an enclosed gas can be measured with a *manometer*, which is a device that measures the *difference* in the pressures of two gases. Generally, the difference in the pressures of the enclosed gas and the atmosphere is the quantity measured. If atmospheric pressure from a barometer is also known, the pressure of the enclosed gas can be calculated.

Discussion Question: Why do your ears pop when in an airplane? Why do they also pop when riding in the elevator of a tall building?

III. The Gas Laws

  A.   Experiment reveals that four variables, temperature, volume, pressure, and amount of material, adequately describe a gas sample under many common conditions. Temperature is generally expressed in Kelvin and amount of material is generally stated in moles. Pressure and volume can be expressed in a variety of units, although atmospheres and liters are the most common.

  B.   If the pressure on a balloon is decreased, the volume of the balloon increases. Conversely, if the pressure on a balloon is increased, the volume of the balloon decreases. The relationship between gas pressure and volume is referred to as *Boyle's law*.
      1.   The volume of a fixed amount of gas at constant temperature is *inversely* proportional to the pressure of the gas, $P \cdot V = $ constant.
      2.   Comparing the gas at two different pressures, $P_1 \cdot V_1 = P_2 \cdot V_2$ or $V_2 = \dfrac{P_1 \cdot V_1}{P_2}$.

  C.   As the temperature of a gas decreases, the volume of the gas decreases. Conversely, as the temperature of a gas increases, the volume of the gas increases. The relationship between the temperature and the volume of a gas is referred to as *Charles's law*.
      1.   The volume of a fixed amount of gas at constant pressure is *directly* proportional to the temperature of the gas, $V = $ constant$\cdot T$.
      2.   Comparing the gas at two different temperatures, $\dfrac{V_1}{T_1} = \dfrac{V_2}{T_2}$ or $V_2 = \dfrac{V_1 \cdot T_2}{T_1}$.

  D.   As the quantity of a gas increases, the volume it occupies increases. Conversely, as the quantity of a gas decreases, the volume it occupies decreases. The relationship between the quantity of gas and the volume it occupies is called *Avogadro's law*.
      1.   The volume of a gas at constant pressure and temperature is *directly* proportional to the amount of gas present, expressed in moles, $V = $ constant$\cdot n$.
      2.   Comparing the gas samples, $\dfrac{V_1}{n_1} = \dfrac{V_2}{n_2}$, $V_2 = \dfrac{V_1 \cdot n_2}{n_1}$, or $n_2 = \dfrac{V_2 \cdot n_1}{V_1}$.

Discussion Question: Qualitatively, what happens to the volume of a gas if both the temperature and the pressure are increased?

Lecture Outline – Chapter 10
Brown, LeMay, & Bursten, *Chemistry: The Central Science*, 6th Edition

IV. The Ideal-Gas Equation

A. The *ideal-gas equation* is $P \cdot V = n \cdot R \cdot T$.
1. An *ideal gas* is one that can be completely described by the ideal-gas equation. This can often be done, but not always.
2. The *ideal gas constant*, R, is the proportionality constant in the ideal-gas equation and has the most common value of 0.08206 L-atm/K-mol. Temperature must be expressed in Kelvin. Units of volume and pressure can be other than liters and atmospheres, but the value of R changes accordingly. The units of volume and pressure *must* match the units of R.

B. The conditions 0.00°C (273.15 K) and 1 atm are referred to as *standard temperature* and *pressure*, *STP*. At STP, the volume of 1 mol of an ideal gas is 22.41 L, the *molar volume* of a gas at STP.

C. All of the individual gas laws can be combined to give the ideal-gas equation. Conversely, the ideal-gas equation can be used to derive the individual gas laws. For example, Boyle's law describes the relationship between pressure and volume when temperature and the amount of material are held constant. By grouping the constants together in the ideal-gas equation, $P \cdot V = (n \cdot R \cdot T) = $ constant. Because n, R, and T are constants, their product must also be a constant. The result, $P \cdot V = $ constant, is the same expression as that arrived at by Boyle. The other gas laws similarly can be gotten from the ideal-gas equation.

D. The relationship between two of the variables of the ideal-gas equation is rather simplistic. In applications, it is useful to be able to relate three variables, P, V, and T. This relationship, called the combined gas law, also comes from the ideal-gas equation with amount of gas held constant. Therefore,

$P \cdot V = (n \cdot R) \cdot T$ or $\frac{P \cdot V}{T} = $ constant. In comparing a gas sample under two different sets of conditions, $\frac{P_1 \cdot V_1}{T_1} = \frac{P_2 \cdot V_2}{T_2}$.

V. Molar Mass and Gas Densities

Density is the ratio of mass to volume for a sample. By noting that the number of moles of substance is given by the mass divided by the molar mass, $n = \frac{mass}{M}$, substitution for n can be performed, yielding $P \cdot V = \frac{mass \cdot R \cdot T}{M}$. Rearranging gives $\frac{mass}{V} = d = \frac{P \cdot M}{R \cdot T}$. The equation can also be rearranged to solve for the molar mass of the gas if it is unknown, $M = \frac{d \cdot R \cdot T}{P}$.

Discussion Question: Why are helium balloons used by the National Weather Service to carry instruments aloft rather than hot air balloons?

VI. Gas Mixtures and Partial Pressures

A. In a gaseous mixture, each component gas exerts a pressure. Clearly, the pressure exerted by each component gas is a fraction of the total pressure and each component gas is being contained in the same volume at the same temperature. The pressure of each component gas is termed its *partial pressure* and the total pressure is the sum of the partial pressures, $P_t = P_1 + P_2 + P_3 ...$ Each partial pressure is given by $P = \frac{n \cdot R \cdot T}{V}$, and so $P_t = \frac{R \cdot T}{V}(n_1 + n_2 + n_3 ...)$. That is, the partial pressure of each component gas is proportional to the number of moles of the gas present and the total pressure is proportional to the total number of moles of gas present.

B. As such, the partial pressures of the gases present in a mixture are given by the *mole fraction* of the gases in the mixture, $\dfrac{P_i}{P_t} = \dfrac{\frac{n_i \cdot R \cdot T}{V}}{\frac{n_t \cdot R \cdot T}{V}} = \dfrac{n_i}{n_t} = X_i$.

VII. Volumes of Gases in Chemical Reactions

A. Gases are often generated in the laboratory, either as a product of a reaction being studied or as a synthesis of the gas to be used in a chemical reaction. A common way to trap and measure the gas formed is to bubble it into an inverted jar full of water, a technique called *collecting a gas over water* or *displacement*.

B. The amount of gas collected over water is a function of volume, pressure, and temperature. Volume and temperature are easily measured. However, the gas that is collected is saturated with *water vapor*. The total pressure inside the jar is then $P_{total} = P_{gas} + P_{water}$. To obtain the pressure of the gas collected, the vapor pressure of water at the particular temperature must be subtracted from the total pressure.

VIII. Kinetic-Molecular Theory

A. The *kinetic-molecular theory* explains why ideal gases behave as they do. To summarize the kinetic-molecular theory:
  1. Gases consist of large numbers of molecules that are in continuous, random motion.
  2. The volume of all the molecules of the gas is negligible compared to the total volume in which the gas is contained.
  3. Attractive and repulsive interactions among gas molecules are negligible.
  4. Energy can be transferred between molecules during collisions, but the *average* kinetic energy does not change with time, as long as the temperature remains constant. The collisions are *elastic*.
  5. The average kinetic energy of the molecules is proportional to the absolute temperature. At any given temperature, the molecules of all gases have the same average kinetic energy.

B. The absolute temperature of a gas is a measure of the *average* kinetic energy of its molecules. Although kinetic energy at a given temperature is the same for all gases, the molecular speeds are different. At a constant kinetic energy, as the molecular mass increases, the molecular speed decreases.

C. The kinetic-molecular theory explains the gas laws.
  1. The effect of a volume increase at constant temperature is such that there are fewer molecules per unit volume and therefore fewer collisions. As a result of fewer collisions, the pressure exerted by the gas decreases.
  2. The effect of a temperature increase at constant volume is such that the molecules have a higher kinetic energy and therefore higher speeds. Because of the increased speeds, more collisions occur in a unit of time and the pressure exerted by the gas increases.

D. The ideal-gas equation can be derived from the kinetic molecular theory.

IX. Molecular Effusion and Diffusion

A. The average molecular speed of a gas, u, is given by $u = \sqrt{\frac{3 \cdot R \cdot T}{M}}$. Consequently, as molecular mass increases, the average speed of the gas decreases.

B. *Effusion* is the ability of a gas to escape a vessel through a tiny hole. *Graham's law of effusion* states that the relative rates of effusion of two gases under identical conditions are inversely proportional to the square roots of their molar masses, $\frac{r_1}{r_2} = \frac{u_1}{u_2} = \sqrt{\frac{3 \cdot R \cdot T/M_1}{3 \cdot R \cdot T/M_2}} = \sqrt{\frac{M_2}{M_1}}$.

C. *Diffusion* is the ability of a gas to disperse itself throughout a vessel. An example of diffusion would be odors spreading throughout a building. *Mean free path* is the average distance traveled by a molecule between collisions.

Discussion Question: Is the relative rate of effusion of two gases affected by changing the temperature at which the experiments are performed?

X. **Deviations from Ideal Behavior**

A. The ideality of gases is an approximation. *Real gases* do not exactly follow the ideal-gas equation. That is, for a mole of an ideal gas, $\frac{P \cdot V}{R \cdot T} = 1$. For a mole of a real gas, $\frac{P \cdot V}{R \cdot T} \neq 1$. At high pressure, gases do not behave ideally because the molecules are forced close together and so the assumption that molecules do not interact is false. At low temperature, gases do not behave ideally because as the container volume decreases, the assumption that the molecular volumes are negligible is false.

B. The behavior of real gases is described by the van der Waals equation,

$(P + \frac{a \cdot n^2}{V^2})(V - n \cdot b) = n \cdot R \cdot T$.

XI. **Chemistry at Work: Gas Separations**

This section discusses the technique of purifying a gas sample by allowing the sample to diffuse. Purification is achieved because the gas of interest will diffuse at a different rate than the impurities. The diffusion often must be repeated many times in order to sufficiently increase the purity of the gas.

Discussion Question: At a particular temperature and pressure, a gas with a high molar mass behaves *less* ideally than a gas with a low molar mass. Why?

Lecture Outline – Chapter 10
Brown, LeMay, & Bursten, *Chemistry: The Central Science, 6th Edition*

SAMPLE QUIZ QUESTIONS

1. If the pressure of 2.0 L of gas is increased from 1 atm to 2 atm while its temperature is increased from 100°C to 200°C, what is its final volume?

2. How many moles of O_2 are there in 1.12 L of this gas at 2 atm and 1092 K?

3. A 10.0 g sample of a gas occupies 5.6 L at STP. Calculate its molecular weight.

4. The total volume of the gas bags of the German dirigible Hindenberg was 2×10^5 m^3. How many moles of H_2 would be required to fill them at 20°C and 1 atm pressure? How much would this hydrogen weigh? What mass of He would be required?

5. If 200 mL of Ar at 400 mm Hg and 25°C is placed in a 300 mL container that already holds He at 250 mm Hg and 25°C, what is the total pressure of the mixture?

6. Why do gases deviate from ideal behavior? Under what experimental conditions are such deviations most noticeable?

7. Show, using a rough graph, how the kinetic energy distribution of a gas changes as its temperature increases.

8. The van der Waals equation, $(P + \frac{n^2a}{V^2})(V - nb) = nRT$, attempts to correct the ideal gas for the properties of real gases. Explain the origins and signs of the $\frac{n^2a}{V^2}$ and nb terms.

9. If oxygen gas is collected over water at 29°C and 676 mm Hg, and the vapor pressure of water at that temperature is 30 mm Hg, what is the pressure of the pure oxygen: (a) 706 mm Hg; (b) 646 mm Hg; (c) 760 mm Hg; (d) 30 mm Hg?

10. At a given temperature: (a) all molecules of $H_2S(g)$ will have the same kinetic energy; (b) the average kinetic energy of $H_2S(g)$ molecules will be greater than that of $H_2(g)$; (c) the average kinetic energy of $H_2(g)$ molecules will be greater than that of $H_2S(g)$; (d) the average kinetic energy of $H_2S(g)$ molecules will be the same as that of $H_2(g)$.

11. Which of the following statements concerning 1 L of H_2 and 1 L of O_2, both at STP, is false: (a) The H_2 molecules have greater average speed than the O_2 molecules; (b) The total number of collisions of molecules with the walls in a unit of time is the same for both gases; (c) Both gases contain the same number of molecules; (d) The average kinetic energy of the O_2 molecules is the same as that of the H_2 molecules?

12. A sample of a gas is confined in a 3.51 L vessel at a pressure of 575 torr. What will the pressure be if the volume is decreased to 2.50 L: (a) 410.0 torr; (b) 0.00124 atm; (c) 807 torr; (d) 1060 torr; (e) 0.539 atm?

13. A sample of a gas occupies 15.3 L at -15.0°C. What volume will the gas occupy if the temperature is increased to 300.0 K: (a) 13.2 L; (b) 306 L; (c) 0.765 L; (d) 17.8 L; (e) 27.5 L?

14. A sample of a gas occupies a volume of 356 L at 522°C. At what temperature will the gas occupy a volume of 999 L: (a) 2230°C; (b) 1460°C; (c) 1960°C; (d) 283 K; (e) 1740 K?

15. A sample of a gas is confined at a pressure of 1.05 atm and a temperature of 28.0°C. What temperature is required to raise the pressure to 5.00 atm: (a) 1160 K; (b) 1430 K; (c) 406 K; (d) 133 K; (e) 1710 K?

16. A sample of 1.50 mol of CO_2 occupies a volume of 40.0 L at a particular pressure and temperature. How many moles of CO_2 must be added to increase the volume to 50.0 L: (a) 0.38 mol; (b) 0.30 mol; (c) 1.88 mol; (d) 1.20 mol; (e) 3.38 mol?

17. A sample of a gas is confined in a 3.00 L vessel at 1.02 atm and 298 K. If the volume of the vessel is reduced to 2.00 L and the temperature is increased to 315 K, what will the resulting pressure be: (a) 0.719 atm; (b) 2.57 atm; (c) 1.45 atm; (d) 1.08 atm; (e) 1.62 atm?

18. A sample of a gas is confined in a 8.03 L vessel at 1.07 atm and 298 K. If the pressure of the vessel is reduced to 0.330 atm and the temperature is increased to 375 K, what will the resulting volume be: (a) 20.7 L; (b) 32.8 L; (c) 0.484 L; (d) 4.08 L; (e) 3.12 L?

19. A sample of a gas was collected in a 1.00 L vessel at 300.0 K and 0.772 atm. At what temperature will the volume be 2.00 L and the pressure be 1.00 atm: (a) 463 K; (b) 389 K; (c) 777 K; (d) 600 K; (e) 194 K?

20. A sample of 30.5 g of N_2 is confined in a 20.0 L vessel at 298 K. What is the pressure of the N_2: (a) 1.22 atm; (b) 37.3 atm; (c) 34.3 atm; (d) 0.112 atm; (e) 1.33 atm?

21. A sample of 2.00 mol of CO_2 is confined at 3.00 atm and 0.0°C. What volume does the gas occupy: (a) 14.9 L; (b) 44.8 L; (c) 7.47 L; (d) 134 L; (e) 33.6 L?

22. A sample of 0.0223 g of a gas is confined in a 1.00 L vessel at 0.00554 atm and 273 K. What is the molar mass of the gas: (a) 0.000247 g/mol; (b) 22.4 g/mol; (c) 4050 g/mol; (d) 0.0111 g/mol; (e) 90.2 g/mol?

23. A sample of 1.00 mol of a gas is confined in a 20.0 L vessel At what temperature will the pressure in the vessel be 2.00 atm: (a) 214 K; (b) 40.0 K; (c) 244 K; (d) 487 K; (e) 273 K?

24. Which of the following is not a valid part of the kinetic molecular theory: (a) Kinetic energy is related to absolute temperature; (b) Gas molecules have significant volumes; (c) Gases are composed of molecules; (d) Collisions of gas molecules are elastic; (e) Molecules are always in motion and therefore possess kinetic energy?

25. A gaseous mixture is analyzed giving the pressures of the components to be 0.205 atm N_2, 75.0 torr CO_2, 5 torr He, and 83 torr CO. What is the total pressure of the mixture: (a) 163 torr; (b) 708 torr; (c) 760 torr; (d) 319 torr; (e) 150 torr?

26. A gaseous mixture of hydrogen, oxygen, and nitrogen exerts a total pressure of 1.00 atm. If the partial pressure of H_2 is 15.0 torr and the partial pressure of oxygen is 152 torr, what is the partial pressure of nitrogen: (a) 593 torr; (b) 760 torr; (c) 168 torr; (d) 608 torr; (e) 927 torr?

27. What is the density of $SO_2(g)$ at STP: (a) 1.00 g/L; (b) 0.0446 g/L; (c) 0.350 g/L; (d) 2.86 g/L; (e) 0.235 g/L?

28. A gas is found to have a density of 2.05 g/L at STP. What is the molar mass of the gas: (a) 10.9 g/mol; (b) 1620 g/mol; (c) 6820 g/mol; (d) 133 g/mol; (e) 45.9 g/mol?

29. Consider the equation, $2\,NO_2(g) \rightarrow N_2O_4(g)$. What volume of N_2O_4 at STP is formed when 100.0 g of NO_2 are used in the reaction: (a) 97.4 L; (b) 24.4 L; (c) 11.2 L; (d) 48.7 L; (e) 100.0 L?

30. Which of the following gases will effuse most rapidly through a given pinhole: (a) H_2S; (b) HCl; (c) Kr; (d) PH_3?

Lecture Outline – Chapter 11
Brown, LeMay, & Bursten, *Chemistry: The Central Science*, 6*th* Edition

Chapter 11 — Intermolecular Forces, Liquids, and Solids

OVERVIEW: Chapter 11 looks at liquids and solids in much the same way that Chapter 10 dealt with gases, focusing on the interactions and energies involved. First, the kinetic-molecular theory is extended to liquids and solids and the observed properties compared to those of gases.

Intermolecular forces are described in detail, with emphasis on the magnitudes of the charges involved and on the strengths of the interactions.

Using intermolecular forces, the properties of liquids are discussed, including freezing point, boiling point, viscosity, surface tension, and vapor pressure

Phase diagrams are also covered, focusing on the typical phase diagram of CO_2 and on the unusual, but important, phase diagram of H_2O.

Solids then make up the remainder of the chapter. Crystal lattices, unit cells, and X-ray diffraction are covered. The bonding in molecular, ionic, covalent network, and metallic crystals is also discussed.

The chapter ends with a discussion of buckminsterfullerene and its relatives, emphasizing the potential of molecular carbon.

LECTURE OUTLINE

I. **The Kinetic-Molecular Description of Liquids and Solids**

  A. *Intermolecular forces* are attractive electrostatic interactions that occur *between* molecules, atoms, or ions of a substance.

  B. In *liquids*, the intermolecular forces are strong enough to hold the molecules together yet weak enough to allow the molecules to move. Liquids are generally denser than gases, are fairly incompressible, have definite volumes, and can flow.

  C. In *solids*, the intermolecular forces are strong enough to prevent the molecules from moving. Hence, solids do not flow nor do they take the shape of their containers. Solids are generally more dense than the liquid of the same substance, are incompressible, and are rigid.
    1. *Crystalline solids* are those whose molecules or ions are arranged in repeating patterns possessing high symmetry.
    2. *Amorphous solids* are those whose molecules are arranged in a random fashion.

II. **Intermolecular Forces**

  A. *Intermolecular forces* are electrostatic interactions between the molecules, atoms, or ions of a sample that act to hold the particles together. At a given temperature, those substances that are gases tend to have weaker intermolecular forces than those that are liquids. Solids have yet stronger intermolecular forces. Many properties, including the melting point and boiling point of a substance, are dependent on the strength of the intermolecular forces. The stronger the intermolecular forces, the higher the melting point and the boiling point.

B. Three kinds of intermolecular forces, the *van der Waals forces*, lead gases to behave nonideally.
1. *Dipole-dipole forces* are those between the δ+ pole of one polar molecule and the δ- pole of another polar molecule.
2. *London-dispersion forces* arise from the *polarizability* of electron clouds. In London forces, the electron clouds distort or become polarized, inducing a dipole. As such, δ+ and δ- poles are created that are smaller in magnitude than the fixed dipoles of a truly polar molecule. Nonetheless, these *induced dipoles* electrostatically attract one another.
3. *Hydrogen bonding* exists when a hydrogen atom attached to N, O, or F on one molecule is attracted to the unbonded electron pairs of an N, O, or F of another molecule.

C. *Ion-dipole forces* are those electrostatic attractions that form between an ion and an oppositely charged pole of a polar molecule.

D. *Dipole-dipole forces* exist between neutral molecules. They are those interactions between the δ+ pole of one polar molecule and the δ- pole of another polar molecule. For molecules of approximately equal mass and size, the intermolecular attraction increases with increasing polarity.

E. *London-dispersion forces* exist between formally nonpolar molecules. All liquids and solids possess London forces. London forces tend to increase with increasing formula weight.

F. *Hydrogen bonding* is a special type of dipole-dipole interaction in which an atom of hydrogen is attached to N, O, or F. In such a case, the concentration of partial positive charge on the hydrogen is enormous, allowing it to interact with the unbonded electron pairs of the negative poles of other molecules in an unusually strong way.

III. A Closer Look: Trends in Hydrogen Bonding

The strength of hydrogen bonding increases in the order H···N < H···O < H···F. The trend is explained in terms of the increasing electronegativities of N, O, and F.

IV. Properties of Liquids: Viscosity and Surface Tension

A. *Viscosity* is the resistance of a liquid to flow. The greater the intermolecular forces, the higher the viscosity. The higher the viscosity, the more slowly a liquid flows.

B. *Surface tension* is the energy required to increase the surface area of a liquid by a unit amount, usually 1 m^2. Qualitatively, the surface tension is the strength of the interaction among surface molecules that attempts to minimize surface area. The stronger the intermolecular forces in the liquid, the higher the surface tension.
1. *Cohesive forces* are those that bind like molecules to one another.
2. *Adhesive forces* are those that bind a substance to a surface.
3. A *meniscus* is the curved surface of a liquid caused by the combination of cohesive forces within the liquid and the adhesive forces between the liquid and the surface of its container. Water adheres strongly to glass, so the meniscus of water on glass is higher at the glass than in the center of the liquid surface.

Discussion Question: Several products at home are liquids. Discuss the viscosities of cooking oil, water, catsup, nail-polish remover, etc. Focus on how well the product runs when it is spilled.

V. Changes of State

A. *Phase changes* or *phase transitions* occur when a substance's structure is altered. Most commonly, this applies to a *change of state*, a change that converts among solid, liquid, and gas. Melting, freezing, vaporizing, and condensing are all changes of state.

B. *Energy changes accompany all changes of state.* Whenever a change produces a less-ordered state, energy must be supplied to overcome the intermolecular forces. Thus, melting (solid to liquid) and vaporization (liquid to gas) are endothermic processes.
   1. The *heat of fusion* is the enthalpy change associated with melting a substance and is usually expressed in kJ/mol.
   2. The *heat of vaporization* is the enthalpy associated with vaporizing a substance and is usually expressed in kJ/mol.

C. Heating a substance obviously requires energy; it is endothermic. As heat is added to a solid sample, the temperature of the sample increases until melting begins. During melting the temperature remains constant, the added heat being used to break the intermolecular forces. Continued heating of the liquid causes a temperature increase until boiling begins. During boiling the temperature remains constant, the added heat being used to break the intermolecular forces. Continued heating of the gas then causes the temperature to increase. The energy required for each of these five processes, warming the solid, melting, warming the liquid, boiling, and warming the gas, is determined by the identity of the substance involved and by the amount of sample present. For the warming processes, $q = s \cdot m \cdot \Delta T$, where the specific heat, s, is different for each physical state. For the phase transitions, $q = (\#moles) \cdot \Delta H$.

D. Each liquefiable gas has an associated critical temperature and critical pressure.
   1. The *critical temperature* of a substance is the highest temperature at which a gas can be liquefied. The stronger the intermolecular forces, the higher the critical temperature.
   2. The *critical pressure* of a substance is the pressure required to bring about liquefaction at the critical temperature.

Discussion Question: Discuss the state changes of common substances like water, dry ice, liquor (if deemed appropriate). Why does dry ice sublime rather than melt? Why does rubbing alcohol evaporate faster than water? Several questions of this type can be asked.

VI. Chemistry at Work: Supercritical Fluid Extraction

A. *Supercritical fluids* are derived from liquids that are held at high temperature and high pressure so that the physical phase is not quite liquid and not quite gaseous, but seems to be a composite of the two. For example, a supercritical fluid will fill its vessel, like a gas, but has a density much closer to that of the liquid phase.

B. Supercritical fluids can be used as solvents, just as liquids at "normal" temperatures and pressures can be. The fluid can be used to selectively dissolve a component of a mixture, thereby *extracting* that component from the mixture.

VII. Vapor Pressure

A. *Vapor pressure* is the pressure exerted by a vapor. As the vaporization process continues in a closed vessel, vapor molecules condense as others vaporize. With increasing concentration of vapors, the rate of condensation increases to the point that vaporization and condensation occur at the same rate. This point is called a

Lecture Outline – Chapter 11
Brown, LeMay, & Bursten, *Chemistry: The Central Science*, 6*th* Edition

dynamic equilibrium because, even though it seems nothing is occurring, molecules are in fact moving rapidly between the liquid and the gas.

B. A liquid that vaporizes readily is said to be *volatile*. The ability to vaporize is increased by increasing the temperature, thereby increasing the vapor pressure.

C. The *boiling point* of a substance is defined as the temperature at which the vapor pressure of the liquid equals the pressure of the environment on the liquid's surface. Consequently, as the external pressure is decreased, the boiling point also decreases. The *normal boiling point* is the boiling point at the standard pressure, 1 atm.

Discussion Question: Why do food instructions often give longer cooking times at higher elevations?

VIII. A Closer Look: The Clausius–Clapeyron Equation

A. The *Clausius–Clapeyron Equation* gives the relationship between temperature and vapor pressure.

B. The equation has the form: $\ln P = \dfrac{-\Delta H_{vap}}{RT} + C$, where P is the equilibrium vapor pressure at temperature T, R is the ideal gas constant in energy units, and ΔH_{vap} is the enthalpy of vaporization of the liquid. Clearly, a graph of $\ln P$ vs. $\dfrac{1}{T}$ should be linear with a slope of $\dfrac{-\Delta H_{vap}}{R}$ and an intercept of C.

IX. Phase Diagrams

A *phase diagram* is a graphical way to show the physical state in which a substance will exist under varying conditions of temperature and pressure and the equilibria which exist among the states. A phase diagram typically has three regions, corresponding to solid, liquid, and gas. In addition, lines separate each region, corresponding to the phase equilibria. The phase diagram, then, can be used to predict the physical state of the substance, and whether it is at equilibrium, given the temperature and pressure.

X. Structures of Solids

A. A *crystalline solid* is one in which the molecules, atoms, or ions are ordered in well-defined, repeating, patterns.

B. An *amorphous solid* is one in which no such order exists. Amorphous solids may have small regions of order, but no long-range order.

C. The patterns of particles in crystalline solids can be described by unit cells. A *unit cell* is the smallest "piece" of the crystal required to show the repeating pattern.
   1. A *crystal lattice* is the long-range pattern shown by the unit cell. We speak of the NaCl lattice; remember that this is a reference to the *pattern* shown by NaCl, not a reference to the chemical NaCl.
   2. The simplest unit cell is the *primitive cubic*. The term cubic simply means that the unit cell is a cube, a box with 90° corners and equal-length sides. The term primitive refers to the simplest; the primitive cubic unit cell consists only of atoms at the corners of the box. Note that these corner atoms simultaneously exist in eight unit cells. Therefore, a corner atom is only 1/8th in a given unit cell.
   3. The *body-centered cubic* unit cell has an atom at the center of the box in addition to the eight atoms in the corners. A body-centered atom contributes to only one unit cell.

4. The *face-centered cubic* unit cell has atoms in the centers of each of the six faces of the box. Note that these face-centered atoms simultaneously exist in two unit cells. Therefore, a face-centered atom is only half in a given unit cell.
5. In some lattices, atoms are also found residing on the edges of the box. These *edge-centered* atoms simultaneously exist in four unit cells. Therefore, an edge-centered atom is only 1/4th in a given unit cell.

D. When atoms of equal size are placed into a lattice, they pack together as closely as possible, much like putting marbles in a box. The arrangements which yield the most efficient use of space are called *close packing*.
1. When stacking atoms, or marbles, the first and second layers are fixed. The addition of the third layer offers two choices. If the third layer is added such that the atoms of the third lie directly over the atoms of the first layer, the arrangement is called *hexagonal close packing*, or ABABAB... Each atom has twelve nearest neighbors, six in its own plane, three in the plane above and three in the plane below.
2. If the third layer is added such that the atoms of the third do not lie directly over the atoms of the first layer, the arrangement is called *cubic close packing*, or ABCABC... Each atom has twelve nearest neighbors, six in its own plane, three in the plane above and three in the plane below.

Discussion Question: List and discuss common solids and try to identify each as crystalline or amorphous.

XI. A Closer Look: X-Ray Diffraction by Crystals

A. X-rays can be diffracted by crystals. The nature of such *X-ray diffraction* is determined by the symmetry of the crystal and by the spacing between adjacent planes of molecules, atoms, or ions. As the crystal diffracts the x rays, the waves of radiation interact with one another, a phenomenon called *interference*.
1. *Constructive interference* occurs when the interacting waves reinforce one another's amplitude. Constructive interference in diffraction is caused by the constant spacing of planes of atoms.
2. *Destructive interference* occurs when the interacting waves diminish one another's amplitude. Diffractions which are not part of the overall pattern in the crystal tend to cancel one another out.

B. The *Bragg equation* describes the spacing of the planes in a crystal in terms of the angle of X-ray radiation, the Bragg angle, below which the radiation is simply reflected. The equation is $n\lambda = 2d \sin\theta$, where $n = 1, 2, 3,...$, d is the distance between planes of atoms, θ is the angle, and λ is the wavelength.

C. Not all crystals are perfect. Often, errors in the pattern in a crystal occur. These errors are called *crystal defects*. The most common type of crystal defect are *vacancy defects* in which one or more atoms are missing from the locations called for by the crystal lattice.

XII. Bonding in Solids

A. *Molecular solids* are those in which particles that comprise the crystal are molecules. Because the particles are neutral molecules, London-dispersion forces, dipole-dipole interactions, and hydrogen-bonding can be used to hold the crystal together.

B. *Covalent network solids* are those in which the crystal itself is a large, covalently bonded, molecule. These crystals tend to be very strong and hard.

- C. *Ionic solids* are those in which the particles in the crystal lattice are ions. Consequently, ionic crystals are held together by the electrostatic attractions among oppositely charged ions. The lattice of an ionic crystal is determined by the charges of the ions and by their relative sizes.

- D. *Metallic solids* are those in which the particles in the lattice are metal atoms. These crystals are generally one of the close packing arrangements or body-centered cubic. Because of the closeness of the atoms and the mobile electrons, the electrons can readily hold the atoms together by interacting with several metals simultaneously, a model viewed as having metal atoms immersed in a sea of valence electrons.

Discussion Question: How would the "electron sea" model of metallic bonding also explain electrical and thermal conductivity?

XIII. A Closer Look: "Buckyball"

The Buckyball or Buckminsterfullerene, C_{60}, was the first molecular form of carbon to be identified. Unlike diamond and graphite, which are infinite arrays of covalently bonded carbon atoms, C_{60} and its relatives are true molecular units, with fixed compositions and geometries. Much research is ongoing into the synthesis and chemistry of this new family of compounds.

SAMPLE QUIZ QUESTIONS

1. What types of attractive forces must be overcome to melt each of the following substances: (a) SiO_2; (b) CO_2; (c) NaCl; (d) K?

2. In which of the following substances is hydrogen bonding the most important: (a) LiH; (b) NH_3; (c) CH_4; (d) H_2S?

3. Which of the following has the highest normal boiling point: (a) CF_4; (b) CCl_4; (c) CBr_4; (d) CI_4?

4. CF_4 boils at -128°C, CCl_4 boils at 76.8°C. Which of the following statements best accounts for the lower boiling point of CF_4: (a) The average kinetic energy of CF_4 molecules is higher at any given temperature than for CCl_4; (b) CCl_4 is more polar than CF_4, thus experiences stronger dipole-dipole interactions in the liquid; (c) CCl_4 has larger dispersion force interactions between molecules; (d) CCl_4 molecules form a less ideal gas than CF_4 molecules.

5. Which of the following exhibits only London dispersion forces: (a) IBr(l); (b) H_2SO_4(l); (c) CCl_4(l); (d) Al_2O_3(s); (e) HF(l)?

6. Which of the following exhibits hydrogen bonding: (a) CH_4(l); (b) PH_3(g); (c) NaH(s); (d) CH_3CH_2OH(l); (e) HI(g)?

7. Liquids are generally less dense than solids because: (a) the molecules in solids are farther apart; (b) the intermolecular forces in liquids are stronger; (c) to organize into a lattice, the distances between molecules increase; (d) molecules in liquids tend to have less kinetic energy than molecules in solids; (e) the molecules in liquids are farther apart.

8. Which of the following liquids should have the highest density (all have about the same molecular weight): (a) C_8H_{18}; (b) GeF_3; (c) BrCl; (d) $SiHCl_3$; (e) C_9H_9?

9. Which of the following would you expect to have the lowest critical temperature: (a) H_2; (b) N_2; (c) O_2; (d) H_2O?

10. Which of the following liquids should have the lowest surface tension: (a) CH_3OH; (b) CH_4; (c) CH_3F; (d) CH_2F_2; (e) CHF_3?

11. Viscosity generally: (a) increases with decreasing molecular weight; (b) is the measure of a liquid's ability to flow; (c) increases with increasing temperature; (d) is unexpectedly low for water; (e) increases with increasing intermolecular forces.

12. A dynamic equilibrium: (a) means that there is no change; (b) exists when opposite processes occur at the same rate; (c) exists between a liquid and its vapor in an open vessel; (d) refers to the production of vapors by a liquid; (e) exists when opposite processes occur simultaneously.

13. Explain briefly how vapor pressure and temperature are related.

14. Which of the following would you expect to have the highest vapor pressure at any given temperature: (a) C_2H_6; (b) C_3H_8; (c) C_4H_{10}; (d) C_5H_{12}?

15. Vapor pressure generally: (a) is affected by the intermolecular forces of the liquid; (b) is the same for most substances at a given temperature; (c) increases as the kinetic energy of the liquid molecules decreases; (d) is independent of temperature; (e) increases with increasing intermolecular forces.

16. Boiling points generally: (a) decrease with increasing strength of the intermolecular forces in a liquid; (b) increase with decreasing atmospheric pressure; (c) decrease with decreasing atmospheric pressure; (d) decrease with increasing viscosity; (e) depend only on the atmospheric pressure.

17. The molar heat of vaporization of chloroform, $CHCl_3$, is 31.9 kJ/mol. How much heat is required to vaporize 97.0 g of chloroform at its boiling point: (a) 3090 kJ; (b) 0.267 kJ; (c) 39.3 kJ; (d) 25.9 kJ; (e) 3.75 kJ?

18. A sample of 1.00 kg of glycerin, $C_3H_8O_3$, requires 828 kJ to vaporize at its boiling point. What is the molar heat of vaporization of glycerin: (a) 9.00×10^3 kJ/mol; (b) 76.2 J/mol; (c) 7.62×10^4 kJ/mol; (d) 1.21 kJ/mol; (e) 76.2 kJ/mol?

19. Given that for H_2O, ΔH_{fus} = 5.98 kJ/mol, ΔH_{vap} = 40.6 kJ/mol, and the specific heat of liquid water is 1.00 cal/g-°C, calculate the amount of heat required to convert 55.5 g of water from ice at 0.00°C to liquid water at 99.9°C: (a) 9950 cal; (b) 18.4 cal; (c) 4410 cal; (d) 5540 cal.

20. A substance has the following properties: normal boiling point = 500 K, normal melting point = 200 K, and triple point = 190 K at 0.1 atm. Draw a rough sketch of the phase diagram for this substance, labeling the points given above as well as the regions in which the solid, liquid, and vapor phases exist.

21. Amorphous solids generally: (a) exhibit an organized packing of atoms within the crystal; (b) melt suddenly at a well defined temperature; (c) result from slow freezing of the liquid; (d) are more stable than crystalline solids; (e) melt over a range of temperatures.

22. The melting points of crystalline substances generally: (a) are lower than for amorphous solids of the same molecular weight; (b) increase with increasing strength of the intermolecular forces in the solid; (c) vary widely with changes in atmospheric pressure; (d) are a broad range of temperatures, not a well defined temperature; (e) increase with decreasing strength of the intermolecular forces in the solid.

23. Molecular solids are generally: (a) good electrical conductors; (b) volatile; (c) quite hard; (d) quite brittle.

24. An example of a covalent solid is: (a) diamond; (b) aluminum; (c) iodine; (d) sodium nitrate.

25. What is the net number of particles in a primitive cubic unit cell: (a) 1; (b) 2; (c) 3; (d) 4?

26. The number of nearest neighbors around an Ar atom in a cubic close-packed Ar lattice is: (a) 4; (b) 6; (c) 8; (d) 12.

27. What is a unit cell? Describe, using sketches, the difference between a primitive cubic and a face-centered cubic cell.

28. The KCl unit cell looks like that of NaCl. If the unit cell has an edge length of 6.29 Å, what is the density of KCl?

29. The principle angle at which X-ray diffraction occurs in a solid is 60° when $\lambda = 1.54$ Å. What is the spacing between lattice planes that gives rise to this diffraction?

Lecture Outline – Chapter 12
Brown, LeMay, & Bursten, *Chemistry: The Central Science*, 6*th* Edition

Chapter 12 — Modern Materials

OVERVIEW: Chapter 12 is concerned with the new materials that have been created in the past 10 to 20 years and that continue to be the focus of a great deal of research.

Liquid-crystalline materials are discussed first, including their basic properties, structures, the essential features of current devices made with them, and the sorts of devices in which their use is being investigated.

Polymers are not as new as liquid crystals, but the recent advances with polymers are noteworthy. Polymers are discussed in terms of the general reactions that form them, their properties, and their structures. In this coverage, fibers, plastics, and crosslinking are described.

Among the modern materials being research most actively, ceramics and ceramic composites perhaps offer the greatest promise for industrial applications. Methods of fabricating ceramic products are discussed, including embedded fibers and mixtures. The discussion of ceramics turns to superconductivity. The criteria for superconductivity, the potential uses for it, and the success at finding these materials are all covered.

Finally, thin films are discussed. While stating that the original thin films were decorative in nature, the discussion focuses on modern applications of these coatings, such as in electronics and in structural materials and tools.

LECTURE OUTLINE

I.   Liquid Crystals

   A.   *Cholesteryl benzoate*, first studied by Frederick Reinitzer, was the first identified liquid-crystalline substance. It melts to a milky liquid at 145°C and to a clear liquid at 179°C, and changes color from red to blue on continued heating. The milky phase is the liquid-crystalline phase.

   B.   Liquid crystals have characteristic structures of only a few types.
   1.   Many liquid crystals have *rodlike shapes*, are long, and are rigid. Intermolecular forces maintain ordering in the liquid-crystalline phase.
   2.   Other liquid crystals have *disklike shapes*, such as Reinitzer's cholesteryl benzoate.

   C.   In addition to their structures, there are different types of liquid-crystalline phases.
   1.   The *nematic liquid-crystalline phase* is the simplest. In this phase, molecules are aligned along their long axes, but there is no ordering with respect to the ends of the molecules.
   2.   *Smectic liquid-crystalline phases* show alignment of the ends of the molecules.
      a.   In the *smectic A phase*, molecules are arranged in sheets, with their long axes parallel and with their ends aligned.
      b.   In the *smectic C phase*, molecules are aligned with their long axes tilted with respect to a line parallel to the plane in which the molecules are stacked and with their ends aligned.
   3.   In the *cholesteric liquid-crystalline phase*, diskshaped molecules are stacked, disk upon disk, in layers. Some twisting of molecules is common to reduce direct intermolecular repulsions.

II. Chemistry at Work: Liquid Crystal Displays

Liquid crystals have found many uses. *Liquid-crystal-displays* consist of a liquid-crystalline layer between two transparent, electrically conductive, plates. When a voltage is applied to a region of the plates, the adjacent liquid crystals align themselves. Using a polarizing filter, light is blocked from passing out of the device when the voltage is applied.

1. Some liquid crystals are *temperature sensitive* materials in that they change *color* as a function of temperature.
2. Liquid-crystalline behavior is seen in many *biological* structures. Cell walls and other boundary membranes appear to have many of the properties of liquid crystals.

Discussion Question: From what you now know about liquid crystals, why does the display on your calculator appear to have all the digits darkened when viewed from a shallow angle?

III. Polymers

A. *Polymers* are molecular substances of high molecular mass that are formed by the joining together of small molecules called monomers.

B. *Monomers* are molecular substances with relatively low molecular masses that can be linked together to form polymers.

C. *Polyethylene* is one of the simplest and most useful manmade polymers. It is formed from ethylene, or ethene, molecules, $H_2C=CH_2$, which "open" their double bonds and link together, end-to-end.

D. *Addition polymerization* is the type of polymerization reaction illustrated by polyethylene. In all addition polymerizations, the double bonds of the monomers "open" and the monomer molecules link together. Several useful polymers are made this way, such as polypropylene, polystyrene, and Teflon®.

E. Along with addition polymerization, there are other types of *polymer processing*.
  1. *Condensation polymerization* is a type of reaction that forms polymers. In condensation polymerization, monomer molecules connect by slitting out a small molecule, such as water. For example, a nylon linkage is formed by reacting an R–NH$_2$ on one molecule with an HO–R' on another molecule, where R and R' are simply carbon-atom chains of organic molecules. The result is the production of water and the connection of the monomers as R–NH–R'.
  2. *Extrusion* is the technique of forcing newly formed polymer through a small hole to form long strands. This is typically done in the manufacture of fibers, such as nylon.
  3. Reaction-injection molding is the technique of conducting the polymerization reaction in a mold so that, upon polymerization, the object has the shape of the mold. This technique is used for solid parts, such as auto parts, telephones, and other plastic objects.

F. There are several *types of polymers*, differentiated by their properties.
  1. *Plastics* are generally those polymeric materials that can be formed into various shapes, usually by application of heat and pressure.
  2. *Thermoplastics* can be reshaped, once formed, and so are candidates for recycling. For example, two-liter soft drink bottles can be melted down and molded into park benches. Such a project is underway (begun summer 1990) between Pepsi® and the City of Chicago.
  3. *Thermosetting plastics*, by contrast, are irreversibly shaped by manufacturing and are difficult to reshape.

Lecture Outline – Chapter 12
Brown, LeMay, & Bursten, *Chemistry: The Central Science*, 6th Edition

        4. *Elastomers* are materials that exhibit rubbery or elastic behavior. As long as its limits of flexibility have not been exceeded, elastomers return to their previous shape after stress is removed.
        5. *Fibers* are long and thin and generally are not elastic.
    G. Different polymers have different structures and physical properties.
        1. Polymers are amorphous materials and so melt over a range of temperatures. However, they do possess some ordering, which we refer to as *crystallinity*. Thermoplastics, for instance, consist of linear chains of polymer molecules. The individual chains are flexible, but can be complicated by side chains that extend off the main chain. The degree of ordering depends on how complicated branching is and on the average chain length.
        2. *Plasticizers* are lower molecular weight substances that are added to polymers to reduce the interactions between polymer chains, making the polymer more flexible.
    H. *Crosslinkages* between polymer chains are bonded connections. Crosslinking of polymers creates more rigid materials than non-crosslinked polymers of the same substance.
        1. *Vulcanization* of natural rubber is the process by which the naturally occurring rubber polymer is crosslinked.
        2. Natural rubber is made from the inner bark of the *Hevea brasiliensis* plant and is a linear polymer of isoprene, $H_2C=C(CH_3)-CH=CH_2$.

Discussion Question: Can you think of something around your home that has *not* been replaced by a polymer? For those things still not made with polymers, what property or properties make the original material more suitable?

IV. Chemistry at Work: Recycling Plastics

Most plastics can be recycled. A challenge for communities who wish to recycle, though, is finding buyers for some of the plastics they collect. Recycling of #1 and #2 plastics is common. The categories of plastics are:
    1   polyethylene terephthalate (PETE)
    2   high density polyethylene (HDPE)
    3   polyvinyl chloride (PVC)
    4   low density polyethylene (LDPE)
    5   polypropylene (PP)
    6   polystyrene (PS)
    7   other plastics

V. Chemistry at Work: Kevlar, An Advanced Material

Kevlar® is an important, fairly new, material with a high heat resistance, a tensile strength greater than steel, and a stiffness greater than glass of the same thickness. The polymer chains are very strong, while hydrogen bonding holds adjacent molecules together.

VI. Ceramics

    A. *Ceramics* are nonmetallic inorganic solid materials that may be crystalline or amorphous. Ceramics can be held together by covalent network bonding, by ionic bonding, or by a combination of the two. They generally are heat resistant but brittle. Some ceramics do contain metals, such as in silicates, oxides, carbides, nitrides, and aluminates. *Engineering ceramics* are ceramic materials fabricated to replace other materials, such as wood, plastic, or metal.

B. Ceramics are often processed to minimize any defects in their structures. For example, some ceramics are first produced as small, uniform beads and then sintered, heated under pressure, to bond the beads together and form the desired object.
   1. The *sol-gel process* is important in forming the small uniform beads. Typically, the sol-gel process begins with a metal alkoxide. That metal alkoxide is then converted to the metal hydroxide in the form of a *sol*, a suspension of extremely small particles. The sol particles are reacted to form a three-dimensional network of metal atoms connected by oxygen atoms, called a *gel*. Heating of the gel then produces uniform, finely divided metal oxide particles.
   2. A *condensation reaction* is a reaction between two molecules that binds them together and splits out a small molecule. In the sol-gel process, the step in which the sol particles are converted to a gel by forming the network of bonds is a condensation reaction.
   3. The sol-gel process is particularly useful for making ceramic coatings and films. Due to shrinkage, it is difficult to form solid objects by this method.

C. *Ceramic composites* are mixtures of two or more ceramic materials.
   1. *Composites* in general are formed by mixing two or more engineering materials to form an object.
   2. *Ceramic fibers* are long, narrow structures (length at least 100 times diameter) made of a ceramic material. The addition of ceramic fibers to another ceramic material to form a composite has been shown to be a useful way to improve the mechanical strength of ceramic parts without compromising the thermal properties.

D. Ceramics, particularly the new ceramic composites, have been used in many applications.
   1. Ceramic composites are used extensively on the blades of cutting tools, on grinding wheels, and as abrasives.
   2. *Piezoelectric ceramics* are ceramics that generate an electrical potential when subjected to mechanical stress. For example, quartz is used to control the frequency in electronic devices, such as quartz watches and ultrasonic generators.
   3. *Thermistors* are ceramic materials with limited electrical conductivity that increases with temperature. As such, thermistors can be used as thermometers and as part of temperature-control devices.

E. Some ceramics are superconducting under certain conditions.
   1. *Superconductivity* is the ability of a substance to conduct electricity without resistance to the flow of electrons.
   2. A substance's *superconducting transition temperature*, T_c, is the maximum temperature at which it is a superconductor. The observed values of T_c are typically very low (20 to 30 K). Recent advances in superconductivity have yielded materials which are superconducting at temperatures as high as 125 K.

F. Theories of superconductivity are rudimentary. The original theory, proposed by Bardeen, Cooper, and Schrieffer, explains superconductivity at low temperatures in terms of electrons moving in pairs, which allows them to more easily escape interactions with atoms.

Discussion Question: Other than dinnerware, what ceramic items do you have around your home or at work?

VII. Thin Films

A. *Thin films* are coatings, generally 0.1 to 0.2 μm or greater in thickness. Paints and varnishes are normally not included because they are much thicker. To be useful, a thin film should possess most or all of these properties:
   1. It should be chemically stable in its intended environment.
   2. It should adhere well to its *substrate*, the structure to which it is applied.
   3. It should have a uniform thickness.
   4. It should be chemically pure or have a controlled composition.
   5. *Dislocations* or other flaws should occur as seldom as possible.

B. Thin films are used extensively in microelectronics, where they are used as conductors, resistors, and capacitors. They are also used as coatings on optics and on tools.

C. Thin films are formed in a variety of ways.
   1. *Vacuum deposition* is used to form thin films of substances that can be vaporized and recondensed without destroying their chemical properties. Such substances include metals, metal alloys, and simple inorganic compounds like oxides.
   2. *Sputtering* involves the use of high voltage to remove material from a source. These atoms are carried through the ionized gas within the chamber, and deposited on the substrate.
   3. *Chemical vapor deposition* is a technique in which the surface of the substrate is coated with a volatile, stable compound. The compound then undergoes a reaction to form an adherent coating.

VIII. Chemistry at Work: Diamond Coatings

A. Diamond remains one of the hardest and most corrosion resistant substances known. As a novel coating, it offers excellent endurance for surfaces, such as glass, plastic, and cutting tools.

B. One method of preparing a diamond surface involves microwave irradiation of the object to be coated in an atmosphere of CH_4 and H_2. The process is far from commercially viable at this point in time.

SAMPLE QUIZ QUESTIONS

1. List the properties typical of a liquid-crystalline material.

2. What are the differences and similarities between nematic and smectic liquid crystals?

3. Define the term *polymer*.

4. Write a chemical equation that describes the formation of polytetrafluoroethylene from tetrafluoroethylene ($F_2C=CF_2$).

5. Why are most polymers amorphous materials?

6. List the common properties of ceramics.

7. Describe the sol-gel process.

8. What is a superconducting material?

9. List the methods used to fabricate thin films.

10. What are composite materials?

11. Describe the process for making thin films called sputtering.

12. Cholesteryl benzoate: (a) melts at a well-defined temperature; (b) is liquid-crystalline between 145°C and 179°C; (c) is white at all temperatures; (d) has rodlike molecules.

13. A smectic liquid crystal: (a) shows no internal ordering of molecules; (b) has its molecules aligned along their long axes, but have no ordering with respect to the ends of the molecules; (c) possesses more order than a nematic liquid crystal; (d) must have its molecules' long axes parallel to one another.

14. In a liquid-crystal-display: (a) an applied voltage changes the orientation of the liquid-crystalline molecules; (b) consists of the liquid-crystalline material and an electrically conducting plate; (c) appears dark in those areas in which there is no applied voltage; (d) light passes through those regions in which there is an applied voltage.

15. An addition polymer is one in which: (a) a small molecule, like water, is usually also produced upon manufacture; (b) only single bonds are modified upon manufacture; (c) there are two or more different monomers; (d) double bonds "open" and monomer molecules link together.

16. A condensation polymer is one in which: (a) a small molecule, like water, is usually also produced upon manufacture; (b) only double bonds are modified upon manufacture; (c) there are two or more different monomers; (d) double bonds "open" and monomer molecules link together.

17. Define the term *crystallinity*.

18. A ceramic composite: (a) is formed by adding a polymer resin to a ceramic material; (b) contains fibers, the length of which is about ten times the diameter; (c) is formed by adding ceramic fibers to a ceramic material; (d) contains embedded fibers that have a random orientation.

19. In general, a thin film: (a) includes coatings such as varnish; (b) can usefully be of random thickness; (c) is prone to dislocations and other imperfections; (d) has a thickness of 0.1 to 0.2 μm or greater.

Chapter 13 — Properties of Solutions

OVERVIEW: Chapter 13 discusses solutions, particularly those in which water is the solvent.

After developing an appreciation for the components of a solution, the physical events that comprise the solubility process are discussed. The emphasis here is on energetics, with enthalpy and entropy being used, although entropy is not covered in detail or by name.

A complete development of units of expressing concentration follows, covering the percentage concentration systems, parts per million, mole fraction, molality, and molarity. Normality is covered as an optional topic.

Solubility, saturated solutions, unsaturated solutions, and supersaturated solutions are covered. As part of this section, the concept of dynamic equilibrium is discussed again (as it was in vapor pressure). Both solid and gaseous solutes in aqueous solutions are covered.

The factors that affect solubility are discussed, such as pressure for gaseous solutes and temperature.

The colligative properties are also covered in this chapter, including vapor pressure lowering, freezing point depression, boiling point elevation, and osmotic pressure. Applications of the colligative properties, such as molecular weight determination, are also discussed.

Finally, in order to put proper emphasis on solutions and the properties of solutes, colloidal dispersions are covered. Particle sizes, the Tyndall effect, and common techniques for removing colloidal particles are also discussed.

LECTURE OUTLINE

I. **Solutions**
   A. *Solutions* are homogeneous mixtures; the components are intimately mixed so that all portions of the mixture have exactly the same characteristics.
   B. The *solvent* in a mixture is that component whose phase is not changed when the solution is formed. If more than one component stays in the same phase, the one which is present in greatest amount is the solvent.
   C. The *solute(s)* in a mixture are all components except the solvent.

II. **The Solution Process**
   A. In the formation of a solution, *solvation* is the process by which intermolecular forces are created between the solute and solvent.
   B. When the solvent is water, the solvation process is called *hydration*.
   C. The formation of a solution is always accompanied by an *energy change*, although it may be small. During the process, several attractive interactions are being broken and others are being formed.
      1. *Solute-solute* intermolecular forces are being broken. This process is always *endothermic*.
      2. *Solvent-solvent* intermolecular forces are being broken. This process is always *endothermic*.
      3. *Solute-solvent* intermolecular forces are being formed. This process is always *exothermic*.

Lecture Outline – Chapter 13
Brown, LeMay, & Bursten, *Chemistry: The Central Science*, 6th Edition

The enthalpy change for the combined process is dependent on the magnitude of each of the three steps. $\Delta H_{soln} = \Delta H_1 + \Delta H_2 + \Delta H_3$

    D. Two principles guide whether the solution process, or any other process, is spontaneous.
        1. Processes in which the *energy content* of the system *decreases* tend to occur spontaneously. This generally also means the process is exothermic.
        2. Processes in which the *disorder* of the system *increases* tend to occur spontaneously.
    E. Care must be taken to differentiate between *physical processes* and *chemical processes*. The formation of a solution is usually considered a physical process because the components can be isolated again in their initial form. Chemical reactions, such as the burning of a fuel, change the identities of the substances.

Discussion Question: Solutions are common in daily life, as are other kinds of mixtures and some pure substances. Name common materials and discuss whether they are pure substances, solutions, or another kind of mixture. Most importantly, discuss how you could test each to decide which it is.

III. A Closer Look: Hydrates

Hydrates are dry compounds that contain molecules of water *attached* through hydration. Consequently, hydrates are not wet crystals, but rather crystals with water molecules incorporated into their structures.

IV. Ways of Expressing Concentration

    A. *Dilute solutions* are those in which the solute concentration is small. As such, it is an arbitrary, qualitative reference.

    B. *Concentrated solutions* are those in which the solute concentration is large. As such, it is an arbitrary, qualitative reference, and complimentary to the term dilute.

    C. *Weight percentage* is a way of expressing concentration in which the concentration of a component is given as a percent of the total mass of the solution.

$$Wt\% = \frac{\text{mass of component in soln}}{\text{total mass of soln}} \times 100$$

    D. *Parts per million* (*ppm*) is a system used for very low concentrations. It is similar to a percentage, except it is based on 1,000,000 parts rather than the 100 parts of a percentage.

$$\text{ppm of component} = \frac{\text{mass of component in soln}}{\text{total mass of soln}} \times 10^6$$

    E. *Mole fraction* (*X*) is a system in which the concentration of a component is given as the fraction of the total number of moles present in the solution.

$$X_{component} = \frac{\text{moles component}}{\text{total moles all components}}$$

    F. *Molarity* (*M*) is the system of expressing concentration which is most important in reaction stoichiometry. In molarity, the concentration of a component is given as the moles of the component per liter of *solution*.

$$M_{component} = \frac{\text{moles component}}{\text{liters soln}}$$

G. *Molality (m)* is a system in which the concentration of a component is given as the moles of the component per kilogram of *solvent*.

$$m_{component} = \frac{\text{moles component}}{\text{kg of solvent}}$$

V. A Closer Look: Normality

Normality (N) is a system in which the concentration of a component is given as the *equivalents* of the component per liter of solution.

$$N_{component} = \frac{\text{equivalents component}}{\text{liters soln}}$$

Equivalents are defined according to the type of reaction being viewed. For *acids*, an equivalent is the number of moles of the acid that supplies 1 mol of H^+. For *bases*, an equivalent is the number of moles of the base that supplies 1 mol of OH^-. In an *oxidation-reduction* reaction, an equivalent is the number of moles of the substance that gains or loses 1 mol of electrons. Consequently, normality is *always* a whole-number multiple of molarity.

VI. Saturated Solutions and Solubility

A. *Crystallization* is the process in which a solid solute converts to its crystalline form. As such, it is the reverse of the solution process for crystalline solutes.

B. A *dynamic equilibrium* is the condition in which a process is proceeding in its forward direction at the same rate as it is proceeding in its reverse direction. In the case of solutions, when the rate of solute dissolving equals the rate of solute crystallizing, the system is in dynamic equilibrium, or simply in equilibrium. At equilibrium, there is no change in the net amounts of dissolved solute and crystalline solute, although each process continues to occur.

C. A *saturated solution* exists when the solubility reaction is in equilibrium. The term signifies that no additional solute can be dissolved; the solution is "at capacity."

D. *Solubility* is the term applied to the amount of solute *dissolved* in an equilibrium solution process. Solubility is often expressed as the number of grams of the solute that can be dissolved into a volume, often 100 mL, of the solvent. The *molar solubility* is solubility expressed in moles of solute that can be dissolved into a liter of solution.

E. By contrast, an *unsaturated solution* is one in which the amount of dissolved solute is *less* that it could be. Unsaturated solutions form when all the crystalline material dissolves; more could dissolve if it was present.

F. A *supersaturated solution* is an unstable condition in which there is *more* dissolved solute than the solution can theoretically hold. These solutions typically result when the solution is formed at high temperature, at which the solubility is higher, and then allowed to cool slowly. Generally, a slight agitation causes the excess dissolved solute to crystallize.

Discussion Question: Can students think of examples of supersaturated solutions? Can they think of any commonly used saturated solutions?

VII. Factors Affecting Solubility

A. In forming a solution, solubility is increased when the solute-solvent interactions created are strong. Generally, solubility is increased when the solvent and the solute engage in similar intermolecular forces. This is often stated as "like

dissolves like." For example, water is polar and can use hydrogen bonding. Consequently, solutes that are polar, or ionic, should be soluble in water. Nonpolar substances should dissolve best in nonpolar solvents.
1. Pairs of liquids that mix in all proportions are said to be *miscible*.
2. Pairs of liquids that do not appreciably mix are called *immiscible*.

B. *Pressure can affect solubility*, particularly of gases because gases are compressible. Henry's law states that the solubility of a gas in liquid is proportional to the pressure of the gas over the liquid, $C_g = kP_g$, where P_g is the pressure of the gas, C_g is the resulting solubility of the gas, and k is the *Henry's law constant*. Each solute-solvent pair has a unique Henry's law constant at a given temperature.

C. *Temperature* also *affects solubility*. In water, the solubility of gases tends to decrease as temperature increases, as would be expected by the kinetic-molecular theory. Also in water, the solubility of ionic salts tends to increase with increasing temperature.

VIII. Chemistry and Life: Fat- and Water-Soluble Vitamins

Vitamins B and C are water soluble, while vitamins A, D, E, and K are fat soluble. Consequently, vitamins B and C need to be replenished by diet on a regular basis because they are carried throughout the body in blood and ultimately expelled in urine. The fat soluble vitamins are concentrated in body fat and so accumulate if we take in substantially more than we need. It is then possible to accumulate fat soluble vitamins to toxic levels.

IX. Chemistry and Life: Blood Gases and Deep-Sea Diving

Due to the high pressures a diver is exposed to on deep dives, the blood concentrations of the gases in the breathing mixture increase by Henry's law. If the diver returns to atmospheric pressure too quickly, the dissolved blood gases bubble out of the blood, causing a condition known as *the bends*. The condition is serious and can be fatal without immediate treatment. Experiments have been conducted in using different breathing gas compositions to minimize the risk. Jacques Cousteau's divers use a helium-enriched mixture, as do NASA astronauts.

X. Colligative Properties

A. Many properties of solutions are due to the amount of solute present, but not its identity. Such properties are called *colligative properties*. The most basic of the colligative properties, by which the others are explained, is *vapor pressure lowering*. The addition of a nonvolatile solute to a solvent decreases the vapor pressure of the solvent and hence the solution. A similar effect is seen with volatile solutes, but it is much harder to quantify these solutions. Consequently, only the effect of nonvolatile solutes is to be discussed.

B. For a solution of a nonvolatile solute dissolved in a liquid solvent, *Raoult's law* quantitatively describes the reduction in the vapor pressure, $P_A = X_A P°_A$, where P_A is the vapor pressure of the solution, $P°_A$ is the vapor pressure of the pure *solvent* at the temperature of interest, and X_A is the mole fraction of the solvent in the mixture. Solutions that obey Raoult's law are said to be *ideal solutions*.

C. *Boiling-point elevation* is a colligative property by which the boiling point of a solution is higher than the boiling point of the pure solvent. The extent to which the boiling point is raised is given by, $\Delta T_b = K_b m$, where K_b is the *molal boiling-point-elevation constant* of the solvent, m is the molality of the solute in the solution, and ΔT_b is the change in the boiling point of the solution relative to the pure solvent. Note that the molality is that of *dissolved particles*; when ionic solutes are used, the molality is a whole-number multiple of the molality of the formula unit. For example, 0.10 *m* CaCl$_2$ dissolves to give 0.30 *m* ions. It is the latter quantity that must be used here.

- D. *Freezing-point depression* is a colligative property by which the freezing point of a solution is lower than the freezing point of the pure solvent. The extent to which the freezing point is lowered is given by, $\Delta T_f = K_f m$, where K_f is the *molal freezing-point-depression constant* of the solvent, *m* is the molality of the solute in the solution, and ΔT_f is the change in the freezing point of the solution relative to the pure solvent. Note that the molality is that of *dissolved particles*; when ionic solutes are used, the molality is a whole-number multiple of the molality of the formula unit. For example, 0.10 *m* CaCl$_2$ dissolves to give 0.30 *m* ions. It is the latter quantity that must be used here.

- E. *Osmosis* is the process by which solvent molecules will cross a semipermeable membrane from a dilute solution to a concentrated solution. The pressure required to stop osmosis is called the *osmotic pressure*, which is given by $\pi = MRT$, where π is the osmotic pressure, *R* is the ideal-gas constant, *T* is the Kelvin temperature, and *M* is the molarity of the solute.
    1. When two solutions of identical osmotic pressure are separated by a semipermeable membrane, no osmosis will occur and the solutions are said to be *isotonic*.
    2. If two solutions of different osmotic pressure are separated by a semipermeable membrane, the solution with the lower osmotic pressure is said to be *hypotonic*.
    3. If two solutions of different osmotic pressure are separated by a semipermeable membrane, the solution with the higher osmotic pressure is said to be *hypertonic*.
    4. Osmosis is crucial to the functioning of living cells. Normal intracellular fluid is isotonic with cell bodies. When cells are placed in a solution that is hypertonic to them, *crenation* occurs, which is the shriveling of the cells due to water leaving them and passing to the more concentrated fluid. Conversely, when cells are placed in a solution that is hypotonic to them, the cells expand as water flows from the solution into the cells and, finally, the cells burst. This phenomenon for red blood cells is called *hemolysis*.

- F. The colligative properties can be used to determine the molecular weight of a solute.

Discussion Question: Discuss how spreading salt can cause ice to melt. Does it make a difference what ionic substance is used?

XI. A Closer Look: Ideal Solutions with Two or More Volatile Components

- A. Raoult's law for two volatile components has the form $P_{total} = P_A + P_B = X_A P_A° + X_B P_B°$. The total vapor pressure is then the sum of the pressure expected from each of the component volatile liquids.

- B. In an ideal mixture, the vapor pressure of the more volatile component is greater than that of the less volatile component; the mole fraction of the more volatile component in the vapor is greater than that of the less volatile component.

- C. The difference in vapor composition is used to separate two or more volatile substances in the technique called *distillation*. In distillation, the mixture is heated and then the vapors are condensed into a separate vessel. The collected liquid now has a much higher concentration of the volatile substance than the original mixture.

XII. A Closer Look: Colligative Properties of Electrolyte Solutions

A. The colligative properties all depend upon the concentration of the solute in the mixture. Further, they depend on the concentration of dissolved solute particles. For example, as 1 M solution of sugar has a dissolved particle concentration of 1 M, because sugar remains as the assembled molecule when dissolved. A 1 M solution of calcium chloride, $CaCl_2$, on the other hand, will have a dissolved particle concentration of 3 M, because it breaks into three ions when dissolved in water.

B. The equations for the colligative properties can then be rewritten to include the *van't Hoff factor, i*, which represents the number of particles into which the solute dissociated when dissolved in water. For example, the boiling point elevation equation becomes $\Delta T_b = i \cdot K_b \cdot m$, and so on. Therefore, $i = 1$ for sugar, while $i = 3$ for calcium chloride.

C. The value of the van't Hoff factor for a given solute can be determined by comparing the experimental value of a colligative property with the value predicted for a nonelectrolyte ($i = 1$).

XIII. Colloids

A. Mixtures in which the lesser component particles are somewhat larger than in solutions are called *colloidal dispersions or simply colloids*. Colloidal particles are in the range of 10 to 2000 Å across, while solutes of solutions are in the molecular size range, typically less than about 10 Å. Because of their large size, colloids reflect light, a phenomenon called the *Tyndall effect*. Colloidal particles are too small to settle out of solution, and are buoyed by collisions with one another.

B. The most important colloids are those in which water is the solvent. Some colloidal particles are attracted to water and form intermolecular attractions with it; these colloidal particles are said to be *hydrophilic*. Other colloidal particles are repelled by water; these colloidal particles are said to be *hydrophobic*.

C. Often, colloidal particles must be removed from their dispersing medium, such as smoke from a chimney. There are two principle techniques for removing colloidal particles.
 1. *Coagulation* is a process in which the colloidal particles are forced to aggregate into larger particles. These enlarged particles can then be removed by filtration or settling.
 2. *Dialysis* is a technique that uses a semipermeable membrane. In this technique, the openings in the semipermeable membrane are large enough for ions and molecular solutes to pass through, but too small for the colloidal particles to pass through. Dialysis is the basis for blood purification by artificial kidney machines.

XIV. Chemistry and Life: Sickle-Cell Anemia

A. Most diseases do not discriminate among ethnic groups or genders. A few diseases do, including sickle-cell anemia, which affects primarily people of African ancestry.

B. In sickle-cell anemia, the hemoglobin molecules in the blood stream are abnormal, especially in shape and solubility. Consequently, the hemoglobin molecules can clog blood vessels easily, which also diminishes oxygen supply to tissues. The result is a painful and generally fatal disease, marked by profound tissue damage.

C. No cause nor cure for sickle-cell anemia has yet been found, although several treatments have been tried with varying degrees of success.

Discussion Question: Many pollutants, such as smoke, are colloidal dispersions. Discuss the impact of technology for removing colloidal particles from a medium. Gas streams, such as from smokestacks, and ground water are logical systems to emphasize.

Lecture Outline – Chapter 13
Brown, LeMay, & Bursten, *Chemistry: The Central Science, 6th Edition*

SAMPLE QUIZ QUESTIONS

1. Calculate the mole fractions of C_2H_5OH and H_2O in a solution containing 100 g of each compound.

2. What is the molality of a solution of 15.75 g NaCl dissolved in 175 g of $CHCl_3$: (a) 0.0900 *m*; (b) 21.2 *m*; (c) 1.54 *m*; (d) 2.76 *m*; (e) 90.0 *m*?

3. What is the molarity of the solute when 15.0 g of $Ba(NO_3)_2$ is dissolved in enough water to make 1450 mL of solution: (a) 2.70 *M*; (b) 0.370 *M*; (c) 0.0396 *M*; (d) 0.0103 *M*; (e) 10.3 *M*?

4. How many grams of LiF are required to make 700.0 mL of a 0.100 *M* solution: (a) 1.43 g; (b) 0.0700 g; (c) 181 g; (d) 1.81 g; (e) 3.70 g?

5. What volume of solution is needed to make a 0.750 *M* solution of CsCl if 16.3 g of solid CsCl is used: (a) 72.6 mL; (b) 0.273 L; (c) 0.129 L; (d) 3921 mL; (e) 7.75 L?

6. What is the concentration of a solution of 3.00 mol of HCl dissolved in enough water to make a 5.00 kg of the solution, expressed in percent by weight: (a) 0.730 %(w/w); (b) 21.8 %(w/w); (c) 36.5 %(w/w); (d) 0.0600 %(w/w); (e) 2.18 %(w/w)?

7. How many grams of H_2SO_4 are required to make 10.0 kg of a 5.00 %(w/w) solution: (a) 5.00×10^2 g; (b) 98.1 g; (c) 1.00×10^2 g; (d) 0.0500 g; (e) 4.90 g?

8. How many grams of $KClO_4$ are required to make 1250.0 mL of a 15.2 %(w/v) solution: (a) 137 g; (b) 1.90×10^2 g; (c) 21.1 g; (d) 169 g; (e) 114 g?

9. What is the percent by volume concentration of 86 proof liquor: (a) 43 %(v/v); (b) 8.6 %(v/v); (c) 86 %(v/v); (d) 95 %(v/v); (e) 170 %(v/v)?

10. What is the normality of a 0.030 *M* solution of H_3PO_4?

11. Clarity: (a) is the same as cloudiness; (b) is the same as transparency; (c) means lack of color; (d) is sometimes observed in true solutions; (e) is the same as colorlessness.

12. Which of the following factors does not affect the initial rate of dissolving a solute in a solvent: (a) the amount of solvent used; (b) temperature; (c) surface area of the undissolved solute; (d) agitation; (e) the amount of solute used?

13. Solubility: (a) is an endothermic process; (b) depends on agitation of the solution; (c) is the maximum amount of solute which can dissolve in a solvent at a given temperature; (d) is an exothermic process; (e) refers to the amount of solute dissolved.

14. Solubility increases: (a) with an increase in the time allowed to dissolve; (b) with increasing temperature if $\Delta H_{solution}$ is a positive value; (c) with agitation; (d) with the amount of solute used; (e) with increasing temperature if $\Delta H_{solution}$ is a negative value.

15. Solubility decreases: (a) with an increase in pressure; (b) with a decrease in temperature; (c) with a solute and solvent which have similar intermolecular forces; (d) with an increase in the value of $\Delta H_{solution}$; (e) with a decrease in pressure.

16. Chemical structure: (a) affects the solubility of a solute in a given solvent; (b) is the sole controlling factor in solubility at a given temperature; (c) is not an important factor in solubility; (d) is the same in the liquid and solid phases of most solvents; (e) refers only to crystalline substances.

17. The vapor pressure of a solution containing a nonvolatile solute is proportional to: (a) the mole fraction of solute; (b) the mole fraction of solvent; (c) the molarity of solvent; (d) the normality of solute.

18. The solubility of O_2 in water at 20°C and $P_{O_2} = 1$ atm is 1.4×10^{-3} M. What is the solubility at the same temperature if the partial pressure of O_2 is 0.20 atm?

19. What is meant by the term "supersaturated solution?" How are such solutions normally prepared?

20. If the freezing point of a 1 m aqueous solution of a soluble salt is -5.5°C, how many ions have resulted from dissociation of each solute particle?

21. Suppose 5.00 g of ethylene glycol, $C_2H_5O_2$, are dissolved in 90.0 g of water. What is the freezing point of the solution($K_f = 1.86$°C/m)?

22. If an aqueous solution of 1 m HX is 10% dissociated, its freezing point is: (a) 1.86°C; (b) -3.72°C; (c) -0.37°C; (d) -2.05°C.

23. The molal freezing point depression constant of water is 1.86°C-kg/mol. A solution of 22.1 g of a substance was dissolved in 750.0 g of water and the freezing point of the solution was -2.35°C. Determine the molar mass of the substance: (a) 129 g/mol; (b) 23.3 g/mol; (c) 3.79 g/mol; (d) 29.5 g/mol; (e) 37.2 g/mol.

24. Exactly 0.705 g of a substance is dissolved in sufficient water to form 100 mL of solution. This solution exhibits an osmotic pressure of 1.05 mm Hg at 25°C. What is the molecular weight of the substance?

25. Colloidal dispersions: (a) are generally clear; (b) are not mixtures; (c) don't reflect light as well as true solutions; (d) don't settle out on sitting; (e) are not stable.

26. Give two ways that colloid particles might be precipitated from a colloidal dispersion.

27. What is the Tyndall effect?

28. Which of the following has the least effect on the solubility of a solid in water: (a) the nature of the solute; (b) the nature of the solvent; (c) the pressure; (d) the temperature?

29. From the following list of substances: $KBr(s)$, $C_2H_5OH(l)$, $CO_2(g)$, $C_6H_{12}(l)$, select the one that best satisfies each of the following descriptions: (a) Likely to be miscible in all proportions with water; (b) Solubility in water is strongly affected by pressure; (c) Produces a conducting solution when dissolved in water; (d) Has the lowest solubility in water.

Lecture Outline – Chapter 14
Brown, LeMay, & Bursten, *Chemistry: The Central Science, 6th Edition*

Chapter 14 — Chemical Kinetics

OVERVIEW: Chapter 14 introduces chemical kinetics. As such, it includes discussions of all the basic issues addressed in kinetics. A complete description of the concept of reaction rate begins the chapter. Through examples, the factors that affect rate are also introduced.

The concentration dependence of reaction rates is covered in terms of reaction order and initial rates. The determination of reaction order and the rate constant from concentration vs. rate data are emphasized.

Following this, the integrated, or time-concentration forms of the rate laws are developed, omitting the calculus, for first and second order reactions. As part of this treatment, the graphic determination of order and the rate constant are emphasized. The concept of half-life is also fully covered.

The temperature dependence of rate is discussed within the context of collision theory, although not named, and quantified using the Arrhenius equation. Again, the graphic determination of the activation energy is emphasized.

Simple mechanisms of chemical equations are also discussed, including the use of the steady-state approximation. Following on mechanisms, the nature and uses of catalysts are described. Both homogeneous and heterogeneous catalysts are covered, with special emphasis given to commercial heterogeneous catalysis and to biological systems.

LECTURE OUTLINE

I. **Factors that affect rate**

  A. *Kinetics* is the study of the *rates*, or speeds, of chemical reactions as well as the study of the sequences of events, or steps, that reactions go through in proceeding from reactants to products. The steps of a chemical reaction are collectively referred to as the reaction's *mechanism*.

  B. The *concentrations of the reactants* used in a reaction usually affect the observed rate. Generally, the more concentrated the reactants, the faster the reaction proceeds.

  C. The *temperature at which the reaction occurs* also affects the rate. Reactions increase in rate as the temperature is increased.

  D. A *catalyst* is a substance that, when added to a chemical reaction, *increases the rate* of the reaction while not being consumed itself. Consequently, catalysts are not present in the balanced chemical equation for the reaction.

  E. For solid or liquid reactants or catalysts, the rate of the reaction increases as the exposed surface area of the substance increases.

Discussion Question: Many reactions are encountered in daily life, from the curing of glues to the spoilage of foods. Discuss chemical processes seen around us and the factors that seem to accelerate or retard these reactions.

II. **Reaction Rates**

  A. *Reaction rate* is expressed as the change in concentration of one of the participants per unit time, much as speed in a car is expressed as the change in position (miles) per unit time (hours). The common units of rate are M/s, although the time is often expressed in any convenient time scale. Note that the units M/s are arithmetically equivalent to the units $M \cdot s^{-1}$.

B. The *average rate* of a chemical reaction is given by the change in concentration over the change in time in all time intervals of the experiment. For the arbitrary reaction A → B, the average rate of loss of "A" is:

$$\text{average rate} = \frac{\text{decrease in concentration of "A"}}{\text{length of time interval}} = -\frac{\Delta[A]}{\Delta t}.$$

The symbolism, [A], is read as "molarity of A."

C. The rate of a reaction expressed in terms of the loss of a reactant is called a *rate of disappearance*.

D. The rate of a reaction expressed in terms of the generation of a product is called a *rate of appearance*.

E. The *instantaneous rate* of a chemical reaction is the most theoretically meaningful description of rate. The instantaneous rate of a chemical reaction is the slope of the curve of the plot of concentration vs. time, taken at any instance in time. As such, the instantaneous rate is a derivative:

$$\text{rate} = \frac{d[\text{concentration}]}{dt}.$$

F. The rate of disappearance of a reactant is related to the rate of appearance of a product through their stoichiometric coefficients. For the hypothetical reaction 2A + 3B → C, the rates are related by: $-\frac{1}{2}\frac{\Delta[A]}{\Delta t} = -\frac{1}{3}\frac{\Delta[B]}{\Delta t} = +\frac{1}{1}\frac{\Delta[C]}{\Delta t}$.

III. The Dependence of Rate on Concentration

A. The *rate law* of a chemical reaction is the algebraic equation that shows the relationship between concentration and rate or time. It has the form: Rate = $k[\text{reactant 1}]^m[\text{reactant 2}]^n\ldots$, where m and n are powers to which the reactant concentrations are taken and k is the *rate constant* for the reaction.

B. The rate constant of a reaction is indeed constant for a given reaction at a given temperature. Different reactions have different values of k, and k for a given reaction generally increases as temperature increases.

C. The powers to which the concentrations are raised in the rate law are called the *orders* of the reaction. Reaction orders are defined for each reactant in the rate law; when the exponent on a given reactant is 1, it is said to be *first order* with respect to that reactant, and similarly for *second order* and *third order*. It is also possible that an exponent is found to equal zero; in this case it is said to be *zeroeth order* in that reactant. In addition, the *overall order* of a reaction is defined as the sum of the exponents. Zeroeth, first, and second order reactions are fairly common, with first order being, by far, the most common.

D. The units of k are derived from the order of the reaction. If Rate = $k[A]^n$, then $k = \frac{\text{Rate}}{[A]^n}$. Looking only at the units, units of $k = \frac{M}{\text{time} \cdot M^n}$. Therefore, for a zeroeth order reaction, k has the units of M/time, for a first order reaction, k has the units of 1/time, for a second order reaction, k has the units of $1/M \cdot$time, and so on.

E. To compare the rates of different reactions or the rates of the same reaction performed under differing conditions, it is necessary to examine each reaction in a consistent way. This is done by using the *initial rates* to determine the rate laws.

Discussion Question: Discuss zeroeth-order processes, such as food spoilage (zeroeth-order at least in regard to amount of food present). What other processes seem to proceed without regard for concentration.

IV. Change of Concentration with Time

A. *First-order reactions* are those in which the exponent on the reactant is 1, i.e., rate = $-\frac{\Delta[A]}{\Delta t} = k[A]^1 = k[A]$. Through integration (not covered here) the rate form of the rate equation can be transformed to give the relationship between concentration and time. For a first-order reaction, the time-concentration form of the rate equation is $\ln[A]_t - \ln[A]_0 = \ln\frac{[A]_t}{[A]_0} = -kt$, where the notation $[A]_t$ is read as "concentration of A at time=t," and $[A]_0$ is read as "concentration of A at time=0." From the rate law, it is clear that a plot of $\ln[A]_t$ vs. t should yield a straight line for a first-order reaction.

B. The *half-life*, $t_{1/2}$, of a reaction is defined as the time required to consume one-half of the reactant present at the beginning of the time interval. For first-order reactions, substituting $[A]_{t_{1/2}} = \frac{1}{2}[A]_0$ into the rate law yields

$t_{1/2} = 0.693/k$.

C. *Second-order reactions* are those in which the exponent on the concentration of the reactant is 2, i.e., rate = $-\frac{\Delta[A]}{\Delta t} = k[A]^2$. Integration yields the time-concentration form of the rate law as $\frac{1}{[A]_t} = kt + \frac{1}{[A]_0}$. From the rate law, it is clear that a plot of $\frac{1}{[A]_t}$ vs. t should yield a straight line for a second-order reaction. The half-life of a second-order reaction is given by $t_{1/2} = \frac{1}{k[A]_0}$.

V. Temperature and Rate

A. The effect of temperature and concentration on the rate of a chemical reaction can be explained by the *collision model* or *collision theory*. According to the collision model reactant molecules must possess certain properties.
   1. The reactant molecules must collide. Only a certain fraction of collisions will lead to reaction due to improper alignment or insufficient energy of collision.
   2. Anything which increases the number of collisions per unit time will also increase the rate of the reaction.

Both an increase in temperature and an increase in reactant concentration will increase the rate of reaction. This is in keeping with the treatment of the concentration dependence given above.

B. A temperature increase provides additional energy to a chemical reaction. In order for a reaction to proceed, a certain amount of energy needs to be possessed by the reactants. Therefore, as temperature increases, a greater fraction of the reactants possess the necessary energy and the reaction proceeds faster. The amount of energy *required* by the reactants for the reaction to proceed is called the *activation energy*, E_a.

C. The particular arrangement of atoms at this maximum energy state is called the *activated complex* or *transition state*. The activated complex generally is an intermediate form between the reactants and the products of the reaction.

D. Svante Arrhenius first described the activated complex and found the quantitative relationship between temperature and the rate constant for a reaction. The expression is called the Arrhenius equation, which has the form, $k = Ae^{-E_a/RT}$, or $\ln k = \ln A - \frac{E_a}{RT}$, where k is the rate constant at a given Kelvin temperature, T, R is the gas constant in energy units (usually 8.314 J/mol·K), and A is the *frequency factor* for the reaction. A is related to the frequency of collisions between reactant molecules and the likelihood that the collisions will lead to reaction. A plot of $\ln k$ vs. $\frac{1}{T}$ should give a straight line with a slope of $-\frac{E_a}{R}$ and an intercept of $\ln A$.

E. Comparison of two reaction trials differing only in temperature, yields the more familiar form of the Arrhenius equation, $\ln \frac{k_1}{k_2} = \frac{E_a}{R}\left[\frac{1}{T_2} - \frac{1}{T_1}\right]$. This form of the equation allows the calculation of the activation energy from values of k at two different temperatures, or the prediction of k at a given temperature from k at another temperature and the activation energy.

Discussion Question: Can students think of any processes that proceed more slowly at high temperature than at low temperature?

VI. A Closer Look: Seeing the Transition State

The study of how molecules collide, transfer energy, and make and break bonds is called *molecular reaction dynamics*. Rapid progress has been made in recent years in making exact measurements of the nature of the transition state.

VII. Reaction Mechanisms

A. A *reaction mechanism* is the sequence of steps a reaction goes through in proceeding from reactants to products.
1. Each step of the reaction is a simple chemical reaction called an *elementary step*. Elementary steps generally involve one, two, or three molecules. The statistical likelihood that the necessary geometry can be attained decreases as the number of involved molecules increases.
2. The *molecularity* of an elementary step is the number of molecules involved in that step.
3. When one molecule is involved in an elementary step, the step is said to be *unimolecular*. When two molecules are involved in an elementary step, the step is said to be *bimolecular*. When three molecules are involved in an elementary step, the step is said to be *termolecular*.
4. A chemical substance that is produced by one elementary step and then consumed by another, so that the substance does not occur as part of the overall products, is called an *intermediate*.

B. The rate law of an overall reaction cannot be reliably gotten from the balanced overall equation. However, the *rate law for an elementary step* does follow from the balanced equation for the step. The rate law for an elementary step is given by the rate constant, k, for the step times the concentration of each substance involved in the step taken to the power of its molecularity.

C. The *rate laws of multistep mechanisms* can be complicated. Clearly, no reaction can proceed faster than its slowest step. The slowest elementary step in a reaction mechanism is called the *rate-determining step*. If the rate-determining step is the first step of the mechanism, the overall rate law is simply the rate law for this step.

D. However, if one or more fast steps precede the rate-determining step, the elementary rate laws for these fast steps and the rate-determining step must be combined to arrive at the rate law for the overall reaction. The *steady-state approximation* states that a fast, reversible reaction step has a product that is consumed at the same rate as it is formed. That is, there is no accumulation of this substance and the rate of its formation equals the rate of its disappearance. Stated another way, it is assumed that fast, reversible steps are *at equilibrium*.

Discussion Question: Choose an example from your area of expertise and discuss its mechanism with the class.

VIII. A Closer Look: Evidence of Mechanism

Often, more than one mechanism is consistent with the experimental rate law. In these cases, other experiments might be performed or extensive statistical treatment of the data might be done. A common experiment is to replace some of the reactant atoms with radioactive isotopes of the atoms. It is then possible to see which reactant atoms go where among the products.

IX. Catalysis

A. Catalysts are, in general, substances that alter the rate of a chemical reaction, although they are themselves not consumed in the reaction. A more specific usage applies the term *catalyst* to a substance that increases the rate of a chemical reaction, and applies the term *inhibitor* to a substance that retards the rate of a chemical reaction. Catalysts speed the reaction by lowering the activation energy of the reaction.

B. In *homogeneous catalysis*, the catalyst is in the same phase as the components of the reaction mixture. Catalysts for homogeneous catalysis are called *homogeneous catalysts*. Homogeneous catalytic systems are fairly efficient, leading to little loss of the catalyst, such loss being called *poisoning*.

C. In *heterogeneous catalysis*, the catalyst is in a different phase than the reactants. Catalysts for heterogeneous catalysis are called *heterogeneous catalysts*. Heterogeneous is the more commercially important of the two types, usually with the catalyst existing as a solid and the reaction mixture being either liquid or gaseous.
   1. In the function of a heterogeneous catalyst, *adsorption* is typically the first step of the reaction. In adsorption, reactant molecules bind to the surface of the catalyst. This step aligns the reactant molecules for the reaction steps to follow.
   2. The places on the surface of the catalyst that are available for binding to reactant molecules are called *active sites*.

D. In living organisms, *enzymes* modify reaction rates. Enzymes are very large carbon-based compounds of the type called *proteins*.
   1. Enzymes have portions of their molecules specifically shaped to accommodate the reactant molecule (the *substrate*). The fitting of substrate to enzyme is often explained in terms of the *lock-and-key model*, in which the substrate is the key and the enzyme is the lock. The substrate must "fit" into the *active site* of the enzyme in order for the reaction to proceed.

2. Other substances may enter the lock and make the enzyme unavailable for reaction. These other substances are referred to as *inhibitors*.
3. Enzymes are very effective catalysts. The *turnover number*, or the number of substrate molecules reacted by a *single* enzyme molecule, is generally in the range of 10^3 to 10^7 reactions per second.

X. Chemistry at Work: Catalytic Converters

A. Catalytic converters have become standard on all newer cars. It is the converter that prevents the owner from using "leaded" gas; the lead destroys the catalyst, sometimes accompanied by fire.

B. The function of the catalytic converter is to oxidize CO and unburned fuel to CO_2 and water, and to reduce the nitrogen oxides to N_2.

Discussion Question: Catalysts are used in many facets of daily life. Discuss some common catalytic systems, such as the catalytic converter on an auto or the production of formaldehyde from H_2 and CO.

Lecture Outline – Chapter 14
Brown, LeMay, & Bursten, *Chemistry: The Central Science*, 6th Edition

SAMPLE QUIZ QUESTIONS

1. The addition of a catalyst to a chemical reaction: (a) affects the rate of the forward reaction; (b) always accelerates the reaction; (c) affects the rate of the forward and reverse reactions; (d) may not affect the reaction; (e) affects the value of the equilibrium constant.

2. In the reaction A + B → C, rate = $k[A]^2$. Doubling [A] will increase the rate by a factor of: (a) 2; (b) 3; (c) 4; (d) 9.

3. If the rate of a reaction is independent of the concentration of all reactants, the order of the reaction is: (a) zeroeth; (b) first; (c) second; (d) third.

4. For the rate law

$$\text{rate} = k[A][B][C]$$

   what are the units of the rate constant: (a) mol^3/L^3-s; (b) mol/L-s; (c) s^{-1}; (d) L^2/mol^2-s; (e) Rate constants have no units?

5. For the reaction $2NO(g) + O_2(g) \rightarrow 2NO_2(g)$, how is the rate of disappearance of NO, $-\Delta[NO]/\Delta t$, related to the rate of disappearance of O_2?

6. What are the units of a second-order rate constant?

7. A transition state: (a) is a low energy interim product of a reaction; (b) is the temperature at which the liquid and vapor phases are at equilibrium; (c) has no dependency on the reactants of a reaction; (d) is a reactive hypothetical chemical which shows characteristics of the reactants and products; (e) has no dependency on the products of a reaction.

8. Explain why all collisions of molecules do not lead to reaction.

9. A reaction A + B → C, occurs in a single step with $\Delta E = -20$ kJ/mol and $E_a = 80$ kJ/mol. Draw a diagram of the energy profile for the reaction. How does the profile change when a catalyst is added?

10. If a first-order reaction has a half-life of 10 minutes, what is the value of the rate constant?

11. The following rate data were obtained for the reaction

$$2NO(g) + 2H_2(g) \rightleftharpoons N_2(g) + 2H_2O(g)$$

[NO]	[H$_2$]	initial rate, M/sec
0.10 M	0.10 M	1.4 x 10^{-4}
0.20 M	0.10 M	5.6 x 10^{-4}
0.10 M	0.20 M	2.8 x 10^{-4}

   (a) Write the rate law for this reaction. (b) Calculate the magnitude of the rate constant. (c) What is the reaction rate when [NO] = 0.15 M and [H$_2$] = 0.25 M?

12. The first-order rate constant for the decomposition of a certain substance in water is 1.50×10^{-2}/min. If a solution that is 0.10 M in this substance is prepared, how long will it take for the concentration to reach 0.01 M?

13. Explain why reaction rates increase with increasing temperature.

14. The activation energy of a reaction: (a) is not related to the rate of the reaction; (b) is the amount of energy required to initiate the reaction; (c) is the amount of energy liberated by initiating the reaction; (d) is a negative quantity; (e) is the difference in energy content between the reactants and products of the reaction.

15. The activation energy of a reaction can be lowered by: (a) raising the temperature; (b) removing products; (c) lowering the temperature; (d) adding a catalyst.

16. The activation energy for a reaction can be determined from a plot of: (a) k vs. T; (b) $\ln k$ vs. T; (c) k vs. $1/T$; (d) $\ln k$ vs. $1/T$.

17. Consider the reaction

$$CH_2CH_2 + HBr \rightarrow CH_3CH_2Br$$

What is the expected rate law for the reaction? (a) rate = $k[CH_2CH_2][HBr]$; (b) rate = $k[HBr]$; (c) rate = $k[CH_3CH_2Br]$; (d) rate = $k[CH_2CH_2]$; (e) rate = $k[CH_2CH_2][HBr]/[CH_3CH_2Br]$

18. A chemical reaction is postulated to occur in two steps, $B + B \rightarrow B_2$ and $B_2 + A \rightarrow AB + B$. (a) What is the overall reaction? (b) Identify the intermediate. (c) If the first step is rate-determining, what is the rate law for this overall reaction? (d) Is the second step unimolecular, bimolecular, or termolecular? Explain briefly.

19. For the bimolecular step

$$CH_3COCH_3 + I_2 \rightarrow CH_3COCH_2I + HI$$

tripling the concentration of CH_3COCH_3: (a) should triple the rate constant of the reaction; (b) should double the rate of the reaction; (c) should have no effect on the rate of the reaction; (d) should drive the reaction to the left; (e) should triple the rate of the reaction.

20. A catalyst: (a) increases the average kinetic energy of the reactants; (b) increases the activation energy of the reaction; (c) alters the mechanism of the reaction; (d) increases the frequency of collisions between reactant molecules.

Lecture Outline – Chapter 15
Brown, LeMay, & Bursten, *Chemistry: The Central Science*, 6th Edition

Chapter 15 — Chemical Equilibrium

OVERVIEW: Equilibrium has been mentioned twice earlier in the text, in the context of vapor pressure and in the context of solubility. Chapter 15 represents the detailed introduction to equilibrium, the properties of equilibrium systems, and the algebraic relationships involving equilibria.

The concept of equilibrium is developed using the Haber process as the historical basis of equilibrium studies. With the law of mass action, the equilibrium constant is introduced and its relationship to the concentrations of species in an equilibrium reaction is discussed. The determination of equilibrium constants from experimental data is emphasized.

Heterogeneous equilibria are introduced from the basis of the concentrations of pure solids and liquids being constant. It is stressed that the approach to evaluating an equilibrium system is the same, regardless of whether it is homogeneous or heterogeneous.

The determination of equilibrium concentrations from initial concentrations and the equilibrium constant is covered in detail. Several different algebraic rearrangements are used to solve these problems, including substitution, square roots, and the quadratic equation.

Finally, using the specific information from the first parts of the chapter, qualitative aspects of Le Châtelier's principle are discussed, citing changes in concentration, pressure, volume, and temperature as the most common forms of stress placed on chemical equilibria.

LECTURE OUTLINE

I. **The Concept of Equilibrium**

   A. In a chemical reaction in the state of *equilibrium*, the concentrations of all reactants and products cease to change, but with some amount of each substance still present. As written in a chemical equation, two single-barbed arrows are used to indicate that the reaction continues to proceed, with the rate of the forward reaction equal to the rate of the reverse reaction, $A \rightleftharpoons B$. Because the rates of the forward and reverse reactions are equal,

   $k_f[A] = k_r[B]$. It then follows that $\frac{[B]}{[A]} = \frac{k_f}{k_r} = \text{constant}$.

   B. An equilibrium is a *dynamic* state. Even though the net concentrations of reactants and products stay the same, individual molecules continue to react, and so reactants and products are continuously formed and consumed.

II. **Chemistry at Work: Nitrogen Fixation and the Haber Process**

   The classic example of equilibrium studies is that of nitrogen fixation and the Haber process.
   1. *Nitrogen fixation* is the conversion of N_2, with its triple bond, into a nitrogen compound with more reactive single bonds, such as NH_3. Nitrogen fixation is important in many ways, from the production of fertilizers to the production of explosives.
   2. The *Haber process*, or the Haber-Bosch process, was developed during World War I in order to produce explosives from materials native to Germany. Much research was performed into the optimum conditions for the reaction, $N_2 + 3H_2 \rightleftharpoons 2NH_3$. Variations in temperature, pressure, and amounts of materials were investigated. Today, the Haber process is

still the principal method for producing ammonia, using atmospheric nitrogen.

III. The Equilibrium Constant

A. In an equilibrium reaction, since the reaction is proceeding in both directions, it makes no difference whether the reaction is begun with all reactants, all products, or a mixture of reactants and products. That is to say, the equilibrium state can be approached either from reactants *or* from products.

B. The *law of mass action* expresses the relative concentrations of reactants and products in terms of a quantity called the equilibrium constant. Also called mass balance, the law states that the mass of reactants consumed must equal the mass of products formed. A balance exists when the ratio of product concentrations to reactant concentrations does not change through further reaction; this is the *equilibrium* condition.

C. The *equilibrium constant* of a reaction mathematically describes the equilibrium condition.
   1. Consider the general reaction, $jA + kB \rightleftharpoons pR + qS$, where A, B, R, and S are hypothetical substances, and $j, k, p,$ and q are their coefficients in the balanced chemical equation. Beginning with some mixture of reactants and/or products, the reaction will proceed toward equilibrium as long as the needed substances are still present.
   2. At equilibrium, the *ratio* of the product concentrations to the reactant concentrations, each taken to the power of its coefficient in the balanced chemical equation, is a constant, $K = \dfrac{[R]^p[S]^q}{[A]^j[B]^k}$, where K is called the *equilibrium constant* for the reaction. As before, the notation, [R], etc., is read to mean "the concentration of R." The value of the equilibrium constant is affected only by temperature; it is independent of the absolute concentrations.

D. Concentrations can be expressed in several unit systems. The most common way to express concentrations, particularly in solution reactions, is to use molarity. For added information, subscripts are commonly placed behind K that denote some relevant fact. When concentrations are expressed in molarity, the symbol, K_c, is used.

E. Equilibrium constants must be positive, although there are no restrictions on how large or how small the value can be. Common values of K_c range from approximately 10^{-30} to 10^{+50}. When the value of K is much *larger* than 1, the equilibrium is said to *lie to the right*; the concentrations of the products are much greater than the concentrations of the reactants. Conversely, when the value of K is much *smaller* than 1, the equilibrium is said to *lie to the left*; the concentrations of the reactants are much greater than the concentrations of the products.

F. The value of K for a reaction is reciprocally related to the value of K for the reverse reaction. Therefore, for $A \rightleftharpoons B$, where the equilibrium constant is K, and for $B \rightleftharpoons A$, where the equilibrium constant is K', the Ks are related by $K = 1/K'$.

IV. Heterogeneous Equilibria

A. Equilibria in which all the participants are in the same phase are called *homogeneous equilibria*.

B. Equilibria in which the participants are in two or more phases are called *heterogeneous equilibria*. The concentration of a *pure* liquid or solid is constant at a given temperature and remains essentially constant as temperature changes. Because of this fact, the equilibrium-constant expression for heterogeneous equilibria are rearranged to group pure solids and liquids with the equilibrium constant, and a net constant is obtained, denoted in the text as K'. The effect on writing equilibrium-constant expressions is that *pure liquids and solids are omitted from the equilibrium-constant expression*.

Discussion Question: The common phenomenon of rusting is a heterogeneous process. Discuss whether rusting is an equilibrium reaction. What are the observations used to support the various arguments?

V. Calculating Equilibrium Constants

A. The value of K is a function of the *equilibrium* concentrations of reactants and products. As an algebraic problem, any one quantity can be determined as long as all other quantities are known or can be easily found. A general approach to *solving for the equilibrium constant* of a reaction is given by:
   1. Tabulate the known initial concentrations and equilibrium concentrations of all species involved in the equilibrium.
   2. For those species for which both the initial and the final concentrations are known, calculate the change in concentration that occurs as the system reaches equilibrium.
   3. Use the stoichiometry of the reaction to calculate the changes in concentration for all the other species in the equilibrium.
   4. From the initial concentrations and the changes in concentration, calculate the remaining equilibrium concentrations. The complete set of equilibrium concentrations is then used to calculate the equilibrium constant.

B. The equilibrium constants used thus far are K_c, equilibrium constants where the amounts of the species are expressed as molarities. For gas-phase reactions, the amounts are commonly expressed as the partial pressures of the gases in atmospheres. When K is determined from partial pressures, it is symbolized as K_p. Gas-phase equilibria can be described by either K_c or K_p, and we can easily convert between the two. From $PV = nRT$, $P = \frac{n}{V}RT = MRT$, and the relationship between molarity and partial pressure is clear. As this relationship applies to a chemical equilibrium, $K_p = K_c(RT)^{\Delta n}$, where Δn is the number of moles of products minus the number of moles of reactants. Equilibrium constants do carry units, although the units are commonly omitted.

VI. Applications of Equilibrium Constants

A. One of the most-asked questions about equilibria is what direction will the reaction proceed from a given set of starting concentrations. The answer is obtained from the *reaction quotient*, Q. The reaction quotient has the same algebraic form as the equilibrium-constant, except that the concentrations used are *initial* concentrations. Only when the system is at equilibrium will $Q = K$. If $Q > K$, the ratio of products to reactants is too high, and the system must react *to the left* in order to reach equilibrium. If $Q < K$, the ratio of products to reactants is too low, and the system must react *to the right* in order to reach equilibrium.

B. The principal use of equilibrium constants is the calculation of equilibrium concentrations. The methods are similar to those outlined above in V.A. The known and easily calculated concentrations are tabulated. Using the known concentrations and the equilibrium constant, an unknown concentration can be calculated. These calculations can become involved, including the use of the quadratic equation, $x = \dfrac{-b \pm \sqrt{b^2 - 4ac}}{2a}$.

VII. Factors Affecting Equilibrium: Le Châtelier's Principle

A. *Le Châtelier's principle* states that if a system at equilibrium is disturbed by a change in temperature, pressure, or the concentration of one of the components, the system will shift its equilibrium position so as to counteract the effect of the disturbance.

B. A *change in reactant or product concentrations* can be an addition or a removal. Consider an addition as creating an "excess" stress; the system will react to reduce that excess by consuming some of the added substance and return to equilibrium. Similarly, consider a removal as a "deficit" stress; the system will react to reduce the deficit by producing more of the substance removed and return to equilibrium.

C. The *effect of a pressure or volume change* is generally only noticeable in a gas-phase reaction. If a system is at equilibrium and the total pressure is increased by the application of external pressure, the system responds by a shift in the equilibrium position in the direction that reduces the pressure. That is, a pressure increase will shift the reaction in the direction that reduces the total number of moles of gas.

D. A *temperature change* is the only form of stress to an equilibrium system that *changes* the value of the *equilibrium constant*. Visualize heat as part of the balanced chemical equation for the equilibrium; if the reaction is endothermic, heat is a reactant and, if the reaction is exothermic, heat is a product. The effect of a temperature change can now be predicted by considering a temperature change as altering the amount of heat present.

E. *Catalysts* increase the *rate* at which equilibrium is achieved, but do not affect the equilibrium position nor the equilibrium constant. In terms of Le Châtelier's principle, the addition of a catalyst has no effect on the equilibrium condition.

Discussion Question: Equilibria are very common, from the competition between CO and O_2 for hemoglobin to the chemical balance in a swimming pool to "pH balanced" shampoos. Discuss some of these phenomena, emphasizing the response of the system to additions of substances.

VIII. Chemistry at Work: Controlling Nitrogen Oxide Emissions

A. In automobiles, the exhaust stream contains several nitrogen oxides, especially NO. The equilibrium for the conversion of NO to O_2 and N_2 is favorable at exhaust temperature.

B. However, the reaction is too slow to proceed before the exhaust cools. A catalyst that could accelerate the reaction *and* not be damaged by the high temperatures would be extremely useful. To date, though, no catalyst has been found that is suitable.

C. Modern automobile catalytic converters accelerate the reaction of NO with H_2 and CO.

Lecture Outline – Chapter 15
Brown, LeMay, & Bursten, *Chemistry: The Central Science*, 6th Edition

SAMPLE QUIZ QUESTIONS

1. Write the expression for the equilibrium constant for each of the following reactions:
   (a) $4NH_3(g) + 3O_2(g) \rightleftharpoons N_2(g) + 3H_2O(g)$; (b) $NH_4Cl(s) \rightleftharpoons HCl(g) + NH_3(g)$.

2. If K_c for $2NO(g) + O_2(g) \rightleftharpoons 2NO_2(g)$ is 2.0×10^3 at a particular temperature, what is the concentration of NO at equilibrium with 1.0×10^{-2} M O_2 and 2.0×10^{-2} M NO_2?

3. Some IBr was placed in a constant temperature 3-liter flask. It partially decomposed and the following equilibrium was established: $2IBr(g) \rightleftharpoons I_2(g) + Br_2(g)$. If 1.00 mol of IBr was present initially and the equilibrium mixture contained 0.86 mol IBr, what is the value of the equilibrium constant?

4. At 600°C, the reaction $N_2(g) + 3H_2(g) \rightleftharpoons 2NH_3(g)$ has an equilibrium constant $K_p = 2.3 \times 10^{-6}$. What is the value of K_c at this temperature?

5. At 700°C, the reaction $2SO_2(g) + O_2(g) \rightleftharpoons 2SO_3(g)$ has an equilibrium constant $K_c = 416$. Is a mixture with $[SO_2] = 0.1$ M, $[O_2] = 8.0 \times 10^{-3}$ M, and $[SO_3] = 3.8 \times 10^{-2}$ M at equilibrium? If not, indicate the direction that the reaction must proceed to reach equilibrium.

6. What effect would each of the following changes have on the concentration of NH_3 in the equilibrium mixture described by the reaction $N_2(g) + 3H_2(g) \rightleftharpoons 2NH_3(g)$ + heat? (a) add N_2; (b) remove H_2; (c) raise the temperature; (d) increase the pressure; (e) decrease the volume; (f) add a catalyst.

7. Provide a brief description or definition of each of the following: (a) reaction quotient; (b) Le Châtelier's principle; (c) Haber process; (d) catalyst.

8. Explain why the heterogeneous catalysis of the decomposition of $H_2(g)$ by palladium is retarded by N_2 gas.

9. In which of the following cases does the reaction $A + B \rightleftharpoons C$ proceed farthest toward completion? (a) $K = 1$; (b) $K = 10^{-2}$; (c) $K = 10^2$; (d) $K = 10$

10. If $K = 0.123$ for $A + 2B \rightleftharpoons 2C$, then for $2C \rightleftharpoons A + 2B$, $K =$: (a) 0.123; (b) -0.123; (c) 6.48; (d) 8.13.

11. The concentration of a pure liquid or solid phase is left out of the equilibrium constant expression because: (a) solids and liquids do not react; (b) solids and liquids react slowly; (c) the concentration of a pure solid or liquid is constant; (d) their concentrations cannot be determined.

12. At a given temperature, 0.300 mol NO, 0.200 mol Cl_2, and 0.500 mol ClNO were placed in a 25.0-liter vessel. The following equilibrium is established: $2NO(g) + Cl_2(g) \rightleftharpoons 2ClNO(g)$. At equilibrium, 0.600 mol ClNO was present. The number of moles of Cl_2 present at equilibrium is: (a) 0.100 mol; (b) 0.150 mol; (c) 0.200 mol; (d) 0.250 mol.

13. At a given temperature, 0.300 mol NO, 0.200 mol Cl_2, and 0.500 mol ClNO were placed in a 25.0-liter vessel. The following equilibrium is established: $2NO(g) + Cl_2(g) \rightleftharpoons 2ClNO(g)$. At equilibrium, 0.600 mol ClNO was present. The value of the equilibrium constant is: (a) 1.50×10^3; (b) 2.50×10^3; (c) 60; (d) none of these.

14. A reversible reaction: (a) is a hypothetical reaction which helps explain activation energy; (b) depends only on temperature; (c) proceeds in both directions simultaneously; (d) can lead to the formation of an equilibrium; (e) can proceed in either direction depending on the conditions placed on it.

15. At equilibrium: (a) the rate constants for the forward and reverse reactions are equal; (b) the reaction ceases; (c) the concentrations of the reactants and products are equal; (d) all of the reactants have been converted to products, barring a limiting reactant; (e) the forward and reverse rates are equal.

16. Consider the reaction

$$2H_2S(g) + 3O_2(g) \rightleftharpoons 2H_2O(g) + 2SO_2(g)$$

Write the expression for the equilibrium constant for the reaction:
(a) $K_c = \frac{[H_2S][O_2]}{[H_2O][SO_2]}$; (b) $K_c = \frac{[H_2O]^2[SO_2]^2}{[H_2S]^2[O_2]^3}$;
(c) $K_c = \frac{[H_2S]^2[O_2]^3}{[H_2O]^2[SO_2]^2}$; (d) $K_c = \frac{[H_2O][SO_2]}{[H_2S][O_2]}$; (e) $K_c = \frac{[H_2O][SO_2]}{[H_2S][O_2]^3}$.

17. Consider the reaction

$$CS_2(g) + 4H_2(g) \rightleftharpoons CH_4(g) + 2H_2S(g)$$

Write the expression for the equilibrium constant for the reaction:
(a) $K_c = \frac{[CS_2][H_2]}{[CH_4][H_2S]}$; (b) $K_c = \frac{[CH_4][H_2S]^2}{[CS_2][H_2]^4}$;
(c) $K_c = \frac{[CH_4][H_2S]}{[CS_2][H_2]^4}$; (d) $K_c = \frac{[CS_2][H_2]^4}{[CH_4][H_2S]^2}$; (e) $K_c = \frac{[CH_4][H_2S]}{[CS_2][H_2]}$.

18. Consider the reaction

$$SnO_2(s) + 2H_2(g) \rightleftharpoons Sn(s) + 2H_2O(g)$$

Write the expression for the equilibrium constant for the reaction:
(a) $K_c = \frac{[Sn][H_2O]}{[SnO_2][H_2]}$; (b) $K_c = \frac{[SnO_2][H_2]}{[Sn][H_2O]}$;
(c) $K_c = \frac{[Sn][H_2O]^2}{[SnO_2][H_2]^2}$; (d) $K_c = \frac{[SnO_2][H_2]^2}{[Sn][H_2O]^2}$; (e) $K_c = \frac{[Sn][H_2O]^2}{[SnO_2][H_2]}$.

19. An equilibrium mixture for the reaction

$$O_2(g) + 4HCl(g) \rightleftharpoons 2H_2O(g) + 2Cl_2(g)$$

had the following concentrations: $[O_2]$ = 0.0223 mol/L, $[HCl]$ = 0.277 mol/L, $[H_2O]$ = 7.03 mol/L, and $[Cl_2]$ = 0.880 mol/L. Calculate the equilibrium constant: (a) 1.00×10^3; (b) 1.00×10^6; (c) 3.43×10^{-6}; (d) 2.92×10^5; (e) 9.98×10^{-4}.

Lecture Outline – Chapter 15
Brown, LeMay, & Bursten, *Chemistry: The Central Science*, 6th Edition

20. An equilibrium mixture for the reaction

$$2NO_2(g) + 7H_2(g) \rightleftharpoons 2NH_3(g) + 4H_2O(g)$$

had the following concentrations: [NO$_2$] = 0.0200 mol/L, [H$_2$] = 0.0100 mol/L, [NH$_3$] = 0.0200 mol/L, and [H$_2$O] = 0.0100 mol/L. Calculate the equilibrium constant: (a) 1.00 x 10^6; (b) 5.00 x 10^3; (c) 2.00 x 10^{-4}; (d) 1.00; (e) 1.00 x 10^{-6}.

21. An equilibrium mixture for the reaction

$$2NOCl(g) \rightleftharpoons 2NO(g) + Cl_2(g)$$

had the following concentrations: [NOCl] = 0.100 mol/L, [NO] = 0.0750 mol/L, and [Cl$_2$] = 0.0250 mol/L. Calculate the equilibrium constant: (a) 71.1; (b) 0.0141; (c) 0.188; (d) 0.0188; (e) 0.00141.

22. The position of equilibrium: (a) describes the relative rates of the forward and reverse reactions; (b) depends on the concentrations of the reactants and products; (c) is independent of temperature; (d) depends on whether the reactants and products are solids, liquids, or gases; (e) gives a measure of completeness of reaction.

23. Consider a reaction for which K_c = 0.798 mol/L. Which of the following statements about the equilibrium is correct? (a) The reaction essentially goes to completion; (b) There are slightly more products than reactants in the mixture; (c) The reaction essentially does not occur; (d) The rates of the forward and reverse reactions are not equal; (e) There are slightly more reactants than products in the mixture.

24. Consider the reaction

$$2N_2O(g) + 3O_2(g) \rightleftharpoons 2N_2O_4(g)$$

which is an endothermic reaction. What is the effect of increasing the concentration of N$_2$O? (a) [O$_2$] will decrease; (b) The reaction will be driven to the left; (c) [N$_2$O$_4$] will decrease; (d) [O$_2$] will increase; (e) There will be no effect.

25. Consider the reaction

$$CO(g) + 2H_2(g) \rightleftharpoons CH_3OH(g)$$

which is an exothermic reaction. What is the effect of increasing the temperature? (a) The reaction will be driven to the right; (b) [CO] will decrease; (c) [CH$_3$OH] will increase; (d) The reaction will be driven to the left; (e) [H$_2$] will decrease.

26. For a given reaction, k_f = 15.01 s^{-1} and k_r = 10.09 L/mol-s. What is the equilibrium constant? (a) 1.488 mol/L; (b) 0.672 mol/L; (c) 1.488 L/mol; (d) 26.10 s^{-1}; (e) 0.672 L/mol

Lecture Outline – Chapter 16
Brown, LeMay, & Bursten, *Chemistry: The Central Science*, *6th* Edition

Chapter 16 — Acid–Base Equilibria

OVERVIEW: Chapter 16 continues the coverage of equilibria into acidic and basic systems. First, the Arrhenius and Brønsted-Lowry definitions of acids and bases are discussed and the interrelationship between acids and bases is described.

pH is introduced in this chapter and is used extensively hereafter.

Strong acids are discussed, including listing the common examples. The discussion of weak acids is really where the differences between strong acids and weak acids are developed. The acid-dissociation constant is introduced and used extensively. Many problems are performed showing algebraic manipulations of acid equilibria.

A similar treatment is given for bases, including common strong bases and manipulations of K_b. After K_b has been explored, the relationship between K_a and K_b for a conjugate acid-base pair is discussed.

Following the complete discussions of acids and bases, the acid-base properties of salt solutions are covered, emphasizing the relative magnitudes of the equilibrium constants. The role of molecular structure in the strength of an acid or base is also described.

Finally, to tie the treatment of acids and bases together and to provide a functional basis for further work in chemistry, the Lewis definitions of acids and bases are given and examples of Lewis acid-base reactions are presented.

LECTURE OUTLINE

I. **The Dissociation of Water**

   A. Water can ionize by the reaction: $H_2O(l) \rightleftharpoons H^+(aq) + OH^-(aq)$. The hydrogen ion does not exist in water, and the reaction: $2H_2O(l) \rightleftharpoons H_3O^+(aq) + OH^-(aq)$ is more descriptive. The *hydronium ion*, $H_3O^+(aq)$, depicts the hydrogen ion as being held by a molecule of water. This *autoionization* is spontaneous.

   B. The autoionization of water is an equilibrium process and can be described by an equilibrium constant, $K_c = \dfrac{[H_3O^+][OH^-]}{[H_2O]^2}$. Because $[H_2O]$ is constant, water being a pure liquid, this is a heterogeneous equilibrium, $K_c \cdot [H_2O]^2 = [H_3O^+][OH^-] = K_w$, where K_w is called the *ion-product constant of water*. K_w has a value of 1.0×10^{-14} at 25°C. Given the autoionization of water, an acidic solution is one in which $[H_3O^+] > [OH^-]$, a basic solution is one in which $[H_3O^+] < [OH^-]$, and a neutral solution is one in which $[H_3O^+] = [OH^-]$.

II. **Brønsted-Lowry Acids and Bases**

   A. Historically, *acids* are substances that taste sour, like the acetic acid in vinegar, and can dissolve active metals. *Bases*, on the other hand, taste bitter and feel slippery to the touch.

   B. Svante *Arrhenius* formally defined *acids* as substances that release H^+ ions in aqueous solution and *bases* as substances that release OH^- ions in aqueous solution.

   C. The hydrogen ion or the proton, H^+, is *hydrated* in water. The hydrated proton consists of the proton surrounded by the oxygen atoms of several water molecules. A variable number of water molecules is bound to the hydrogen, with the attachment of one water, the hydronium ion, H_3O^+, being perhaps the best representation. The Arrhenius definition of acids is then modified to state that an acidic solution is formed by a chemical reaction in which an acid transfers a hydrogen ion to water.

- D. Brønsted and Lowry independently proposed similar definitions of acids and bases. In the *Brønsted-Lowry* concept, an *acid* is a substance capable of donating a proton, and a *base* is a substance capable of accepting a proton. Note that there is no requirement of a solvent in the Brønsted-Lowry definition.

- E. In the Brønsted-Lowry definition, acid-base reactions involve the transfer of a proton from the acid to the base. As such, the ion or molecule left over after an acid donates its proton is a base. The acid and the ion or molecule left after donation are referred to as a *conjugate acid-base pair*. For example, HCl and Cl$^-$, HCN and CN$^-$, H_2O and OH$^-$, and H_3O^+ and H_2O are conjugate pairs.

- F. Substances, such as water, that can behave both as a Brønsted acid and as a Brønsted base, are called *amphoteric*.

- G. The relative strengths of an acid and its conjugate base are reciprocally related; the stronger an acid, the weaker its conjugate base, and the weaker an acid, the stronger its conjugate base.

Discussion Question: Discuss the warning labels placed on household products that are acids or bases.

III. The pH Scale

- A. The *pH scale* offers a convenient method of expressing hydrogen ion concentration. By definition, $pH = -\log[H^+] = \log\frac{1}{[H^+]}$, where "log" is the common or base-10 logarithm. Because of the sign change, higher pH indicates lower $[H^+]$, while lower pH indicates higher $[H^+]$. Further, acidity can now be restated in terms of pH; acidic solutions are those for which pH < 7.00, basic solutions are those for which pH > 7.00, and a neutral solution is one for which pH = 7.00 at 25°C.

- B. The logarithmic scale is also common for expressing other quantities, such as [OH$^-$] or K_c. In general, $pX = -\log(X)$, where "X" is the quantity of interest. Thus, $pOH = -\log[OH^-]$ or $pK_c = -\log K_c$.

- C. pH can be measured by several means.
  1. The use of *acid-base indicators* has historically been the principle way of determining pH. Some substances, such as litmus and phenolphthalein, possess different colors within different ranges of pH. For example, litmus is red in acidic solution and blue in basic solution, while phenolphthalein is colorless below a pH of about 9 and pink above a pH of about 9. The pH of a solution can be determined by observing the colors of several indicators in portions of the solution. Many companies sell papers impregnated with several indicators; observing the color of the paper after dipping it in the solution gives the solution's pH.
  2. About 60 years ago, an electronic device was invented to measure pH. The *pH meter* measures the concentration of hydrogen ions in the test solution by determining the solution's ability to conduct electricity through a thin glass sheet. Under these conditions, only hydrogen ions are small enough to pass through the pores in the glass. The conductivity is converted to pH electronically.

Discussion Question: The autoionization constant of water increases with temperature. Discuss the effect this phenomenon has on the pH scale.

IV. Strong Acids and Bases

A. *Strong acids* are acids that are strong electrolytes. That is, the acidic compound converts virtually all of its molecules into its ions when placed in solution. For strong acids, the observed [H$^+$] equals the formal concentration of the acid. The term *formal concentration* in this context means the total concentration of the solute in the solution. For example, we can talk about a solution being formally 0.10 M HCl, without regard for what fraction of the solute exists as HCl or as H$^+$ and Cl$^-$.

B. The *common strong acids* are HCl, HBr, HI, HNO$_3$, HClO$_4$, and H$_2$SO$_4$ (1st H$^+$ lost only).

C. *Strong bases* are bases that are strong electrolytes. That is, the basic compound converts virtually all of its molecules into its ions when placed in solution. For strong bases, the observed [OH$^-$] equals the formal concentration of the base.

D. The common strong bases are the hydroxides of the group 1A and heavier group 2A metals, such as NaOH, KOH, and Ca(OH)$_2$.

E. The pH of a solution of a base is found by first determining the hydroxide ion concentration, [OH$^-$], calculating [H$^+$] from [H$^+$] = K_w/[OH$^-$], and then solving for pH using pH = -log[H$^+$].

V. Weak Acids

A. *Weak acids* are those acids that are weak electrolytes. That is, weak acids are those that do not convert all their molecules into ions when dissolved in water. Weak acids ionize to produce ions in an equilibrium reaction, HX*(aq)* $\rightleftharpoons$ H$^+$*(aq)* + X$^-$*(aq)*. The equilibrium constant for this important reaction type is given the symbol K_a and is called the *acid-dissociation constant*. The value of K_a can be calculated using the same methods covered previously for K_c. The fact that the methods used in working with acid or base equilibria are the same as those discussed for other equilibria should be emphasized.

B. *Calculating pH* for solutions of weak acids is an important problem type in working with acids. First, [H$^+$] must be determined using the methods covered previously for K_c. Then the [H$^+$] is converted to pH using the expression pH = -log[H$^+$].

C. *Polyprotic acids* are those acids which can release more than one proton. The dissociation of each proton is a separate Brønsted-Lowry acid-base reaction, and each dissociation is described by a separate value of K_a, often referred to as K_{a1}, K_{a2}, and so forth. K_a values for successive losses of protons from a polyprotic acid usually decrease by at least a factor of 10^3.

Discussion Question: Qualitatively discuss how H$_2$CO$_3$/HCO$_3^-$/CO$_3^{2-}$ help the body maintain a fairly constant pH.

VI. Weak Bases

A. *Weak bases* are those bases that are weak electrolytes. That is, weak bases are those that do not convert all their molecules into ions when dissolved in water. Weak bases react to produce ions in an equilibrium reaction, B*(aq)* + H$_2$O*(l)* $\rightleftharpoons$ HB$^+$*(aq)* + OH$^-$*(aq)* or B$^-$*(aq)* + H$_2$O*(l)* $\rightleftharpoons$ HB*(aq)* + OH$^-$*(aq)*. The equilibrium constant for this important reaction type is given the symbol K_b and is called the *base-dissociation constant*. The value of K_b can be calculated using the same methods covered previously for K_c. The fact that the methods used in working with acid or base equilibria are the same as those discussed for other equilibria should be emphasized.

- B. Many nitrogen-containing compounds in which the nitrogen is bonded to three different atoms are bases. These compounds, derived from ammonia, are called *amines* and have the general formula NR_3, where R is a carbon-containing group or a hydrogen atom. Regardless of the identities of the R groups, amines are basic due to the unbonded electron pair on the nitrogen, which can abstract a hydrogen ion from water.

- C. *Anions of weak acids* also are bases. Thus, the anion of HCN, CN^-, is a base, and so on.

VII. Chemistry at Work: Amines and Amine Hydrochlorides

Amines can abstract a hydrogen ion from an acid other than water. When an amine is treated with an acid, an *acid salt* is formed. If hydrochloric acid is used, the amine consumes the hydrogen ion and the chloride balances the charge of the resulting cation, $A + HCl \rightleftharpoons AH^+Cl^-$ or $A \cdot HCl$. Some amines, notably drugs, form acid salts with sulfuric acid, $2A + H_2SO_4 \rightleftharpoons (AH^+)_2 \cdot SO_4^{2-}$.

VIII. Relation Between K_a and K_b

- A. From Hess's law, if two reactions are combined to yield a third reaction, the equilibrium constant of this third reaction is given by the product of the equilibrium constant of the two reactions being combined. In this context, the addition of the acid-dissociation reaction and the base-dissociation reaction yields the autoionization reaction of water:

  $HA \rightleftharpoons H^+ + A^-$

  $A^- + H_2O \rightleftharpoons HA + OH^-$

  $H_2O \rightleftharpoons H^+ + OH^-$.

  Multiplying K_a and K_b yields: $K_a \cdot K_b = \frac{[H^+][A^-]}{[HA]} \cdot \frac{[HA][OH^-]}{[A^-]} = K_w$, as expected. The usefulness of this relationship is that for any base, the K_b value can be found from the K_a of its conjugate acid. Conversely, the K_a value of any acid can be found from K_b of its conjugate base.

- B. Using the pX notation, we commonly speak of the pK_a of an acid and the pK_b of a base, where pX = -logX. Therefore, $pK_a = -\log K_a$, $pK_b = -\log K_b$, and $pK_w = -\log K_w$. Because $K_a \cdot K_b = K_w$, it follows that $pK_a + pK_b = pK_w$. Further, the value of K_w at 25°C is 1.0×10^{-14} and so $pK_w = 14.00$ at 25°C.

IX. Acid–Base Properties of Salt Solutions

- A. *Salt solutions* are solutions of ionic compounds that contain neither H^+ nor OH^-. However, anions are the conjugate bases of acids and hydrogen-releasing cations are the conjugate acids of bases. Therefore, it is common for the ions in salt solutions to react with water to form acidic or basic solutions. For example, solutions of NaF yield Na^+ and F^- and, as the conjugate base of a weak acid, F^- attacks water to produce OH^-. From the relationship $K_a \cdot K_b = K_w$, it is clear that the larger the value of K_a, the lower the value of K_b and visa versa.

- B. Anions of polyprotic acids that still contain at least one ionizable proton can act as either an acid or as a base.

C. Metal cations, except those of group 1A and the heavier members of group 2A, act as weak acids in aqueous solution, bonding with OH⁻ from water and releasing H⁺.

D. The pH of a salt solution can be qualitatively predicted.
   1. A salt solution derived from a *strong acid* and a *strong base* is neutral. That is the pH of the solution is 7.
   2. A salt solution derived from a *weak acid* and a *strong base* has a pH greater than 7.
   3. A salt solution derived from a *strong acid* and a *weak base* has a pH less than 7.
   4. A salt solution derived from a *weak acid* and a *weak base* has a pH that depends on the relative values of K_a and K_b for the cation and anion. If the base is stronger than the acid, the pH will be greater than 7. If the acid is stronger than the base, the pH will be less than 7.

X. Acid–Base Behavior and Chemical Structure

A. In order for a hydrogen-containing compound to be an acid, the hydrogen must be bonded to a more electronegative element. That is, the hydrogen must possess δ+ character in the compound to be acidic. If the hydrogen is negative, as in NaH, the hydride anion will consume a proton, or act as a base. If the bond to hydrogen is covalent, the hydrogen will possess no appreciable acid-base character. In general, metal hydrides are either basic or show no pronounced acid-base character, while nonmetal hydrides are either acidic or show no pronounced acid-base character. Further, an acidic hydrogen is more easily lost when the bond strength is small.

B. Many common acids, called *oxyacids*, contain oxygen in addition to hydrogen and another element, usually a nonmetal. In oxyacids, the ionizable hydrogen atoms are bonded to the oxygen, not to the other element. This other element is generally a nonmetal because, if the oxygen draws too much electron density from this element, H⁺ will not released, but rather an OH⁻ will be released. Trends as to the strengths of oxyacids can also be seen.
   1. For acids that have the same structure but differ in the electronegativity of the central atom, acid strength increases with increasing electronegativity.
   2. In a series of acids that have the same central atom, but differ in the number of attached oxygens, acid strength increases with increasing oxidation state of the central atom.

C. *Carboxylic acids* constitute a large class of oxyacids. All carboxylic acids contain the same *functional group*, $\overset{\overset{\displaystyle O}{\|}}{C}-OH$, in which the carbon is also attached to another carbon atom or to a hydrogen atom.

XI. Chemistry and Life: The Amphoteric Behavior of Amino Acids

Most amino acids, the building blocks of proteins, are amphoteric. As their name implies, amino acids contain both an amine group (basic) and a carboxylic acid group (obviously acidic). At physiological pH, most amino acids lose their acidic proton, which is consumed by the basic amine nitrogen. As a result, the "normal" form for the amino acids is for the acidic part to be anionic and the basic part to be cationic. An ion comprised of a cationic site and an anionic site is referred to as a *zwitterion*.

XII. Lewis Acids and Bases

A. G.N. Lewis proposed a theory of acids and bases in which an *acid* is defined as an electron-pair acceptor and a *base* is defined as an electron-pair donor. As such, any compound identified as an acid prior to this point is also an acid in the Lewis theory, and any compound identified as a base prior to this point is a base in the Lewis theory.

B. In the Lewis theory, *metal ions* that react with water to undergo *hydrolysis* are acids. During the dissolution process, metal-containing ionic compounds are *hydrated*, the water molecules separating and stabilizing the ionic charges. These waters of hydration on the metal are acidic, releasing protons to the solution. The extent of dissociation increases as the charge of the metal ion increases and as the size of the metal ion decreases.

SAMPLE QUIZ QUESTIONS

1. In general, acids: (a) have a sour taste; (b) have a bitter taste; (c) have a soapy feel; (d) are insoluble in water; (e) turn red litmus to blue.

2. In general, bases: (a) turn blue litmus to red; (b) react with metals to produce hydrogen gas; (c) are insoluble in water; (d) have a sour taste; (e) have a bitter taste.

3. Classify each of the following as being a strong acid, strong base, weak acid, or weak base: (a) HNO_2; (b) NH_4^+; (c) HF; (d) HCl; (e) KOH; (f) F^-.

4. The ion-product constant for water: (a) accounts for the high conductivity of water; (b) has the symbol K_a; (c) is equivalent to $K_c \cdot [H_2O]$; (d) has the value 1.00×10^{-7} at 25°C; (e) shows that water is slightly acidic.

5. Which of the following most properly illustrates the reaction of an acid with water?

    (a) $PbCrO_4(s) \rightarrow Pb^{2+}(aq) + CrO_4(aq)$;
    (b) $HClO_2(aq) + H_2O(l) \rightarrow ClO_2^-(aq) + H^+(aq)$;
    (c) $Mg(OH)_2(aq) \rightarrow Mg^{2+}(aq) + 2\,OH^-(aq)$;
    (d) $HClO(aq) + H_2O(l) \rightarrow ClO^-(aq) + H_3O^+(aq)$;
    (e) $BaO(s) + H_2O(l) \rightarrow Ba(OH)_2(aq)$

6. What is $[H_3O^+]$ in a 3.00 M solution of $HClO_4$: (a) 1.50 M; (b) 3.00 M; (c) 6.00 M; (d) It depends on the K_a of $HClO_4$; (e) It depends on the volume of solution being considered.

7. The pH of a solution is 3.88. What are $[H_3O^+]$ and $[OH^-]$: (a) $[H_3O^+] = 7.6 \times 10^{-11}\,M$ and $[OH^-] = 1.3 \times 10^{-4}\,M$; (b) $[H_3O^+] = 1.3 \times 10^{-4}\,M$ and $[OH^-] = 7.6 \times 10^{-11}\,M$; (c) $[H_3O^+] = 1.3 \times 10^{-4}\,M$ and $[OH^-] = 7.6 \times 10^{-4}\,M$; (d) $[H_3O^+] = 1.3 \times 10^{-11}\,M$ and $[OH^-] = 7.6 \times 10^{-11}\,M$; (e) $[H_3O^+] = 1.0 \times 10^{-7}\,M$ and $[OH^-] = 1.0 \times 10^{-7}\,M$?

8. If $[H_3O^+]$ is $2.50 \times 10^{-2}\,M$ in an aqueous solution, what is the pOH: (a) 12.398; (b) 2.50; (c) 1.602; (d) 4.00×10^{-13}; (e) 1.00?

9. If $[OH^-] = 0.500\,M$ in an aqueous solution, what is the pH: (a) 0.301; (b) 4.00; (c) 2.00×10^{-14}; (d) 13.500; (e) 13.699?

10. Nitrous acid, HNO_2, has $K_a = 4.5 \times 10^{-4}$. What is the pH of a solution of nitrous acid in which the equilibrium concentration of HNO_2 is 0.200 M: (a) 2.02; (b) 3.35; (c) 9.49×10^{-3}; (d) 0.70; (e) 4.05?

11. Carbonic acid, H_2CO_3, has $K_a = 4.3 \times 10^{-7}$. If the pH of a solution of carbonic acid is 3.000, what is the equilibrium concentration of H_2CO_3: (a) 0.43 M; (b) 0.18 M; (c) 2.3 M; (d) $1.0 \times 10^{-6}\,M$; (e) $1.0 \times 10^{-3}\,M$?

12. Consider the reaction

    $$HClO(aq) + H_2O(l) \rightleftharpoons H_3O^+(aq) + ClO^-(aq)$$

K_a for the reaction is 3.2×10^{-8}. If $[H_3O^+] = 0.000300\ M$ and $[HClO] = 0.210\ M$, what is $[ClO^-]$? (a) $4.6 \times 10^{-11}\ M$; (b) $6.7 \times 10^{-9}\ M$; (c) $1.1 \times 10^{-4}\ M$; (d) $4.3 \times 10^{-7}\ M$; (e) $2.2 \times 10^{-5}\ M$

13. An aqueous solution of formic acid, HCOOH, has pH = 1.888. The equilibrium concentration of HCOOH is found to be $0.987\ M$. What is the value of K_a for formic acid? (a) 1.31×10^{-2}; (b) 1.00×10^{-7}; (c) 1.65×10^{-4}; (d) 7.83×10^{-13}; (e) 1.70×10^{-4}

14. For polyprotic acids, K_a values for the ionization of successive H^+ ions: (a) decrease due to an increase of negative charge on the central atom; (b) decrease due to the fewer number of remaining H^+ atoms; (c) remain fairly constant; (d) increase due to the increasing polarity of the O–H bonds; (e) increase due to repulsions between the H^+ ions.

15. The acid dissociation constants for phosphoric acid vary in the order $H_3PO_4 > H_2PO_4^- > HPO_4^{2-}$. Explain this order of decreasing values for successive K_a.

16. Show that for an acid, HA, and its conjugate base, A^-, the product $K_a \times K_b = K_w$.

17. Describe the relationship between the relative strengths of a series of bases and the relative strengths of their conjugate acids.

18. The conjugate base of an acid: (a) is the molecule or ion left after the acid accepts a proton; (b) is not stable under most conditions; (c) is a stronger acid than the original acid; (d) is the molecule or ion left after the acid donates a proton; (e) cannot itself be an acid.

19. Which of the following is the strongest Brønsted base? (a) NO_3^-; (b) Cl^-; (c) HSO_4^-; (d) CN^-

20. K_b for CO_3^{2-} is 2.1×10^{-4}. If the concentration of CO_3^{2-} is $0.075\ M$, what is $[OH^-]$? (a) $1.6 \times 10^{-5}\ M$; (b) $0.075\ M$; (c) $1.4 \times 10^{-2}\ M$; (d) $4.0 \times 10^{-3}\ M$; (e) $2.5 \times 10^{-10}\ M$

21. K_b for triethylamine, $N(C_2H_5)_3$, is 5.2×10^{-4}. $[OH^-]$ and $[HN(C_2H_5)_3^+]$ are known to be $0.000103\ M$. Calculate $[N(C_2H_5)_3]$: (a) $0.20\ M$; (b) $2.0 \times 10^{-5}\ M$; (c) $4.9 \times 10^4\ M$; (d) $0.40\ M$; (e) $1.1 \times 10^{-8}\ M$.

22. Consider the reaction

$$N(CH_3)_3(aq) + H_2O(l) \rightleftharpoons HN(CH_3)_3^+(aq) + OH^-(aq)$$

for which $K_b = 6.3 \times 10^{-5}$. If the following equilibrium concentrations exist: $[OH^-] = 0.0126\ M$ and $[N(CH_3)_3] = 0.100\ M$, calculate $[HN(CH_3)_3^+]$: (a) $6.3 \times 10^{-6}\ M$; (b) $5.0 \times 10^{-4}\ M$; (c) $7.9 \times 10^{-6}\ M$; (d) $5.0 \times 10^{-3}\ M$; (e) $7.9 \times 10^{-7}\ M$.

23. Carbonate, CO_3^{2-}, has $K_b = 2.1 \times 10^{-4}$. What is the pH of a solution of carbonate in which the equilibrium concentration of CO_3^{2-} is $0.500\ M$: (a) 1.05×10^{-4}; (b) 1.989; (c) 0.0102; (d) 12.011; (e) 9.76×10^{-13}.

24. Cyanide, CN^-, has $K_b = 1.6 \times 10^{-5}$. If the pH of a solution of cyanide is 12.000, what is the equilibrium concentration of CN^-? (a) $1.2\ M$; (b) $6.2\ M$; (c) $0.16\ M$; (d) $1.6 \times 10^{-9}\ M$; (e) $6.2 \times 10^{-8}\ M$

25. An acid anhydride: (a) generally is composed of a metal and oxygen; (b) is formed by reacting an acid with water; (c) forms a base when reacted with water; (d) is formed by reacting a base with water; (e) forms an acid when dissolved in water.

26. The compound formed by Cl_2O_7 when it is reacted with water is: (a) $HClO_4$; (b) $HClO_3$; (c) $HClO_2$; (d) $HClO$; (e) H_2ClO_4.

27. Which of the oxyacids of the halogens would you expect to be the strongest acid? Explain.

28. Indicate whether the following salts will give acidic, basic, or neutral solutions: (a) NH_4Cl; (b) $AlCl_3$; (c) $NaCN$; (d) KBr.

29. Which of the following ions will hydrolyze the most? (a) Al^{3+}; (b) Ga^{3+}; (c) In^{3+}; (d) Tl^{3+}

30. The degree of hydrolysis of a cation is highest for an ion with: (a) high charge and small radius; (b) high charge and large radius; (c) low charge and large radius; (d) low charge and small radius.

Lecture Outline – Chapter 17
Brown, LeMay, & Bursten, *Chemistry: The Central Science*, 6th Edition

Chapter 17 — Additional Aspects of Aqueous Equilibria

OVERVIEW: Chapter 17 completes the coverage of equilibria. The common-ion effect, buffers, titrations, ionic solubility, and qualitative analysis are described here.

The chapter is opened with a thorough discussion of the common-ion effect. It is introduced in the context of acid-base reactions but is extended throughout with the reactions under study.

Buffers are discussed within this framework because the effect of adding small amounts of acids or bases to a buffer represents an application of the common-ion effect.

With the common-ion effect fully covered in acid-base systems, it is the ideal place to introduce acid-base titrations. This is especially true in terms of explaining the regions of the titration curve.

The chapter then moves to solubility equilibria. The relationship between solubility and K_{sp} is fully detailed, which allows for the topics of the effect of added ions, the effect of pH, and the effect of complex ion formation on solubility to be quantitatively covered within the context of the common-ion effect.

As a predictive tool, the criteria for dissolution or precipitation of an aqueous solution of ions is covered. This material is again approached from the common-ion effect.

Finally, qualitative analysis is covered briefly. The emphasis here is that the traditional qualitative analysis scheme is an application of comparative solubilities.

LECTURE OUTLINE

I. **The Common-Ion Effect**

　　A. *Le Châtelier's principle* states that when a stress is placed on a chemical reaction the equilibrium position will shift to reduce that stress.

　　B. *Strong electrolytes* are used to introduce ions into solution. When added this way, the concentrations of the ions can be easily calculated and no complicating equilibria are introduced.

　　C. In solutions of *weak electrolytes*, strong electrolytes can be added to increase the concentration of one of the ions of the weak electrolyte. In general, the dissociation of a weak electrolyte is decreased by the addition of a strong electrolyte that has an ion in common with the weak electrolyte. This phenomenon is called the *common-ion effect*. The common ion can be an anion or a cation.

　　D. Reactions between (1) strong acids and strong bases, (2) strong acids and weak bases, and (3) weak acids and strong bases proceed essentially to completion. However, the products of these reactions may enter into equilibria with water.
　　　　1. If the solution consists only of a conjugate acid-base pair, you need only consider the acid-dissociation equilibrium between them.
　　　　2. If the solution contains a strong acid and a weak base or a strong base and a weak acid, first consider the stoichiometry of the acid-base reaction. If a conjugate acid-base pair is present after the neutralization reaction, you must subsequently consider the acid-dissociation equilibrium between them.

II. Acid-Base Titrations

A. *Titration* is a common technique in which one reactant is added in small, measured increments to a fixed amount of the other reactant. A *titration curve* is a graph of the amount of added *titrant* vs. a measured property of the resulting solution. In acid-base titrations, the titration curve is a graph of the amount of titrant added vs. the pH of the reaction solution.

B. With a strong acid in the receiving flask and with a strong base being added, the titration curve begins at the pH of the initial acid solution and rises with added base. The *equivalence point* is that point on the graph at which the number of moles of added OH^- equals the number of moles of H^+ originally present to be titrated. Continued addition of the base represents an excess amount of the base and pH continues to rise to approximately the pH of the original base solution, ignoring its dilution. For a strong acid-strong base titration, the pH at the equivalence point is 7.00 because neither resulting ion reacts with water.

C. *Titrations involving a weak acid or weak base* are qualitatively similar to those of a strong acid with a strong base. Quantitatively, they differ in that the pH of the equivalence point is generally not 7.00.

D. In the titration of a weak acid with a strong base, the resulting conjugate of the weak acid is a strong base, and it will hydrolyze yielding a basic solution at the equivalence point. To calculate the pH:
   1. Stoichiometry: Assume that the strong base reacts completely with the weak acid, producing a solution that contains the weak acid and its conjugate base.
   2. Equilibrium: Use the value of K_a and the equilibrium expression to calculate the equilibrium concentrations.

E. In the titration between a weak base and a strong acid, the opposite condition exists:
   1. Stoichiometry: Assume that the strong acid reacts completely with the weak base, producing a solution that contains the weak base and its conjugate acid.
   2. Equilibrium: Use the value of K_b and the equilibrium expression to calculate the equilibrium concentrations.

F. If a titration is performed between a weak acid and a weak base, the pH of the equivalence point will depend on the relative strengths of the resulting conjugate acid and base.

G. The *titration of a polyprotic acid* with a base results in the ionizable protons being sequentially consumed by the added OH^-. When the neutralization steps are sufficiently separated in terms of K_a, the acid exhibits a titration curve with multiple equivalence points.

Discussion Question: Revisit pH indicators. What are the criteria for a good choice of an indicator? What are the consequences of a poor choice of an indicator?

III. Buffered Solutions

A. *Buffers* are solutions that resist a change in pH upon addition of small amounts of an acid or a base.

B. Buffered solutions contain both members of a conjugate weak acid-base pair. The acidic member consumes small amounts of added base, while the basic member consumes small amounts of added acid. Considering the acid-dissociation reaction of HX, $HX(aq) \rightleftharpoons H^+(aq) + X^-(aq)$, it is clear that additions of acid react with X⁻ to shift the equilibrium to the left and that additions of base react with H⁺ also to shift the reaction to the left. Rearranging the expression for K_a yields $[H^+] = K_a \frac{[HX]}{[X^-]}$; the pH of the buffer depends on the K_a of the acid and on the ratio of the acid to the conjugate base.

C. *Buffer capacity* is the amount of acid or base that can be added to a buffer before the pH begins to change appreciably. Buffering capacity increases with the concentrations of HX and X⁻; if the ratio $\frac{[HX]}{[X^-]}$ remains constant, the pH of the buffer remains constant.

D. The *Henderson-Hasselbalch equation* is a form of the acid-dissociation equilibrium expression commonly used by biochemists. In the Henderson-Hasselbalch equation, K_a and $[H^+]$ are converted to pK_a and pH by taking the negative logarithm of each term, giving $pH = pK_a + \log\left(\frac{[base]}{[acid]}\right)$. Generally, buffers are constructed such that the amount of dissociation is negligibly small compared to the formal concentrations of the acid and the base, and the formal concentrations of the acid and base can be used rather than calculating equilibrium concentrations.

E. The *addition of acids or bases to buffers* represents a specialized application of Le Châtelier's principle. To determine the effect of an addition of an acid or a base to a buffer:
1. perform the stoichiometric calculation to find the effect of the addition on the formal concentrations, and
2. then perform the equilibrium calculation to find the effect of dissociation of the acid's H⁺.

Discussion Question: In what other algebraic forms can the Henderson-Hasselbalch equations be stated? (Students historically have a poor understanding of the properties of logarithms.)

IV. Chemistry and Life: Blood as a Buffered Solution

The blood is, in part, a complex buffer system. The most important buffer system within blood is the *carbonic acid-bicarbonate buffer system*. This system both buffers the blood and provides a carrier system for carbon dioxide:

$H^+(aq) + HCO_3^-(aq) \rightleftharpoons H_2CO_3(aq)$

$H_2CO_3(aq) \rightleftharpoons H_2O(l) + CO_2(g)$

V. Solubility Equilibria

A. The *solubility* of an ionic solid in water is a *heterogeneous* process. Solubility was considered qualitatively in Chapter 4. Here, solubility is treated quantitatively as an equilibrium process. In this regard, it is no longer necessary to simply state that a substance is soluble or insoluble; it is possible to describe the extent to which it is soluble. The solubility equilibrium of an ionic solid in water is described by the *solubility-product constant*, K_{sp}, where K_{sp} is simply the equilibrium constant for the general reaction $MX(s) \rightleftharpoons M^+(aq) + X^-(aq)$. The associated equilibrium constant expression or K_{sp}-expression is $K_{sp} = [M^+][X^-]$. As in other equilibria, each ion's concentration is taken to the power of its stoichiometric coefficient.

B. The solubility and the K_{sp} of a given substance are related.

1. The *solubility* of a substance is the amount of the substance that dissolves to form a saturated solution, usually expressed in grams per liter of solution, g/L.
2. The *molar solubility* is the number of moles of the solute that dissolves to form a liter of saturated solution, mol/L.
3. The relationship becomes apparent when viewing an example, such as $Cu(CN)_2$. The chemical equation is $Cu(CN)_2(s) \rightleftharpoons Cu^{2+}(aq) + 2CN^-(aq)$. The concentrations of ions produced on dissolution are $[Cu^{2+}] = x$ mol/L and $[CN^-] = 2x$ mol/L. The value x is the molar solubility.

C. As with other equilibria, the *common-ion effect* is important in solubility equilibria. The consequence of adding a strong electrolyte that contains one of the ions of the equilibria is that the equilibrium is shifted to the left; solubility is decreased.

VI. A Closer Look: Limitations of Solubility Products

For three principal reasons, there are limitations on the accuracy of solubility products.

1. Electrostatic interactions among dissolved ions tend to make ion solutions "behave" as though they are less concentrated than they actually are; saturated solutions generally have higher concentrations of ions than predicted from K_{sp}. These "ion-strength" effects can be dealt with mathematically but are beyond this coverage.
2. It is assumed that all the dissolved salt has separated into its constituent ions, an assumption that is sometimes faulty.
3. In the treatment of K_{sp}, any additional acid-base equilibria are ignored. Whenever one of the ions released in the solubility is an acid or a base, that ion will be consumed in an acid-base reaction. By Le Châtelier's principle, this will shift the solubility reaction to the right, increasing solubility.

Discussion Question: Discuss products sold to remove "bathroom scale." What do these products do and how do they achieve the result?

VII. Criteria of Precipitation or Dissolution

A. One of the most commonly asked questions about solubility reactions is whether a given solution is expected to show *precipitation*. Precipitation is the formation of a solid product in a metathesis reaction. In this context, precipitation is the reverse of solubility. As defined in Chapter 15, the reaction quotient, Q, is used. In solubility reactions, however, the equilibrium-constant expression is simply the product of the ion concentrations. Consequently, we speak of Q as the *ion product*. As before, if $Q > K_{sp}$, the reaction shifts to the left and precipitation occurs. If $Q < K_{sp}$, the reaction shifts to the right and dissolution occurs as long as there is additional solid material present to dissolve. If $Q = K_{sp}$, the reaction is at equilibrium and there will be no *net* dissolution or precipitation.

B. The solubility of any solute whose anion is basic will be affected by the pH of the solution. This is an application of the common-ion effect. If the anion is OH^-, there is a direct affect by pH in that the pH gives an initial concentration of the hydroxide ion. If the anion is the conjugate base of a weak acid, the anion will enter into an equilibrium with the conjugate acid. In either case, the equilibria act to decrease the concentration of the dissolved ion, shifting the solubility reaction to the right or increasing solubility, with increasing acidity.

C. Ions can be separated from one another by *selective precipitation* of the ions based on the solubilities of their salts.

Lecture Outline – Chapter 17
Brown, LeMay, & Bursten, *Chemistry: The Central Science*, 6th Edition

    D. If the cation involved in a solubility equilibrium reacts with a substance in solution to form a coordination complex, the solubility of the salt increases.
        1. A *complex ion* is an assembly of a metal ion and Lewis bases bonded to it. The bonds are coordinate covalent bonds and so these assemblies are often called coordination complexes. The Lewis bases involved are generally anions or small neutral molecules and are referred to as *ligands*. It is most common for four or six ligands to bond to the metal. Because the formation of a complex ion decreases the concentration of the metal ion, the solubility of the salt containing the metal increases as the concentration of the ligand increases.
        2. The reaction that describes the formation of the complex ion is an equilibrium reaction, with an equilibrium constant called a *formation constant*, K_f. As such, this is another example of the common-ion effect.
    E. Many metal hydroxides and oxides that are relatively insoluble in water dissolve in strongly acidic *and* strongly basic media. Such substances are called *amphoteric*. The property of amphoterism is used in many commercial applications, including the separation of aluminum in bauxite ore from iron impurities. Although $Fe(OH)_3$, like most hydroxides, is soluble in acid, it is insoluble in base. The aluminum in bauxite is present as hydrated Al_2O_3, which dissolves in both basic and acidic media. Therefore, by treating the ore with base, the aluminum oxide dissolves but the iron(III) hydroxide does not.

Discussion Question: Discuss film developing. The common processes involve reduction of silver halides to silver metal and complexation to dissolve the coating off the unexposed areas.

VIII. Chemistry and Life: Tooth Decay and Fluoridation
    A. Fluoride can help prevent tooth decay by replacing the hydroxides in hydroxyapatite, $Ca_{10}(PO_4)_6(OH)_2$.
    B. Because fluoride is effective in preventing cavities, and because it is *generally* considered to be safe, fluoride is added to the public water supply in many cities and has become a common ingredient in most toothpastes.

IX. Qualitative Analyses for Metallic Elements
    A. *Qualitative analysis* is a series of chemical or "wet" tests to confirm the presence or absence of ions. Such analyses proceed in three stages. (1) The ions are separated into broad groups based on solubility properties. (2) The ions within each group are then separated by selectively dissolving members of the group. (3) The ions are then identified using specific tests. The cations are divided into five groups.
        1. Cations that form insoluble chlorides are precipitated with dilute HCl. These cations are Ag^+, Hg_2^{2+}, and Pb^{2+}.
        2. Cations that form acid-insoluble sulfides are precipitated from the supernatant from #1 with H_2S. These cations are Cu^{2+}, Bi^{3+}, Cd^{2+}, Hg^{2+}, As^{3+}, Sb^{3+}, and Sn^{4+}.
        3. Cations that form base-insoluble sulfides are precipitated from the supernatant from #2 with OH^- and additional S^{2-}. These cations are Al^{3+}, Cr^{3+}, Fe^{3+}, Zn^{2+}, Ni^{2+}, Co^{2+}, and Mn^{2+}.
        4. Cations that form insoluble phosphates are precipitated from the supernatant from #3 with $(NH_4)_2HPO_4$. These cations are Mg^{2+}, Ca^{2+}; Sn^{2+}, and Ba^{2+}.
        5. The alkali metals and NH_4^+ are individually tested for in the solution left from #4.

B. *Quantitative analysis* is the measurement of the amount of a substance present in a sample.

Lecture Outline – Chapter 17
Brown, LeMay, & Bursten, *Chemistry: The Central Science*, 6th Edition

SAMPLE QUIZ QUESTIONS

1. What is the hydrogen ion concentration in a solution that initially contains 0.010 M NaCNO and 0.025 M HCNO if the acid dissociation constant for HCNO is 1.4 x 10^{-4}?

2. A solution prepared from 0.050 mol of a weak acid, HX, is diluted to 200 mL and has a resulting pH of 3.0. What is the pH after 0.080 mol of solid NaX is added?

3. What is the pH at the equivalence point in the titration of 0.10 M NH$_3$ (K_b = 1.8 x 10^{-5}) with 0.10 M HCl?

4. What is the pH of a solution formed by adding 10.0 mL of 0.10 M NaOH to 90.0 mL of 0.200 M HCl?

5. A buffer pair: (a) is composed of a weak acid and a strong base; (b) is a buffer system using two acids and their conjugate bases; (c) is used to alter the pH of a solution; (d) is composed of a weak acid and its conjugate base or a weak base and its conjugate acid; (e) is composed of a weak base and a strong acid.

6. Which of the following would constitute a buffer? (a) HCN and KCN; (b) HCl and NaCl; (c) HC$_2$H$_3$O$_2$ and NaOH; (d) H$_2$SO$_4$ and Li$_3$PO$_4$; (e) H$_2$CO$_3$ and NH$_4$NO$_3$

7. Explain briefly how you would go about making a buffer solution. What determines the pH that is buffered? What determines the capacity of the buffer?

8. A buffer system is made to be 0.35 M NH$_4$Cl and 0.15 M NH$_3$. What is the pH of the system? For ammonia, K_b = 1.8 x 10^{-5}.

9. Explain how a mixture of 0.10 M benzoic acid (K_a = 6.5 x 10^{-5}) and 0.10 M sodium benzoate is capable of functioning as a buffer. What pH does this mixture have?

10. What is the pH of a liter solution composed of 0.050 M HC$_2$H$_3$O$_3$ (K_a = 8.4 x 10^{-4}) and 0.040 M NaC$_2$H$_3$O$_3$, after 1.00 x 10^{-3} mol of NaOH is added?

11. 100.0 mL of a 0.100 M solution of nitrous acid reacts with 100.0 mL of a 0.100 M solution of sodium hydroxide to form the salt, NaNO$_2$. What is the pH at equilibrium? (At 25°C, K_a of nitrous acid is 4.5 x 10^{-4}.)

12. When 50.00 mL of 0.100 M HCN is titrated with 0.0500 M potassium hydroxide, what is the pH at the equivalence point? (K_a for HCN at 25°C is 6.2 x 10^{-10} M.)

13. Which of the following substances will be more soluble in acidic than in basic solution? (a) Na$_2$SO$_3$; (b) Fe(OH)$_3$; (c) BaCl$_2$

14. BaSO$_4$ is used in radiographic studies of the gastrointestinal tract because it is opaque to X rays and only slightly soluble (K_{sp} = 1.6 x 10^{-9}). What is the concentration of Ba^{2+} in a saturated solution of BaSO$_4$?

15. Why does FeS precipitate from a solution saturated with H$_2$S when [H$^+$] is low but not when [H$^+$] is high?

16. K_{sp} for Zn(OH)$_2$ is 1.8 x 10^{-14}. What is the solubility of Zn(OH)$_2$ in mol/L? (a) 2.6 x 10^{-5} mol/L; (b) 6.6 x 10^{-6} mol/L; (c) 1.3 x 10^{-7} mol/L; (d) 6.7 x 10^{-8} mol/L; (e) 1.7 x 10^{-5} mol/L

17. K_{sp} for Ag_2S is 1.8×10^{-49}. If $[S^{2-}] = 0.0010\ M$, what is $[Ag^+]$? (a) $1.3 \times 10^{-23}\ M$; (b) $1.8 \times 10^{-43}\ M$; (c) $4.2 \times 10^{-25}\ M$; (d) $3.6 \times 10^{-17}\ M$; (e) $1.8 \times 10^{-46}\ M$

18. How many moles of $Ag_2C_2O_4$ will dissolve in 100 mL of $0.050\ M\ Na_2C_2O_4$ if K_{sp} for $Ag_2C_2O_4$ is 1.1×10^{-11}?

19. The solubility of $Mg(OH)_2$ in pure water at 25°C is 1.8×10^{-4} mol/L. What is the: (a) K_{sp} for $Mg(OH)_2$ at pH = 5.0; (b) molar solubility for $Mg(OH)_2$ at pH = 5.0?

20. K_{sp} for CaF_2 is 3.9×10^{-11}. Will CaF_2 precipitate when 50 mL of $2.0 \times 10^{-4}\ M$ $Ca(NO_3)_2$ is mixed with 50 mL $1.0 \times 10^{-4}\ M$ NaF?

21. $Cr(OH)_3$ is amphoteric. Show, using balanced chemical equations, the behavior of this substance in both acidic and basic solutions.

22. Calculate the concentration of silver ions in a solution that is $0.10\ M$ with respect to $Ag(NH_3)_2^+$. ($K_f = 1.7 \times 10^7$ for $Ag(NH_3)_2^+$)

Lecture Outline – Chapter 18
Brown, LeMay, & Bursten, *Chemistry: The Central Science*, 6th Edition

Chapter 18 — Chemistry of the Environment

OVERVIEW: Chapter 18 focuses on the environment. The chapter opens with an introduction to the atmosphere, detailing the regions of the atmosphere, their distinguishing characteristics, and their principal chemistries. The chemistries of the outer regions of the atmosphere are also included.

A section is included in the chapter covering the ozone layer. After developing the "good" functions of the ozone layer, attention is turned to the depletion of the ozone layer and our ability to influence the growth of a hole in it.

Next, the chemistry of the troposphere, the lowest atmospheric layer, is covered. The emphasis here is on pollution due to sulfur-containing and nitrogen-containing compounds. The controversial topics of acid rain and photochemical smog are dealt with, as is the greenhouse effect.

The focus then shifts to the earth itself. A section is included that discusses the oceans of the earth, noting that they are all interconnected and not isolated bodies of water. Pollution of the oceans and methods of extracting useful materials from the oceans are also discussed.

Finally, the chapter covers fresh water. The emphasis is again on pollution of fresh water. The related topics of municipal water treatment and sewage treatment are also discussed.

LECTURE OUTLINE

I. **The Hydrosphere**

    The *hydrosphere* is that part of the earth, beneath the atmosphere, in which the world's liquid and solid water exist. As such, it includes the oceans, the polar ice caps, and the water in the earth's interior.

II. **Earth's Atmosphere**

    A. The earth's atmosphere surrounds the entire globe. It extends upward from the surface of the planet to an altitude of over 100 km. In a somewhat arbitrary way the atmosphere is divided into layers for ease of reference and to denote particular features of each region.
        1. The *troposphere* is the layer that lies just above the surface of the earth; it extends to an altitude of about 12 km. Life exists only within this layer and the hydrosphere. The upper limit of the troposphere is called the *tropopause*. Traveling up from the earth's surface, atmospheric temperature reaches a minimum at the tropopause.
        2. The *stratosphere* lies above the tropopause and extends to an altitude of about 50 km. The upper limit of the stratosphere is called the *stratopause*. Atmospheric temperature increases with altitude through the stratosphere and reaches a maximum at the stratopause.
        3. The *mesosphere* lies above the stratopause and extends to an altitude of about 85 km. The upper limit of the mesosphere is called the *mesopause*. Atmospheric temperature decreases with altitude through the mesosphere and reaches a minimum at the mesopause.
        4. The *thermosphere* lies above the mesosphere and is the highest layer, extending to about 110 km in altitude. Atmospheric temperature increases with altitude through the thermosphere.

B. The *composition of the atmosphere* varies in amount with altitude but the proportion of each gas in the atmosphere stays fairly constant with altitude. The most abundant gases in the atmosphere, in mole fraction, are nitrogen, 0.78084; oxygen, 0.20948; argon, 0.00934; and carbon dioxide, 0.000355. The most common way to express gas concentrations is in *parts per million, ppm*. That is, we express concentration as parts in every 1,000,000 parts. The conversion from mole fraction to ppm is achieved by simply multiplying by 10^6. As commonly applied to gases in low concentration, the concentration of argon is 0.00934×10^6 or 934 ppm.

III. The Outer Regions of the Atmosphere

A. Several crucial chemical reactions occur high in the atmosphere, above the mesosphere. *Photodissociation* occurs when a chemical bond ruptures due to absorption of a photon. Of environmental importance, the reaction $O_2(g) + h\nu \rightarrow 2O(g)$ occurs above 120 km. This reaction consumes ultraviolet radiation with wavelengths less than 242 nm.

B. *Photoionization* is a process in which absorption of radiation by a molecule results in the ejection of an electron from the molecule, leaving it a cation. At altitudes above about 90 km, photoionization occurs for N_2, O_2, O, and NO. These reactions consume radiation with wavelengths less than about 135 nm.

Discussion Question: Discuss the implications of allowing short-wavelength ultraviolet radiation to reach earth. There is little suspicion that the photolytic processes of nitrogen are being diminished by pollution, but the impact of such an event could add a new dimension to the students' views of the atmosphere.

IV. Ozone in the Upper Atmosphere

A. *Ozone*, O_3, is an allotrope of oxygen. At altitudes down to about 30 km, the short-wavelength solar radiation is strong enough to photodissociate O_2, allowing the reaction $O(g) + O_2(g) \rightleftharpoons O_3^*(g)$ to proceed, where the "*" denotes that the ozone molecule has an excessive amount of energy. Ozone can dissipate the energy by performing the reverse reaction or by passing it to another molecule through collision. Stable ozone can react further or absorb additional radiation. The net effects are that ozone is continuously formed and destroyed and it absorbs radiation in the 200-310 nm range.

B. The current concern over the *depletion of the ozone layer* is based on measurements of the past 15 years which indicate that industrial pollutants have created a hole in the ozone layer near the South Pole. Clearly, if ozone is depleted to a significant extent, the incidence of high-energy radiation at the earth's surface will increase. Few organisms can function well in this high-energy radiation. *Chlorofluorocarbons (CFCs)* are organic substances that contain chlorine and fluorine in place of several hydrogen atoms. These substances are believed to be responsible for the vast majority of ozone depletion. The principal members of this group are $CFCl_3$, Freon-11, and CF_2Cl_2, Freon-12, which have been used widely as aerosol propellants, refrigerants, and foaming agents in plastics. These compounds are virtually unreactive in the lower atmosphere and are insoluble in water. Therefore, the chlorofluorocarbons persist in the atmosphere and rise into the ozone layer. At that altitude, the CFCs are subject to high-energy radiation which causes *photolysis* or the light-induced rupture of a carbon-chlorine bond. The atomic chlorine thus produced catalyzes the conversion of ozone to O_2.

Discussion Question: Discuss the recent discoveries about the Antarctic ozone hole. Also mention the discovery of an Arctic ozone hole. In particular, focus on the length of time required to reverse the suspected damage and the current products that would need to be curtailed, modified, or abandoned.

V. A Closer Look: Stratospheric Clouds and Ozone Depletion

A. The depletion of the ozone in the stratosphere has been of increasing concern. The depletion is most obvious over Antarctica, with the worst period being in the Antarctic winter and early spring.

B. It had been previously assumed that Cl was tied-up as HCl and ClONO$_2$, meaning that Cl was unavailable for reaction with ozone and that chlorine releases into the atmosphere in the form of CFCs had a minor impact on ozone depletion.

C. A new theory holds that in the cold of winter, clouds form in the Antarctic stratosphere. Cloud formation removes NO$_2$ from the atmosphere, leaving the Cl free to react with ozone. In spring, the clouds dissipate and the "ozone hole" stops growing, and in fact ozone-depleted air mixes with other air in the Southern Hemisphere.

VI. The Chemistry of the Troposphere

A. The chemical reactions that proceed in the troposphere are much more immediately susceptible to the actions of humans. In the troposphere, virtually all radiation is of wavelengths longer than 300 nm. The constituents of the troposphere do not react with radiation of these longer wavelengths, and photodissociation and photoionization do not occur to an appreciable extent. The chemistry of the troposphere is a study of *air pollution.*

B. *Sulfur compounds* are released into the atmosphere through volcanic gases, bacterial decay, and, most importantly, fossil fuel combustion. Approximately 80% of the sulfur, mainly in the form of SO$_2$, released into the air in the United States is from coal and oil combustion. SO$_2$ can be converted to SO$_3$, which is the major source of *acid rain.* Acid rain refers to precipitation and fog that is noticeably acidic. Acid rain has been found to affect fish populations in lakes, the fullness of vegetation in forest regions, and human health. The corrosiveness of acid rain has caused much damage to stone and metal buildings and other structures.

C. *Carbon monoxide* is a subtle pollutant in that it is fairly unreactive and so does not affect materials or vegetation to any great extent. However, CO does pose a risk to animals, particularly humans. It is produced as a result of incomplete combustion of carbon-based fuels such as coal and oil. The effect CO has on animals is that it has the ability to bind to hemoglobin in the blood.
  1. *Hemoglobin* is the iron-containing protein responsible for carrying oxygen throughout the bloodstream. The *heme* groups in each molecule consist of an iron atom surrounded by four nitrogen atoms; it is a complex capable of bonding to small molecules, such as oxygen or carbon monoxide.
  2. *Oxyhemoglobin* is the form of hemoglobin in which an oxygen molecule is attached to the iron for transport. In normal respiration, the oxyhemoglobin carries the oxygen to tissues and the hemoglobin returns to the lungs with carbon dioxide.
  3. Carbon monoxide binds strongly to hemoglobin to form *carboxyhemoglobin.* Carbon monoxide has an affinity for hemoglobin approximately 210 times that of oxygen. Consequently, carbon monoxide

in inhaled air can seriously reduce the amount of oxygen reaching body tissues. The effects range from fatigue to chronic health problems to death.

D. Recent work has shown that smoking is the most potent source of exposure to CO. Smokers have blood CO levels 5 to 10 times higher than nonsmokers.

E. *Nitrogen oxides* are also produced in combustion reactions. Their source, however, is the air used in combustion, not the fuel. The NO formed is rapidly oxidized in air to NO_2, which can undergo photodissociation to NO and O using visible light with a wavelength less than 393 nm. The atomic oxygen is highly reactive, forming principally O_3, which is reactive, corrosive, and toxic. These gases and others collectively form *photochemical smog*. The other gases in smog include carbon monoxide and a variety of organic compounds. The organic compounds involved include hydrocarbons, both saturated hydrocarbons and olefins, and aldehydes, which have a C(=O)H group on at least one terminus.

F. *Water vapor and carbon dioxide* are crucial in maintaining the *earth's temperature balance*. The earth is in overall temperature balance with its surroundings; the planet radiates energy into space at the same rate as it absorbs radiation from the sun. Carbon dioxide and water vapor absorb much of the infrared (heat) radiation given off from the earth's surface. The concentration of water vapor in the atmosphere varies with location and weather, but man's activities have little long-term effect on it because excess amounts of water vapor condense. Carbon dioxide, however is gaseous over all normal conditions and so it can accumulate in the atmosphere and is fairly uniformly distributed. As carbon dioxide accumulates, more irradiated heat from the earth is absorbed before it can escape, leading to an overall warming of the earth. This warming effect is called the *greenhouse effect*. We now know that the greenhouse effect is real; the atmosphere is warming. What is frightening to many people is the unknown consequences on weather, crops, and the polar ice caps.

Discussion Question: Just for fun, ask the smokers and the nonsmokers to compare their typical states of health and physical well-being. This is an emotional issue, so "refereeing" may be required.

VII. The World Ocean

A. Roughly 72% of the earth's surface is covered by water. The largest portion of this layer is comprised of the world's oceans. *Sea water* is therefore an important and inescapable part of our environment. The *salinity* of water is variable. Salinity is defined as the mass in grams of dry salts present in 1 kg of sea water. In the ocean, salinity ranges from about 33 to 37 (3.3% to 3.7%) but can be much higher in brackish backwaters and land-locked salt lakes. The most prevalent dissolved ions in sea water are Cl^-, Na^+, SO_4^{2-}, Mg^{2+}, Ca^{2+}, and K^+. There are vast amounts of these substances present, but little commercial use of them.

B. The principal use of the substances in sea water is the recovery of water, itself. *Desalination* is the removal of dissolved solutes from salt water so that the water can be used. Two major techniques are used in desalination.
1. *Distillation* can be used. In this technique, the salt water is boiled, vaporizing the water but leaving the salts behind. The water is recondensed for use.
2. *Reverse osmosis* can also be used. The water in the sea water will cross a semipermeable membrane, under pressure, from the more concentrated side to the less concentrated side. The net effect is the separation of the water from the dissolved solutes.

Lecture Outline – Chapter 18
Brown, LeMay, & Bursten, *Chemistry: The Central Science*, 6th Edition

    C.   Of increasing concern is *ocean pollution*. Because so many countries and industries have access to the oceans but the oceans belong to no one, they have become a convenient part of the environment to ignore. The pollution problems seen mirror the problems seen in lakes and streams, including the accumulation of toxic chemicals and disruption of the balance between aquatic life and oxygen-demanding wastes.

Discussion Question: Discuss the *Exxon Valdez* accident in Alaska. Focus on the economic and environmental impact of the accident in terms of the local population, the consuming public, and the company.

VIII. Fresh Water

    A.   Fresh water contains many dissolved solutes, mainly Na^+, K^+, Mg^{2+}, Ca^{2+}, Fe^{2+}, Cl^-, SO_4^{2-}, HCO_3^-, O_2, N_2, and CO_2, but in much smaller quantities than in sea water. Through our use of water, additional solutes are placed into it. *Dissolved oxygen* is an important indicator of *water quality*. Dissolved oxygen is required for all aquatic life and for the function of aerobic bacteria that consume organic wastes. Those wastes that can be consumed by bacteria are called *biodegradable*. As the amount of biodegradable waste increases, so does the *biological oxygen-demand*, or BOD. Therefore, the measure of dissolved oxygen indicates the amount of biodegradable waste present in the water.

    B.   Most *municipal water supplies* require treatment before the water is distributed to consumers. Municipal water treatment usually consists of five steps.
        1.   *Coarse filtration* removes large objects from the incoming stream. This step is minor when the water is pumped from wells, but is a major step when the water is taken from a lake or river.
        2.   *Sedimentation* is a step in which the water is allowed to stand so that finely divided particles can settle out. Sometimes, chemicals are added to promote the aggregation of the particles.
        3.   *Sand filtration* is a final filtration through a bed of sand to remove minute particles.
        4.   *Aeration* is generally a step in which the water is sprayed into the air to hasten the oxidation of dissolved organic substances
        5.   *Sterilization* is the final step. Usually, a substance such as Cl_2 is added to kill bacteria and to prevent their growth.

IX. A Closer Look: Water Softening

    In many locales, the natural waters contained dissolved ions that cause problems such as scale build-up on pipes and fixtures and diminished performance of detergents. Depending on the level of dissolved ions, *water softening* may be used.
        1.   *Water hardness* is a measure of dissolved ions, mostly Ca^{2+} and Mg^{2+}, with increasing levels of dissolved ions representing harder water.
        2.   The *lime-soda process* is the chief method used for softening water. In this process, lime, CaO, or slaked lime, $Ca(OH)_2$, and soda ash, Na_2CO_3, are added to the water. These additions cause the precipitation of $CaCO_3$ and $Mg(OH)_2$ from the water stream. The problems associated with the lime-soda process focus on two issues. First, the precipitate particles formed are often too small to settle out well. Second, the process leaves the water too basic for use. Subsequent treatment is required to remedy both problems.

Discussion Question: What are students willing to dump down a drain in laboratory? What impact do they perceive such dumping to have on the environment?

SAMPLE QUIZ QUESTIONS

1. The ionization potential of O_2 is 1205 kJ/mol. What is the minimum wavelength of radiation needed to photoionize an electron from O_2?

2. Explain, using equations, the presence of O_3 in the mesosphere and the stratosphere.

3. What is meant by a stabilizing collision in a reaction between molecules?

4. Explain the origin of sulfate in rain in regions of high industrial activity.

5. Write complete, balanced equations for the following processes: (a) Combustion of sulfur, forming sulfur dioxide; (b) Reaction between sulfur dioxide and calcium oxide; (c) Atom transfer between O^+ and N_2.

6. The most abundant noble gas in the atmosphere is: (a) He; (b) Ne; (c) Ar; (d) Kr.

7. The air pollutant present in greatest quantity in most polluted air is: (a) SO_2; (b) NO_2; (c) CH_4; (d) CO.

8. Which of the following is a precursor to photochemical smog? (a) CO; (b) CO_2; (c) SO_2; (d) NO_2

9. 99% of the total atmosphere is in the region below: (a) 5 km; (b) 50 km; (c) 500 km; (d) 5000 km.

10. The region just above the Earth's surface is known as the: (a) stratosphere; (b) mesopause; (c) troposphere; (d) thermosphere.

11. The second most abundant gas in the Earth's atmosphere is: (a) He; (b) O_2; (c) CO_2; (d) Ar.

12. Which of the following components of the upper atmosphere is responsible for preventing ultraviolet light in the 200-300 nm range from reaching the surface of the Earth? (a) O^+; (b) NO; (c) O_3; (d) CF_2Cl_2

13. Explain, using chemical equations, how a water supply containing Ca^{2+} and HCO_3^- is softened by the soda-lime process.

14. What type of treatment does water receive in a primary sewage plant?

15. Which of the following ions is the most abundant in sea water? (a) Na^+; (b) Ca^{2+}; (c) Cl^-; (d) HCO_3^-

16. The salinity of sea water is defined as: (a) the mass in grams of NaCl in a liter of sea water; (b) the mass in grams of dry salts in a liter of sea water; (c) the mass in grams of dry salts in a kilogram of sea water; (d) None of these is correct.

17. Which two of the following ions are responsible for water hardness? (a) Ca^{2+}; (b) Na^+; (c) Mg^{2+}; (d) H^+

18. Which of the following is not a product of aerobic decay of biodegradable substances?
   (a) H_2O; (b) H_2S; (c) NO_3^-; (d) CO_2

Lecture Outline – Chapter 19
Brown, LeMay, & Bursten, *Chemistry: The Central Science, 6th Edition*

Chapter 19 — Chemical Thermodynamics

OVERVIEW: Chapter 19 contains a detailed treatment of thermodynamics. As such, it builds on the thermochemistry chapter, Chapter 5. The emphasis here is on entropy, free energy, and work.

The chapter begins by detailing logically what is meant by a spontaneous process and how to qualitatively identify one. The idea of entropy is introduced, although not addressed by name.

The second section of the chapter describes entropy and discusses the theoretical work behind the concept. This is followed by a discussion of entropy on the molecular level, where qualitative comparisons in entropies are explored. Finally, numerical values of entropy are covered, as well as determining entropy changes for processes.

With entropy covered and with numerous references to enthalpy, Gibbs free energy is now presented as the "master indicator" of spontaneity. The calculation of $\Delta G°_{rxn}$ from free energies of formation is covered, as is the relationship among free energy, enthalpy, and entropy.

The relationship between free energy change and temperature and between free energy change and the equilibrium constant of a reaction are covered in detail.

Finally, the relationship between ΔG and the maximum obtainable work is covered for spontaneous reactions, as is the relationship between ΔG and minimum applied work for nonspontaneous reactions.

LECTURE OUTLINE

I. **Thermodynamics**

   A. *Thermodynamics* is the area of chemistry dealing with energy relationships.

   B. The *first law of thermodynamics* is the law of conservation of energy. The *law of conservation of energy* states that energy is neither created nor destroyed; it can only be transformed, such as from potential energy to kinetic energy, or exchanged between the system and the surroundings.

   C. *Enthalpy* is defined in terms of *P–V* work as follows. If pressure is constant, $P\Delta V \neq 0$, and $\Delta E = q_p - P\Delta V$. Therefore, $q_p = \Delta E + P\Delta V$. The quantity, q_p, is defined as the enthalpy change of the process, ΔH.

   D. Although the spontaneity of most reactions can be predicted by ΔH, it is clear that enthalpy is not the only quantity needed to reliably predict spontaneity. The second factor in spontaneity is the change in *disorder* or *randomness* over the course of the process.

Discussion Question: Discuss events in everyday life. Can students think of any process that does not have an entropy increase?

II. **Spontaneous Processes**

   The *second law of thermodynamics* can be stated in several ways. In one sense, it says that processes that are spontaneous in one direction are nonspontaneous in the reverse direction.

Lecture Outline – Chapter 19
Brown, LeMay, & Bursten, *Chemistry: The Central Science, 6th Edition*

Discussion Question: Discuss the astronomical theory that the universe is expanding. How do the observations of the Doppler effect red-shift support the second law?

III. Spontaneity, Enthalpy, and Entropy

A. The amount of disorder a system exhibits is called its *entropy*, S. The more disordered a system, the higher its entropy.

B. The *change in entropy* of a system, due to a process, is given by $\Delta S = S_{final} - S_{initial}$. Entropy change is a *state function*. Positive values of ΔS indicates an increase in disorder, while a negative value of ΔS indicates a decrease in disorder.

C. Within the context of entropy, the second law of thermodynamics states that, in any spontaneous process, there is always an increase in entropy somewhere in the universe.

IV. Chemistry and Life: Entropy and Life

In our daily lives, we strive to produce order from disorder. Raw materials are made into engineered materials such as steel, plastic, and concrete, for example. The second law simply states that our activities must lead to greater disorder elsewhere. *Thermal pollution* is one major price we pay for these activities.

V. A Molecular Interpretation of Entropy

A. Entropy is related to (1) the number of particles present and (2) the amount of mobility these particles have. The forms of motion are given below.
   1. *Translational motion* is movement that covers distance, such as a molecule traveling from one side of its container to the other.
   2. *Vibrational motion* is motion due to the chemical bond acting like a spring. Vibrations can be bond lengthening, bond compressing, or opening and closing a bond angle, like a scissors.
   3. *Rotational motion* is motion due to the molecule turning like a top.

B. The *third law of thermodynamics* states that the entropy of a *pure* crystal at 0 K is zero. The substance must be crystalline in order to possess maximum order. It must also be pure and without crystal defects.

C. In general, entropy increases are expected in processes in which
   1. Liquids or solutions are formed from solids.
   2. Gases are formed from either solids or liquids.
   3. The number of molecules of gas increases during a chemical reaction.
   4. The temperature of a substance is increased.

VI. A Closer Look: Entropy, Randomness, and Ludwig Boltzmann

Ludwig Boltzmann found a method to calculate entropies of substances based on the number of possible arrangements of the particles in a system. With practical sample sizes, it is a tedious task, but it does give a statistical basis for entropy.

VII. Calculation of Entropy Changes

A. The entropy values at 298 K and 1 atm pressure are called the *standard entropies*, denoted $S°$. The units of entropy are most commonly expressed in the units J/mol-K.

B. The entropy change for a reaction is found in a way analogous to the way enthalpy changes are determined. Considering the general reaction $aA + bB \rightleftharpoons$

pP + qQ, the entropy change is given by $\Delta S° = [pS°(P)+qS°(Q)+...] - [aS°(A)+bS°(B)+...]$; it follows the saying "products minus reactants."

VIII. Gibbs Free Energy

A. The *Gibbs free energy* of a substance combines its enthalpy and entropy into a single quantity that describes the total amount of energy available for use. Arithmetically, $G = H - TS$, where G is the free energy, H is the enthalpy, S is the entropy, and T is the Kelvin temperature. For a process at constant temperature, $\Delta G = \Delta H - T\Delta S$. The free energy is a "master predictor" of spontaneity:
   1. If ΔG is negative, the reaction is spontaneous in the forward direction.
   2. If ΔG is zero, the reaction is at equilibrium. There is no driving force tending to make the reaction go in either direction.
   3. If ΔG is positive, the reaction is nonspontaneous in the forward direction. However, the reverse direction of the process is spontaneous.

B. In a manner analogous to enthalpies, we tabulate *standard free energies of formation*, $\Delta G_f°$, the free energy change associated with the manufacture of a substance from the elements in their standard states. The standard conditions for free energy are the same as those for any other thermodynamic quantity, 298 K, 1 atm pressure, and 1 M concentrations of substances in solution.

C. The Gibbs free energy change for a process is determined as before. Considering the general reaction aA + bB $\rightleftharpoons$ pP + qQ, the free energy change is given by $\Delta G_{rxn}° = [p\Delta G_f°(P)+q\Delta G_f°(Q)+...] - [a\Delta G_f°(A)+b\Delta G_f°(B)+...]$.

IX. Chemistry at Work: Free Energy and Work

A. The manner in which a reaction is performed determines how much of the work available from the reaction is actually obtained.

B. The change in free energy, ΔG, equals the *maximum useful work* that can be done by a system on its surroundings in a spontaneous process occurring at constant temperature and pressure, $\Delta G = w_{max}$.

C. For nonspontaneous processes, ΔG is the measure of the *minimum* amount of work that must be applied *to the system* to cause the process to occur.
   Discussion Question: Think of examples in which two different types of energy are released if a reaction is performed two different ways. To get started, a battery produces mostly work when wired to a load, while it produces mostly heat if it is shorted.

X. Free Energy and Temperature

A. Assuming that ΔH and ΔS are independent of temperature, the free energy change for a process varies with temperature according to the term $-T\Delta S$. Qualitatively, there are four combinations of signs.
   1. If $\Delta H > 0$ and $\Delta S < 0$, the reaction is nonspontaneous.
   2. If $\Delta H < 0$ and $\Delta S > 0$, the reaction is spontaneous.
   3. If $\Delta H > 0$ and $\Delta S > 0$, the spontaneity of the reaction depends on which term is larger.
   4. If $\Delta H < 0$ and $\Delta S < 0$, the spontaneity of the reaction depends on which term is larger.

B. Only when ΔH and ΔS have the same sign do the relative magnitudes of the two terms come into play. If $|\Delta H| > |T\Delta S|$, then the ΔH will prevail and the reaction will be spontaneous if enthalpy indicates it. Conversely, if $|\Delta H| < |T\Delta S|$, then the ΔS will prevail and the reaction will be spontaneous if entropy indicates it.

Lecture Outline – Chapter 19
Brown, LeMay, & Bursten, *Chemistry: The Central Science*, 6th Edition

XI. Free Energy and the Equilibrium Constant

A. $\Delta G°$ is the free energy change at standard conditions, while ΔG is the free energy change under nonstandard conditions.

B. ΔG is given in terms of $\Delta G°$ and the reaction quotient, Q: $\Delta G = \Delta G° + RT\ln Q$, where T is the Kelvin temperature and R is the gas constant in energy units, usually 8.314 J/mol-K. Therefore, ΔG is seen as the driving force in a reaction due to deviation from equilibrium.

C. At equilibrium, there is no driving force and $\Delta G = 0$ and $Q = K$. The above expression then becomes $\Delta G° = -RT\ln K$; it provides a direct link between free energy and equilibrium. Clearly, when $K > 1$ then $\Delta G°$ is negative; when $K < 1$ then $\Delta G°$ is positive; and when $K = 1$ then $\Delta G°$ is zero.

XII. Chemistry at Work: Driving Nonspontaneous Reactions

Nonspontaneous reactions can be "driven" by the input of energy. Sometimes this can be accomplished by heating the reaction or by irradiating it. The necessary energy can also be obtained from another, highly spontaneous, reaction that can "drive" the nonspontaneous reaction. The text uses the example of purifying Cu_2S ore. The reaction is nonspontaneous, but the combustion of sulfur is spontaneous. The two reactions combine as:

$Cu_2S(s) \rightarrow 2Cu(s) + S(s) \qquad \Delta G° = +86.2$ kJ

A. $+ O_2(g) \rightarrow SO_2(g) \quad \Delta G° = -300.1$ kJ

$Cu_2S(s) + O_2(g) \rightarrow 2Cu(s) + SO_2(g) \quad \Delta G° = -213.9$ kJ

In this case, the second reaction not only provides energy for the first reaction, but also takes away the sulfur product. By Le Châtelier's principle, that too should drive the first reaction.

Lecture Outline – Chapter 19
Brown, LeMay, & Bursten, *Chemistry: The Central Science*, 6th Edition

SAMPLE QUIZ QUESTIONS

1. Account for the fact that KCl dissolves spontaneously in spite of the fact that this process is endothermic.

2. Predict the sign of ΔS for the following processes: (a) $CCl_4(l) \rightarrow CCl_4(g)$; (b) $2NO(g) + O_2(g) \rightarrow 2NO_2(g)$; (c) the combustion of glucose; (d) the freezing of water.

3. For each of the following reactions, state whether you expect it to be spontaneous or nonspontaneous based on entropy. The equations are balanced:

    (a) $MgCO_3(s) \rightarrow MgO(s) + CO_2(g)$;
    (b) $2Al(s) + 3ZnO(s) \rightarrow Al_2O_3(s) + 3Zn(s)$;
    (c) $CH_4(g) + 2O_2(g) \rightarrow CO_2(g) + 2H_2O(l)$;
    (d) $C_2H_2(g) + 5N_2O(g) \rightarrow 2CO_2(g) + H_2O(g) + 5N_2(g)$;
    (e) $Ca(OH)_2(s) \rightarrow CaO(s) + H_2O(l)$;
    (f) $CaO(s) + CO_2(g) \rightarrow CaCO_3(s)$.

4. Which of the following would you expect to possess the larger absolute entropy: (a) 1 mol O_2 at 298 K or 1 mol O_3 at 298 K; (b) 1 mol $He(g)$ at 25 K or 1 mol $He(g)$ at 25°C? Explain briefly.

5. Given $\Delta G°_f(H_2O) = -236.8$ kJ/mol and $\Delta H°_f(H_2O) = -285.8$ kJ/mol, calculate $\Delta S°$ (at 25°C) for the reaction $2H_2(g) + O_2(g) \rightarrow 2H_2O(l)$.

6. Consider the reaction, $2N_2O(g) + 3O_2(g) \rightarrow 2N_2O_4(g)$. Given $\Delta H_f°(N_2O) = 81.6$ kJ/mol, $\Delta H_f°(N_2O_4) = 9.66$ kJ/mol, $S°(N_2O) = 220.0$ J/mol-K, $S°(N_2O_4) = 304.3$ J/mol-K, and $S°(O_2) = 205.0$ J/mol-K, determine the Gibbs free energy per mole of N_2O for the reaction at 800.0°C.

7. Consider the reaction, $C_2H_4(g) + H_2O(g) \rightarrow C_2H_5OH(l)$. Given $\Delta G_f°(C_2H_4) = 68.11$ kJ/mol, $\Delta G_f°(H_2O) = -228.61$ kJ/mol, and $\Delta G_f°(C_2H_5OH) = -174.76$ kJ/mol, determine the Gibbs free energy for the reaction at standard conditions.

8. The free energy change of the reaction, $H_2O(l) \rightleftharpoons H_2O(g)$, at 25.0°C is 8.58 kJ/mol. Determine the value of the equilibrium constant for the reaction.

9. The reaction, $N_2(g) + 3H_2(g) \rightleftharpoons 2NH_3(g)$, has a standard Gibbs free energy change of -33.2 kJ at 25.0°C. Determine K_p for the reaction.

10. The equilibrium constant of the reaction, $H_2O(l) \rightleftharpoons H_2O(g)$, at 25.0°C is 3.1×10^{-2}. Determine the value of the free energy change for the reaction.

11. At 25°C, K_p equals 6.7×10^5 for the equilibrium $N_2(g) + 3H_2(g) \rightleftharpoons 2NH_3(g)$. Given that $R = 8.314$ J/K-mol, calculate $\Delta G°$ for this reaction.

12. Calculate $\Delta S°$ for the reaction $H_2(g) + Cl_2(g) \rightarrow 2HCl(g)$, given the following absolute entropies: H_2 (130.6 J/K-mol), Cl_2 (223.0 J/K-mol) and HCl (186.7 J/K-mol).

13. Is the reaction, $CaCO_3(s) \rightarrow CaO(s) + CO_2(g)$, spontaneous at 25°C? $\Delta G°_f(CaCO_3) = -1129$ kJ/mol; $\Delta H°_f(CaCO_3) = -1207$ kJ/mol; $\Delta G°_f(CaO) = -604$ kJ/mol; $\Delta H°_f(CaO) = -636$ kJ/mol; $\Delta G°_f(CO_2) = -394$ kJ/mol; $\Delta H°_f(CO_2) = -394$ kJ/mol. What is the minimum temperature at which the reaction is spontaneous?

Lecture Outline – Chapter 19
Brown, LeMay, & Bursten, *Chemistry: The Central Science*, 6th Edition

14. For any system at equilibrium at constant T and P: (a) $\Delta S = 0$; (b) $\Delta G = 0$; (c) $\Delta H = 0$; (d) $K = 0$.

15. When a liquid is vaporized: (a) $\Delta S < 0$; (b) $\Delta S > 0$; (c) $\Delta T > 0$; (d) $\Delta T < 0$.

16. If an endothermic process occurs spontaneously at constant T and P, the following must be true: (a) $\Delta G > 0$; (b) $\Delta S > 0$; (c) $\Delta H < 0$; (d) $\Delta G = 0$.

17. Calculate ΔG at 298 K for the reaction, $4NH_3(g, 0.10 \text{ atm}) + SO_2(g, 5.0 \text{ atm}) \rightarrow 4NO(g, 1.0 \text{ atm}) + 6H_2O(g, 0.10 \text{ atm})$, given that $\Delta G°$ is -958.3 kJ.

18. For any system at equilibrium at constant T and P: (a) $\Delta H = \Delta S$; (b) $\Delta H = T\Delta S$; (c) $\Delta G = T\Delta S$; (d) $\Delta H = -T\Delta S$.

19. If a reaction is exothermic and proceeds with a decrease in disorder in the system, what effect will increasing temperature have on the reaction spontaneity? (a) increase spontaneity; (b) decrease spontaneity; (c) no effect

Lecture Outline – Chapter 20
Brown, LeMay, & Bursten, *Chemistry: The Central Science*, 6th Edition

Chapter 20 — Electrochemistry

OVERVIEW: Chapter 20 discusses the many facets of electrochemical reactions. The chapter opens with a treatment describing oxidation and reduction, assigning oxidation numbers, and determining the change in oxidation state during a reaction.

Oxidation numbers are then put to use in the coverage of balancing oxidation-reduction, or redox, reactions. Both the oxidation number method and the ion-electron or half-reaction method are used. The balancing of solution redox reactions is stressed.

The authors chose to cover voltaic or spontaneous electrochemical reactions first. The construction of typical voltaic cells is presented, including the construction of gas electrodes. Cell potentials are covered in detail, beginning with the concept of driving force and the difference in electron potential energy in the two electrodes.

The coverage of voltaic cells includes half-reaction potentials, overvoltages, the effect of concentration and temperature on cell potential, and the relationships among cell potential, ΔG, and the equilibrium constant.

A separate section fully describes commercially important voltaic cells, including dry cells, the lead storage battery, alkaline batteries, and fuel cells.

Electrolysis or nonspontaneous electrochemical reactions are then covered. The events occurring during electrolysis are described, as is the typical electrolytic cell. The idea of the potential required for electrolysis is also discussed. The quantitative aspects of electrolysis are described in detail, culminating in the calculation of metal masses from current-time data.

Finally, the chapter ends with a discussion of corrosion, its physical and economic effects, and measures employed to prevent corrosion.

LECTURE OUTLINE

I. Oxidation and Reduction

A. *Oxidation state* or *oxidation number* is the "apparent" charge on an atom in a molecule or an ion. The oxidation state of an atom is determined by assigning bonding electrons to the more electronegative element in the bond. The electrons assigned in this way are added to the number of unbonded electrons on the atom and this total subtracted from the number of valence electrons seen in the neutral atom.

B. *Oxidation* is a process that removes electrons from an atom, molecule, or ion.

C. *Reduction* is a process that adds electrons to an atom, molecule, or ion.

D. *Oxidation-reduction reactions*, or *redox reactions*, are those in which one or more species undergoes oxidation, while one or more different species undergo reduction.

E. The field of chemistry that studies the relationship between electricity and chemical reactions is called *electrochemistry*.

Lecture Outline – Chapter 20
Brown, LeMay, & Bursten, *Chemistry: The Central Science*, 6th Edition

II. Oxidation-Reduction Reactions

A. We determine whether a reaction is a redox reaction by monitoring the oxidation numbers of the elements involved. This indicates which atoms, if any, are being oxidized, and which, if any, are being reduced.

B. In a redox reaction, electrons are being transferred; they are being removed from one substance by oxidation, and added to another substance by reduction. Because of conservation, oxidation and reduction *must* occur simultaneously. Further, the number of electrons removed by oxidation must equal the number of electrons added by reduction.

C. In a redox reaction, the substance being oxidized is called a *reducing agent* or *reductant*.

D. In a redox reaction, the substance being reduced is called an *oxidizing agent* or *oxidant*.

III. Balancing Oxidation-Reduction Reactions

A. In balancing redox reactions, the number atoms of each element must be the same on both sides of the equation. This requirement must be met for all chemical equations. Additionally, in oxidation-reduction reactions the total *charge* must be the same on both sides of the equation.

B. In the *method of half-reactions* or the *ion-electron method* for balancing redox reactions, the reaction is split into two parts. One *half-reaction* describes the substances undergoing oxidation and the other half-reaction describes those substances undergoing reduction. Each half-reaction is balanced separately and then they are recombined to give the net balanced equation. The procedure as applied to *acidic solutions* can be summarized as follows.
   1. Write the unbalanced equation for the reaction.
   2. Divide the overall reaction into two unbalanced half-reactions, one for oxidation and the other for reduction.
   3. In each half-reaction, balance the elements other than H and O.
   4. In each half-reaction, balance the O atoms by adding H_2O, then balance the H atoms by adding H^+.
   5. In each half-reaction, balance the charge on each side of the half-reaction by adding e^- to the side with the greater positive charge.
   6. Multiply the two half-reactions by coefficients such that the overall electron loss equals the overall electron gain.
   7. Add the two half-reactions; simplify where possible by eliminating terms appearing on both sides of the equation.

C. For reactions in *basic solution*, balance the reactions as if occurring in acidic solution. Cancel any H^+ ions by adding an equal number of OH^- ions to both sides of the equation, and simplify, if necessary, the number of water molecules in the reaction.

IV. Voltaic Cells

A. *Voltaic cells* or *galvanic cells* are devices for performing a spontaneous electrochemical reaction in which the electron transfer is forced to occur through an external pathway.

B. In a voltaic cell, the two half-reactions are contained in separate vessels; each vessel and its contents are referred to as *half-cells*. The two half-cells are connected by (1) an external circuit between the *electrodes*, generally containing a load, such as a light or a voltage meter, and (2) a *salt bridge* between the two solutions.

C. The electrode at which oxidation occurs is made of the substance being oxidized or an inert material coated with the substance being oxidized. This electrode is called the *anode*. In a voltaic cell, the anode has a negative charge.

D. The electrode at which reduction occurs is often, but not necessarily, made of the substance which is the product of the reduction. This electrode is called the *cathode*. In a voltaic cell, the cathode has a positive charge.

E. The salt bridge is generally a gel that contains an ionic substance that does not interfere with the redox reaction. $NaNO_3$ is commonly used because Na^+ is difficult to reduce and NO_3^- is difficult to oxidize. The purpose of the salt bridge is to create a path for ion migration, without mixing the contents of the two half-cells. The migration of ions results in electrical conductivity or a completed electrical circuit.

V. Cell EMF

A. In spontaneous electrochemical reactions, the electrons flow from anode to cathode because it is energetically favorable for them to do so. The potential energy of the electrons in the anode is higher that the potential energy of the electrons in the cathode; electrons in the anode lower their potential energy by traveling through the external circuit to the cathode. The potential *difference* of an electron in the anode and an electron in the cathode is measured in *volts*, where 1 V = 1 joule/coulomb = 1 J/C.

B. The potential energy difference is the driving force of a voltaic reaction. Consequently, the voltage is commonly referred to as the *electromotive force*, *emf*, or *cell potential*.

C. Under standard conditions, the voltage is referred to as the *standard emf* or the *standard cell potential*, $E°_{cell}$. The standard conditions of 25°C, 1 atm pressure, and 1 M solutions are assumed unless noted otherwise.

D. The cell potential can be thought of as the combination of half-cell potentials, just as the redox reaction can be thought of as the combination of two half-reactions.
   1. The half-cell potential for the half-cell undergoing oxidation is the *oxidation potential*, E_{ox}.
   2. The half-cell potential for the half-cell undergoing reduction is the *reduction potential*, E_{red}.
   3. The cell potential is the sum of the half-cell potentials, $E_{cell} = E_{ox} + E_{red}$.
   4. If the reaction is performed under standard conditions, the *standard emf* is given by the sum of the *standard oxidation potential* and the *standard reduction potential*, $E°_{cell} = E°_{ox} + E°_{red}$.

E. Oxidation or reduction potentials cannot be directly measured for an isolated half-reaction. Therefore, one half-reaction is arbitrarily chosen to be 0.00 V, and other half-reactions are then measured relative to it. The reference half-reaction is the reduction of hydrogen ion, $2H^+(aq)$ (1 M) + 2e⁻ $\rightleftharpoons$ $H_2(g)$ (1 atm), for which $E°_{red} = E°_{ox} = 0.00$ V at 25°C. The half-cell is constructed in the normal way, except that the electrode consists of an unreactive platinum plate mounted at the lip of an inverted tube so that hydrogen gas can be bubbled over the plate's surface.

F. The half-cell potential for a reduction is equal in magnitude but opposite in sign to the same half reaction written as an oxidation, $E°_{red} = -E°_{ox}$.

G. Because the oxidation potential and reduction potential for a given half-reaction are so easily related, we tabulate only the *standard reduction potentials* or *standard electrode potentials* for half-reactions.

Lecture Outline – Chapter 20
Brown, LeMay, & Bursten, *Chemistry: The Central Science, 6th Edition*

H. The more positive the $E°$ value for a half-reaction, the greater the tendency for that reaction to occur as written. Therefore, the half-reactions with the most positive reduction potentials are the most powerful oxidizing agents, while the half-reactions with the most positive oxidation potentials are the most powerful reducing agents.

VI. Spontaneity of Redox Reactions

A. A positive emf indicates a spontaneous process, while a negative emf indicates a nonspontaneous process.

B. The emf of a redox reaction describes spontaneity, as does ΔG. Not surprisingly, the two are related, $\Delta G = -n\mathcal{F}E$, where n is the number of moles of electrons transferred in the reaction and $\mathcal{F}$ is Faraday's constant. *Faraday's constant* is the electrical charge on 1 mol of electrons, $1\,\mathcal{F} = 96{,}500\,\frac{C}{mol\,e^-} = 96{,}500\,\frac{J}{V\text{-mol}\,e^-}$. When both half-reactions are at standard conditions, $\Delta G° = -n\mathcal{F}E°$.

VII. Effect of Concentration on Cell EMF

A. Cell emf varies with the concentrations of the participating species. Substituting $\Delta G = -n\mathcal{F}E$ into $\Delta G = \Delta G° + 2.303RT\log Q$ gives $E = E° - \frac{2.303RT}{n\mathcal{F}}\log Q$, an expression known as the *Nernst equation*. At 25°C, the Nernst equation simplifies to $E = E° - \frac{0.0591\,V}{n}\log Q$. It is predicted by the Nernst equation that cell emf increases as reactant concentrations increase and as product concentrations decrease; this is verified experimentally. It is also clear that as the cell operates, reactant concentrations decrease and product concentrations increase, and so the cell potential decreases as the reaction proceeds.

B. EMF is also logically related to the equilibrium constant of a reaction. Because $\Delta G° = -n\mathcal{F}E°$ and $\Delta G° = -RT\ln K$, $-n\mathcal{F}E° = -RT\ln K$, and $E° = \frac{RT}{n\mathcal{F}}\ln K = \frac{2.303RT}{n\mathcal{F}}\log K$. At 25°C, combining the constants, $E° = \frac{0.0591\,V}{n}\log K$. It is clear that as the equilibrium constant increases, so does the cell potential.

VIII. Chemistry and Life: Heartbeats and Electrocardiography

A. The human heart pumps more than 7000 L of blood per day through the body. The heart rate and rhythm are controlled electrochemically.

B. The cell walls are *semipermeable* to many ions, most importantly here being Ca^{2+}, Na^+, and K^+. The concentrations of these ions are different inside and outside the cell, in the *intracellular fluid* and the *extracellular fluid*, respectively. The differences in concentration lead to differences in cell potential across the membrane.

C. The *pacemaker cells*, monitor and govern the ion concentrations, leading to a rhythmic fluctuation of concentrations and rhythmic contraction of the heart muscle.

D. Although there are many kinds of heart malfunctions, indicating different kinds of treatments, damage to the pacemaker cells can often be treated with an artificial pacemaker, which emits the triggering electrical signals in place of the pacemaker cells.

IX. Commercial Voltaic Cells

A. *Batteries* are voltaic cells of commercial value. A typical battery can consist of a single voltaic cell or several connected in series, in which the emf of the battery is the sum of the emf values of the individual voltaic cells.

B. The *lead storage battery* is the type commonly used in automobiles; a 12-V battery consists of six 2-V voltaic cells connected in series. The anode is lead, while the cathode is PbO_2 on a metal support, and both are dipped into sulfuric acid.

C. The *dry cell* is the common flashlight battery. In the acidic dry cell, the anode is a zinc can in which there is a paste of MnO_2, NH_4Cl, and C. The cathode, made of a carbon rod, is inserted in the center of the paste. In the alkaline dry cell, or simply *alkaline battery*, the species are almost the same as in the acidic cell. In the alkaline cell, KOH is used in place of NH_4Cl and the zinc is present as a powdered gel.

D. *Nickel-cadmium*, or *Ni-Cad*, *batteries*, are rechargeable. That is, the reaction can be reversed to restore the battery's composition. In these cells, cadmium is the anode and NiO_2 is the cathode, both immersed in a basic medium.

E. *Fuel cells* are batteries in which conventional fuels, such as H_2 or CH_4, are used. For example, H_2–O_2 fuel cells are used on spacecraft because they give a tremendous amount of power per pound launched. Since the reactants, H_2 and O_2 are supplied continuously, the fuel cells do not become depleted; the product water is used by the crew.

F. Recent advances in battery technology have led to the *solid electrolyte* or *fast-ion conductor* battery. In these cells, the charge-carrying medium is a solid, not an aqueous paste. Therefore, these cells have shown promise in research of providing more power with less weight than is available in current batteries.

Discussion Question: Discuss the various sizes and shapes of batteries commonly available. Does the physical size of the battery affect, within reason, the voltage output of the battery?

X. Electrolysis

A. An electrochemical reaction that is nonspontaneous can be driven to proceed by applying an external electrical supply. Such reactions are called *electrolysis reactions* and they are performed in *electrolytic cells*. Electrolysis can be performed in aqueous solution or in systems called *melts* in which the solid electrolyte reactants are molten. As in voltaic cells, oxidation occurs at the anode, while reduction occurs at the cathode. The electrode charges are the reverse of those in voltaic cells; in an electrolytic cell, the anode is positive and the cathode is negative.

B. *Electrolysis of molten salts* is a straightforward process, although it generally requires high temperatures to melt the salts. However, because only the ions of the molten salt are present, the products are formed in high purity.

C. The *electrolysis of aqueous solutions* is attractive in that high temperatures need not be attained; the species being electrolyzed must be soluble in aqueous solution. However, electrolysis in aqueous solution does not work for all reactions; water can be oxidized or reduced. If water is more easily reduced than the solute ($E°_{red}(H_2O) = -0.83$ V), it is the water that will be reduced. Similarly, if water is more easily oxidized than the solute ($E°_{ox}(H_2O) = -1.23$ V), it is the water that will be oxidized. Due to an *overvoltage*, or greater than theoretical voltage, needed to oxidize water, the strict comparison of $E°$ values often cannot predict whether water or the solute will react.

Lecture Outline – Chapter 20
Brown, LeMay, & Bursten, *Chemistry: The Central Science*, 6th Edition

- D. Electricity has three describing characteristics, the *emf*, E, the *current*, i, and the *resistance*, R, related by $E = iR$. Of these, we are interested in the emf and the current in discussing electrolysis.
    1. The emf describes the driving force of the process. In order for electrolysis to occur, the applied electricity must have an emf greater than or equal to $E°_{cell}$. Because of internal resistance and the overvoltage phenomenon, the actual applied emf is invariably greater than $E°_{cell}$.
    2. The current describes the number of electrons flowing past a point per second.
- E. The electrodes of an electrolysis can be *inert* or they can be *active*. Inert electrodes are those that are not modified during the electrolysis; they are not made of materials shown in the chemical equation and products do not adhere to them. Active electrodes do participate in the electrolysis, either by being degraded or by electrolysis products adhering to them. For example, in *electroplating*, the reduction product, often silver, nickel, or gold, is deposited in a uniform layer on the cathode. Active electrodes are also used in many metal purification processes.

Discussion Question: There are many commercial applications of electrolysis, many of which are described in the text. Can you or the students propose other materials that could be produced by electrolysis? Is the popular technique of hair electrolysis consistent with students' understanding of electrolysis?

XI. Quantitative Aspects of Electrolysis

- A. Stoichiometrically, the number of ions of a metal that are reduced by an applied electrical current is proportional to the number of electrons delivered to the cell by the current. Recall that $1\ \Im = 96,500$ coulombs = charge of 1 mol of electrons and the associated conversion factor is $\dfrac{96,500\ C}{mol\ e^-}$.
- B. The amount of charge passing through an electrical circuit is generally measured in *coulombs*. A coulomb is the amount of charge passing a point in a circuit operating at 1 ampere (A) for 1 second. Therefore, the total charge passing a point in a circuit in a specified period of time is given by charge = (current)(time), or coulombs = (amperes)(seconds).
- C. The amount of charge passed through a circuit is related through the Faraday constant to the number of moles of electrons passed by $\#\ \Im = \dfrac{(amperes)(seconds)}{96,500\ C}$ and to the number of moles of metal ion reduced by $\#\ moles\ metal = \dfrac{(amperes)(seconds)}{(96,500\ C)}\dfrac{(1\ mol\ metal)}{(\#\ mol\ e^-)}$ and to the mass of metal reduced by $mass\ metal = \dfrac{(amperes)(seconds)}{(96,500\ C)}\dfrac{(1\ mol\ metal)}{(\#\ mol\ e^-)}\dfrac{(g\ metal)}{(mol\ metal)}$. This expression can, of course, be rearranged to solve for any one missing quantity.
- D. Because $\Delta G = w_{max}$ and $\Delta G = -n\Im E$, $w_{max} = -n\Im E$, which defines the maximum amount of work that can be derived *from* a galvanic cell. Similarly, the minimum amount of work that must be performed *on* an electrolytic system in order for the reaction to proceed is given by $w_{min} = -n\Im E$, although a higher potential is generally required due to overvoltage.
- E. Electrical work, in joules, is also given by the *power*, in *watts*, times time, where 1 J = 1 watt-second. A kilowatt-hour is then 3.6×10^6 J.

Discussion Question: Discuss the amount of electricity used by a plant that produces aluminum electrolytically. Propose an amount of aluminum to be produced in a day and apply your local electric rates.

XII. Corrosion

A. *Corrosion* is a redox reaction in which a metal is attacked by its environment to form an unwanted substance. All metals except gold and platinum, are thermodynamically capable of undergoing oxidation at room temperature. Uninhibited corrosion can, of course, be very destructive. Some metals form protective coatings in the initial stages of corrosion which protect them from further corrosion; this is the case for aluminum and magnesium in their air oxidations.

B. Of commercial and governmental concern is the *corrosion of iron*. This process, commonly referred to as *rusting*, requires the presence of oxygen and water. The pH and the presence of salts and other metals more resistant to oxidation than iron can accelerate rusting.

C. Iron and other metals are often coated with paint or another metal, such as tin, zinc, or chromium, to protect its surface from corrosion. Tin coatings are used in food cans (tin cans), but protect only as long as the tin surface remains intact. Coating of iron with zinc results in *galvanized iron*. Zinc protects the iron, even if the zinc surface is broken. Because zinc is oxidized more readily than iron, corrosion attacks the zinc, making zinc the anode in an electrochemical reaction and iron the cathode. This type of protection, in which the metal being protected serves as a cathode, is called *cathodic protection.*

Discussion Question: Modern materials, like plastics and ceramics, do not corrode. Why haven't these materials replaced structural metals? What typical applications of metals have been taken over by plastics and ceramics?

Lecture Outline – Chapter 20
Brown, LeMay, & Bursten, *Chemistry: The Central Science*, 6th Edition

SAMPLE QUIZ QUESTIONS

1. Oxidation: (a) occurs at the cathode; (b) also means a decrease in the oxidation number; (c) is the addition of electrons to an atom; (d) is the removal of electrons from an atom; (e) is the process undergone by the oxidizing agent.

2. Reducing agents: (a) cause another reactant to be oxidized; (b) are reduced in redox reactions; (c) cause another reactant to be reduced; (d) cause a precipitation reaction; (e) usually have positive oxidation numbers.

3. Which of the following reactions is a redox reaction?

    (a) $3Se(s) + 4HNO_3(aq) \rightarrow 3SeO_2(s) + 4NO(g) + 2H_2O(l)$;
    (b) $CaCl_2(aq) + 2AgNO_3(aq) \rightarrow Ca(NO_3)_2(aq) + 2AgCl(s)$;
    (c) $BaCO_3(s) \rightarrow BaO(s) + CO_2(g)$;
    (d) $3Mg(OH)_2(s) + 2NH_3(g) \rightarrow Mg_3N_2(s) + 6H_2O(l)$;
    (e) $NH_3(g) + H_2O(l) \rightarrow NH_4^+(aq) + OH^-(aq)$

4. Consider the reaction

    $$ClO_3^-(aq) + 3S^{2-}(aq) + 3H_2O(l) \rightarrow Cl^-(aq) + 3S(s) + 6OH^-(aq).$$

    Which substance is the reducing agent? (a) $S(s)$; (b) $ClO_3^-(aq)$; (c) $H_2O(l)$; (d) $OH^-(aq)$; (e) $S^{2-}(aq)$

5. Balance the following redox reaction in acidic solution.

    $$Fe^{3+}(aq) + I^-(aq) \rightarrow Fe^{2+}(aq) + I_2(s)$$

    (a) $Fe^{3+}(aq) + 2I^-(aq) \rightarrow 2Fe^{2+}(aq) + I_2(s)$;
    (b) $2Fe^{3+}(aq) + 2I^-(aq) \rightarrow 2Fe^{2+}(aq) + I_2(s)$;
    (c) $2Fe^{3+}(aq) + 2I^-(aq) \rightarrow Fe^{2+}(aq) + I_2(s)$;
    (d) $Fe^{3+}(aq) + 2I^-(aq) \rightarrow Fe^{2+}(aq) + I_2(s)$;
    (e) $2Fe^{3+}(aq) + 2I^-(aq) \rightarrow 2Fe^{2+}(aq) + I_2(s) + 2H^+(aq)$.

6. Balance the following redox reaction in acidic solution.

    $$NO_3^-(aq) + H^+(aq) + I_2(s) \rightarrow NO_2(g) + H_2O(l) + IO_3^-(aq)$$

    (a) $10NO_3^-(aq) + 8H^+(aq) + I_2(s) \rightarrow 10NO_2(g) + 2H_2O(l) + 2IO_3^-(aq)$;
    (b) $5NO_3^-(aq) + 8H^+(aq) + I_2(s) \rightarrow 5NO_2(g) + 4H_2O(l) + 2IO_3^-(aq)$;
    (c) $10NO_3^-(aq) + 4H^+(aq) + I_2(s) \rightarrow 10NO_2(g) + 2H_2O(l) + 2IO_3^-(aq)$;
    (d) $10NO_3^-(aq) + 8H^+(aq) + I_2(s) \rightarrow 10NO_2(g) + 4H_2O(l) + 2IO_3^-(aq)$;
    (e) $10NO_3^-(aq) + 8H^+(aq) + I_2(s) \rightarrow 10NO_2(g) + 4H_2O(l) + IO_3^-(aq)$.

7. Balance the following redox reaction in acidic solution.

    $$H_2O_2(aq) + Cr_2O_7^{2-}(aq) \rightarrow Cr^{3+}(aq) + O_2(g)$$

    (a) $3H_2O_2(aq) + Cr_2O_7^{2-}(aq) + 8H^+(aq) \rightarrow 2Cr^{3+}(aq) + 3O_2(g) + 7H_2O(l)$;
    (b) $H_2O_2(aq) + Cr_2O_7^{2-}(aq) \rightarrow 2Cr^{3+}(aq) + O_2(g)$;
    (c) $H_2O_2(aq) + Cr_2O_7^{2-}(aq) \rightarrow 2Cr^{3+}(aq) + 9O_2(g)$;

(d) $3H_2O_2(aq) + Cr_2O_7^{2-}(aq) \rightarrow 2Cr^{3+}(aq) + 3O_2(g) + 3H_2(g)$;
(e) $3H_2O_2(aq) + Cr_2O_7^{2-}(aq) \rightarrow 2Cr^{3+}(aq) + 3O_2(g)$.

8. Balance the following redox reaction in basic solution.

$$Cl_2(g) \rightarrow Cl^-(aq) + ClO^-(aq)$$

(a) $2Cl_2(g) + 2OH^-(aq) \rightarrow 2Cl^-(aq) + 2ClO^-(aq) + H_2O(l)$;
(b) $Cl_2(g) + 4OH^-(aq) \rightarrow Cl^-(aq) + ClO^-(aq) + 2H_2O(l)$;
(c) $2Cl_2(g) + OH^-(aq) \rightarrow 2Cl^-(aq) + 2ClO^-(aq)$;
(d) $2Cl_2(g) + 4OH^-(aq) \rightarrow 2Cl^-(aq) + 2ClO^-(aq) + 2H_2O(l)$;
(e) $Cl_2(g) + 4OH^-(aq) \rightarrow Cl^-(aq) + ClO^-(aq) + 2H_2O(l)$.

9. Balance the following redox reaction in basic solution.

$$CN^-(aq) + MnO_4^-(aq) \rightarrow MnO_2(s) + CNO^-(aq)$$

(a) $CN^-(aq) + MnO_4^-(aq) + H_2O(l) \rightarrow MnO_2(s) + CNO^-(aq) + 2OH^-(aq)$;
(b) $CN^-(aq) + 2MnO_4^-(aq) + 3H_2O(l) \rightarrow 2MnO_2(s) + CNO^-(aq) + 6OH^-(aq)$;
(c) $3CN^-(aq) + 2MnO_4^-(aq) + H_2O(l) \rightarrow 2MnO_2(s) + 3CNO^-(aq) + 2OH^-(aq)$;
(d) $3CN^-(aq) + MnO_4^-(aq) + 2OH^-(aq) \rightarrow MnO_2(s) + 3CNO^-(aq) + H_2O(l)$;
(e) $3CN^-(aq) + 2MnO_4^-(aq) \rightarrow 2MnO_2(s) + 3CNO^-(aq) + OH^-(aq)$.

10. Complete and balance each of the following:

(a) $NCl_3(g) + H_2O(l) \rightarrow NH_3(g) + HOCl(aq)$;
(b) $FeO(s) + O_2(g) \rightarrow Fe_3O_4(s)$;
(c) $PbO_2(s) + N_2H_4(g) \rightarrow Pb(s) + N_2(g) + H_2O(g)$.

11. Complete and balance each of the following:

(a) $S_2O_3^{2-}(aq) + Cl_2(g) \rightarrow SO_4^{2-}(aq) + Cl^-(aq)$ (acidic);
(b) $S_8(s) + OH^-(aq) \rightarrow S^{2-}(aq) + S_2O_3^{2-}(aq)$ (basic);
(c) $MnO_4^-(aq) + NO_2^-(aq) \rightarrow Mn^{2+}(aq) + NO_3^-(aq)$ (acidic).

12. Complete and balance each of the following:

(a) $H_2O_2(aq) + H_2SO_3(aq) \rightarrow$;
(b) $Cd(s) + NO_3^-(aq) + H^+(aq) \rightarrow$;
(c) $Sn^{2+}(aq) + O_2(aq) + H^+(aq) \rightarrow$.

13. A voltaic cell: (a) is usually housed in a single vessel; (b) cannot be reversed; (c) requires the application of external electricity; (d) must prevent the half-cells from physically mixing; (e) is comprised of a nonspontaneous reaction.

14. Voltaic cells: (a) are of little commercial value; (b) are generally large, bulky devices; (c) generate electricity; (d) are reversible; (e) require only an oxidation half-cell.

15. The standard potentials for the reduction of the following ions are given in parentheses: Mg^{2+} (-2.37 V); Fe^{2+} (-0.44 V); Cu^{2+} (0.34 V); Ag^+ (0.80 V). (a) Which of the following is the best reducing agent? Mg, Cu, Ag^+, or Mg^{2+}. Explain briefly. (b) What emf is generated by a standard cell utilizing the following reaction: $2Ag^+ + Fe \rightleftharpoons 2Ag +$

Fe²⁺. (c) Will the following reaction be spontaneous? Mg + Cu²⁺ ⇌ Mg²⁺ + Cu. Explain briefly.

16. Consider the following electrode potentials: Fe²⁺ + 2e⁻ ⇌ Fe E° = -0.44 V   Cu²⁺ + 2e⁻ ⇌ Cu E° = 0.34 V (a) What voltage would be produced by a standard voltaic cell that utilizes these half-reactions? (b) Draw the cell, labeling the anode, the cathode, and the directions of the ion and electron motions. (c) What effect will increasing the concentration of Cu²⁺ have on the cell potential?

17. The voltage produced in the reaction Zn + Cu²⁺ ⇌ Cu + Zn²⁺ is independent of: (a) the size of the cathode; (b) the metal used for the anode; (c) the temperature; (d) the concentration of Cu²⁺ in the cathode compartment.

18. If the reaction Zn + Cu²⁺ ⇌ Cu + Zn²⁺ produces a standard potential of 1.10 V, what voltage is produced when [Cu²⁺] = 0.50 *M* and [Zn²⁺] = 0.005 *M*?

19. If *E°* = 0.76 V for Zn + 2H⁺ ⇌ Zn²⁺ + H₂, what is the free energy change and equilibrium constant for this reaction?

20. A battery: (a) produces the same amount of electricity, regardless of composition; (b) is rechargeable; (c) generally has no liquid components; (d) produces a constant amount of electricity for the duration of its lifetime; (e) spontaneously produces electricity.

21. When a lead-storage battery is being charged, it is operating as: (a) a voltaic cell; (b) a dry cell; (c) a fuel cell; (d) an electrolytic cell.

22. The primary reason for the development of the fuel cell is its: (a) low cost; (b) high theoretical efficiency; (c) high voltage; (d) low operating temperature.

23. Batteries are: (a) generally transportable; (b) always based on metal redox reactions; (c) all rechargeable; (d) only used to operate small devices; (e) generally expensive.

24. Electrolytic cells: (a) generally are composed of two vessels; (b) generate electricity; (c) are spontaneous cells; (d) can be used for electroplating; (e) have little commercial value.

25. When aqueous HCl is electrolyzed: (a) Cl₂ is produced at the anode; (b) H₂ is produced at the anode; (c) O₂ is produced at the anode; (d) O₂ is produced at the cathode.

26. Silver can be plated onto nickel: (a) by applying electricity to a nickel anode in a solution of silver ions; (b) by applying electricity to a silver anode in a solution of nickel ions; (c) by applying electricity to a nickel cathode in a solution of silver ions; (d) by allowing a solution of nickel ions to react with a piece of silver; (e) by dipping the nickel object into a solution of silver ions.

27. If 6.00 amps are passed through a bath of aqueous CuSO₄ for 2.00 hrs, how much copper metal would plate out?

28. What is the minimum energy required to produce 1.0 mol of Cu by electrolysis if the process requires an emf of 1.0 V and is 100% efficient?

29. Describe and explain the relative merits of paint, tin, and zinc as protective coatings on iron.

30. Explain the distinction between an electrolytic cell and a voltaic cell.

Lecture Outline – Chapter 21
Brown, LeMay, & Bursten, *Chemistry: The Central Science, 6th Edition*

Chapter 21 — Nuclear Chemistry

OVERVIEW: Chapter 21 discusses nuclear reactions and their impact on our lives. The chapter opens with a review of the subatomic particles and the forms of radioactivity commonly seen. Attention is also given to the production of new nuclei through bombardment of otherwise stable nuclei.

The concept of half-life is covered in detail, emphasizing the idea that varying rates of decay can explain why some radioisotopes are found in nature while others are not. Calculations involving half-life are covered, including dating.

A description is given of methods for detecting radiation, especially the Geiger counter and the scintillation counter.

Nuclear binding energy is covered and extended to the energies of nuclear reactions. Calculations are performed based on the mass-energy relationship.

Nuclear fission and fusion are described in depth. These two topics invoke more curiosity than most others in general chemistry. Real events and accidents are covered, adding to the relevance of the coverage. Peaceful and destructive uses of nuclear energy, as well as the hazards of each are discussed.

The chapter closes with a discussion of the ways of expressing radiation dosages and the biological effects of radiation. Unintentional exposures are covered, as well as controlled exposures for the treatment of disease, are covered.

LECTURE OUTLINE

I. Radioactivity

    A. *Nuclear chemistry* studies the reactions of the nucleus of the atom. Recall from Chapter 2 that there are two principal nuclear particles, *protons* and *neutrons*. Protons and neutrons are collectively referred to as *nucleons*.

    B. All atoms of a given element have the same *atomic number* or the same number of protons in their nuclei.

    C. An atom's *mass number* is the sum of the number of protons and the number of neutrons in its nucleus. Atoms of a given element can have a different number of neutrons; they can have a different mass number.

    D. Atoms of the same element that have different mass numbers are called *isotopes* of one another.

    E. Specific isotopes are referred to in nuclear chemistry by listing the mass number with the atomic symbol, as in ^{235}U or $^{235}_{92}U$, where the subscript is the redundant atomic number.

    F. *Radioactivity* is the *spontaneous* emission of particles or radiation from the nucleus. Those isotopes that are radioactive are called *radioisotopes*. The nuclei of radioisotopes are called *radionuclides*.

    G. When a radioactive isotope undergoes decomposition, the process is referred to as *radioactive decay* or simply *decay*.

    H. Some nuclear reactions occur only after the nucleus has been struck or *bombarded* by another particle. Nuclear reactions that require this type of induction are called *nuclear transmutations*. It is important to remember that decay is spontaneous, while transmutation needs initiation.

I. There are many types, or modes, of radioactive decay.
1. *Alpha particles* (α) are helium-4 nuclei, ^4_2He. Alpha emission consists of a stream of alpha particles released from the reacting nuclei. Alpha emission has the effect of decreasing both the atomic number and the mass number.
2. *Beta particles* (β) are electrons originating in the nucleus. They are often symbolized as $^{\,\,0}_{-1}e$, in order to emphasize their charge and relative mass. Beta emission consists of a stream of nuclear electrons released from the reacting nuclei. Beta particles are produced by the decomposition of a neutron, $^1_0n \rightarrow\ ^1_1p +\ ^{\,\,0}_{-1}e$. Beta emission has the effect of increasing the atomic number by one, but leaves the mass number unchanged.
3. *Gamma radiation* (γ) consists of very short-wavelength radiation. Gamma rays are often represented as $^0_0\gamma$. Gamma radiation does not affect either the mass number or the atomic number; it is a mechanism for dispersing energy.
4. A *positron* (^0_1e) is a subatomic particle that has the same mass as an electron, but carries a positive charge. Positrons are produced by the decomposition of a proton, $^1_1p \rightarrow\ ^1_0n +\ ^0_1e$. Positron emission has the effect of decreasing the atomic number by one, while leaving the mass number unchanged. If a positron collides with an electron, an *annihilation* reaction occurs, $^0_1e +\ ^{\,\,0}_{-1}e \rightarrow 2^0_0\gamma$.
5. In *electron capture*, is the consumption of an inner-shell *orbital* electron by a proton in the nucleus, $^1_1p +\ ^{\,\,0}_{-1}e \rightarrow\ ^1_0n$. Electron capture has the effect of reducing the atomic number by one, while leaving the mass number unchanged.

II. **Patterns of Nuclear Stability**
A. It is observed that stability can be correlated to some degree to the *neutron-to-proton ratio*. For light elements, a ratio of 1:1 indicates stability, while for heavier elements, the ratio that indicates stability increases to about 1.5:1. In a graph of number of neutrons vs. number of protons, nuclei falling near the ratios that imply stability are said to be within the *belt of stability*.
1. Nuclei above the belt of stability (high neutron-to-proton ratios) generally decompose by beta emission, which decreases the number of neutrons and increases the number of protons.
2. Nuclei below the belt of stability (low neutron-to-proton ratios) generally decompose by positron emission or electron capture. Either event decreases the number of protons and increases the number of neutrons.
3. Nuclei with atomic numbers ≥ 84 are all unstable. They tend to decompose by alpha emission.

B. Some nuclei do not produce a stable product nucleus with a single nuclear reaction, but rather by a sequence of nuclear reactions. Such a sequence of nuclear reactions leading from an unstable nucleus to a stable one are referred to as a *radioactive series* or *nuclear disintegration series*.

C. Although not foolproof, there are some general statements for predicting nuclear stability.

1. Nuclei with 2, 8, 20, 28, 50, or 82 protons are likely to be stable. Similarly, nuclei with 2, 8, 20, 28, 50, 82, or 126 neutrons are likely to be stable. These values are called *magic numbers*, and are analogous to the electron closed shells of noble gases.
2. Nuclei with *even numbers* of both protons and neutrons are likely to be stable.

III. Nuclear Transmutations

A. Rutherford and his coworkers were the first to artificially produce a nucleus from a different nucleus. They bombarded ^{14}N with alpha particles, producing ^{17}O, $^{14}_{7}N(\alpha,p)^{17}_{8}O$, where the notation in parentheses lists first the bombarding particle and second the emitted particle. In this reaction, then, a proton was emitted as part of the reaction. In equation form, $^{14}_{7}N + ^{4}_{2}He \rightarrow ^{17}_{8}O + ^{1}_{1}p$.

B. Charged bombarding particles, such as alpha particles, required acceleration to high velocities in order to overcome the repulsion encountered as they approach a target nucleus. Charged particles can be accelerated using strong electrical and magnetic fields. The three most common accelerators are *cyclotrons*, *synchrotrons*, and *linear accelerators*. The particles to be accelerated are injected into the vacuum of the apparatus, where strong fields act to accelerate them. In a cyclotron, the particle path is a spiral beginning at the center of the device and enlarges as the velocity increases. In a synchrotron, increasing field strengths are used to maintain a fixed path for the particles. Linear accelerators use alternating fields to boost the particle velocity as the particle proceeds along the device. Neutrons, which carry no charge, cannot be accelerated; neutron bombardment generally is performed in a nuclear reactor, where neutrons abound.

C. Artificial transmutations are used to produce the elements beyond uranium in the periodic table. These elements, atomic numbers 93 to 109, are referred to as the *transuranium elements*.

IV. Rates of Radioactive Decay

A. The *half-life* of a radioisotope is the time required for half the sample to decay. Nuclear decays are invariably first-order. Therefore, 50% of the sample decays in 1 half-life, 50% of the remaining amount decays within the second half-life (for a total of 75% consumed), and so on.

B. The characteristic rates of decay of radioisotopes make them useful in determining the age of objects, a technique called *dating*. Carbon-14 is produced in the atmosphere and is incorporated into all living things. Upon the death of the organism, no additional carbon-14 is accumulated. By knowing the concentration of carbon-14 in an organism at the time of death and measuring its current concentration, time elapsed since its death can be estimated. The half-life of ^{14}C is 5730 years. The decay of other isotopes, such as $^{238}_{92}U$ or $^{40}_{19}K$, can be used to date rocks and minerals.

C. Using the first-order rate law, calculations based on half-life can be performed.
1. Recall that the first-order rate law has the form, rate = kN, where N is the number of nuclei of the isotope undergoing decay and k is the rate constant.
2. The concentration-time form of the rate law then is $\ln\frac{N_t}{N_0} = -kt$, where N_0 is the number of nuclei of the isotope undergoing decay present in the sample at time = 0 and N_t is the number of nuclei of the isotope

undergoing decay present in the sample at the present time, time = t. Thus, given N_0 and k, and measuring N_t, one can calculate the age, t. Indeed, the amount of the radioisotope need not be expressed as number of nuclei; amounts can be expressed in mole fraction, in molarity, or in number of decays per second. N_0 and N_t need only be expressed in identical units.

3. Recall also that, if $t_{1/2}$ is known, the value of the rate constant can be easily found, $k = 0.693/t_{1/2}$.

Discussion Question: Discuss the reliability of carbon dating. Some people believe it to be perfectly accurate, others believe it to be fundamentally flawed. Under what circumstances might everyone consider it accurate?

V. Detection of Radioactivity

A. Several methods have been devised to detect emissions from radioactive sources. The mostly commonly known method uses a *Geiger counter*. The Geiger counter contains a gas that is ionized if struck by radiation. Two electrodes inside the device conduct electricity only when ionized gas molecules are generated. Consequently, the voltage registered on the counter is proportional to the number of ionized gas molecules present and thus necessarily to the intensity of the radiation. The counter generally translates the voltage into audible "clicks," the frequency of which is proportional to the radiation intensity.

B. Some chemical substances fluoresce, or emit visible light, when struck by radiation. Instruments that detect this fluorescence are called *scintillation counters*. Scintillation counters are commonly used in biochemistry to determine the extent to which radioisotopes have been incorporated into biologically important molecules.

C. Radiotracers have become increasingly important in medical diagnostics and the treatment of illness. For example, thyroid function can be tested by having a patient ingest ^{131}I and then measuring the fraction of the dosage that accumulates in the thyroid gland. In positron emission tomography, or PET, a positron emitter is injected into the body and its travels throughout the body measured; these measurements are converted with the aid of computers into high-resolution pictures of internal body structures. Different chemical compounds are concentrated in specific organs of the body. As a treatment for cancer, work has been progressing in which compounds that are specifically concentrated in the diseased organ are synthesized with radioisotopes. Upon injection, the compound, along with the radioisotope it contains, moves to the affected organ. In this way, radiation therapy can be given only to the diseased organ, allowing the healthy tissues to avoid large doses of radiation.

VI. Chemistry and Life: Medical Applications of Radiotracers

Radiotracers are radioisotopes used to "trace" the path of an element in a system. As such, the radioisotope can be traced through a chemical reaction to determine what position it occupies in the structure of the product, or it can be placed in a pipe system where any locations in which the radioisotope accumulates can be taken as a leak in the plumbing.

VII. Energy Changes in Nuclear Reactions

A. Judging from the energy released in a nuclear blast, it is apparent that the energies holding nuclei together are enormous. From Einstein's equation, $E = mc^2$, or ΔE

Lecture Outline – Chapter 21
Brown, LeMay, & Bursten, *Chemistry: The Central Science, 6th Edition*

$= \Delta mc^2$, we can see that the energy consumed or produced in a reaction, chemical or nuclear, has an associated mass change. In chemical reactions, the mass changes are so small that the law of conservation of mass holds true. However, the energies of nuclear reactions are so large that the associated mass changes are readily observed.

B. The apparent change in mass during a nuclear reaction is related to the *nuclear binding energies* of the isotopes involved in the reaction. The binding energy of a nucleus is the amount of energy needed to break the nucleus into its constituent nucleons.
  1. The difference in the mass of a nucleus and the sum of the masses of its individual nucleons, Δm, is its *mass defect*.
  2. The *binding energy* is then given by Einstein's equation, $\Delta E = \Delta mc^2$. With mass expressed in kg, and the speed of light expressed in m/s, the energy calculated is in units of joules. This energy value can then be multiplied by Avogadro's number to arrive at the binding energy per mole of nuclei, or it can be divided by the number of nucleons in the nucleus to arrive at the binding energy per nucleon. This method can be used to determine the energy change for *any* reaction.
  3. Heavy atoms release energy and gain stability if they are fragmented, a process called *fission*. Fission is the mode of decay that occurs in the atom bomb and in nuclear power plants.
  4. Light atoms release energy and gain stability if they are fused together, a process called *fusion*. Fusion occurs as the principal reaction in the sun and in the hydrogen bomb.

Discussion Question: The amount of U-235 needed for a nuclear detonation is slightly in excess of 1 kg. Discuss how much energy released by the fission of that amount of U-235. By contrast, discuss the amount of energy released by the fusion of hydrogen-3 with hydrogen-2.

VIII. Nuclear Fission

A. *Chain reactions* are those in which released particles initiate the reaction of other nuclei. For example, the fission of $^{235}_{92}U$ follows the reactions, $^{1}_{0}n + ^{235}_{92}U \rightarrow ^{137}_{52}Te + ^{97}_{40}Zr + 2^{1}_{0}n$ and $^{1}_{0}n + ^{235}_{92}U \rightarrow ^{142}_{56}Ba + ^{91}_{36}Kr + 3^{1}_{0}n$. In both cases, the reaction produces more neutrons than consumed. Therefore, the reaction can accelerate as more neutrons are formed unless it is controlled.
  1. In order for a fission process to maintain a chain reaction, a certain minimum amount of fissionable material must be present. This minimum mass is called the *critical mass*. At the critical mass, there is sufficient material present that many neutrons escape the sample, but enough collide with and initiate other nuclei that the fission reaction continues at a constant rate.
  2. When the amount of fissionable material present is less than the critical mass, a *subcritical mass*, enough neutrons are lost that the reaction slows as it proceeds.
  3. When the amount of fissionable material present exceeds the critical mass, a *supercritical mass*, enough neutrons are held within the sample, reacting with more nuclei, so that the reaction quickly accelerates. The outcome of this acceleration could be a nuclear explosion.

B. A controversial form of energy is the use of *nuclear reactors*. These power plants operate by using pellets consisting of UO_2, with approximately 3% of the uranium being ^{235}U. The pellets are packed into stainless-steel or zirconium tubes, forming *fuel rods*. The mass of fissionable material is below the critical mass so there is no danger of a nuclear detonation. However, should the plant overheat, structural damage can be done to the plant, and hot gases can collect within the structure. It is feared by some that these accumulated gases could cause a gas-pressure explosion, rupturing the structure and spreading radioactivity.

    1. The fate of neutrons within the nuclear fuel is determined by use of *control rods*. Control rods are made of a substance that absorbs neutrons, such as cadmium or boron. The greater distance to which the control rods are inserted within bundles of fuel rods, the slower the nuclear reaction. The nuclear reaction can be accelerated by withdrawing the control rods.

    2. Also present in the core of the reactor is a *moderator*, which is a substance, such as water or carbon, that slows the neutrons to an extent that other fuel pellets can absorb them.

    3. A *cooling liquid* circulates around the reactor core. The function of the cooling liquid is to remove heat from the core, avoiding structural damage due to high temperatures. The cooling liquid, upon vaporization due to the heat of the reactor, can be used to run generators for the generation of electricity. In practice, the heat of the cooling liquid is transferred to a secondary cooling liquid which is then directed to the generators; the purpose of this is to isolate the liquid that actually circulates in the core from the rest of the facility.

C. A *breeder reactor* is a type of nuclear reactor that uses a nonfissionable or slowly fissionable isotope and, through bombardment, converts it to a more useful fissionable isotope, usually ^{239}Pu or ^{233}U. Breeder reactors are controversial because ^{239}Pu can be used to make atomic bombs; the fear is the ease with which the technology or the materials might be acquired by a group not afraid to use them for bomb manufacture. There are also profound waste disposal issues involved with breeder reactors.

IX. A Closer Look: The Dawning of the Nuclear Age

The *atom bomb*, or nuclear fission bomb, makes use of the acceleration of the fission of a supercritical mass to the point of explosion. Many of these weapons have been built, and two of them have been used on "real" targets. The bomb called "Little Boy" was dropped on Hiroshima on August 6, 1945. The bomb was shipped such that the uranium-235 it contained was held in two separate masses, each subcritical. Detonation was achieved by using conventional explosives to ram the two masses together, resulting in a supercritical mass. On August 9, 1945, the atom bomb called "Fat Man" was exploded over Nagasaki. This bomb used a spherical array of conventional explosives to compress a core of fissionable plutonium-239, resulting in the nuclear detonation. Combined, the estimated dead and missing in Japan numbered 132,000 people.

X. Nuclear Fusion

A. *Fusion* is the process of fusing light nuclei together to form heavier ones. Because extremely high temperatures are required for fusion reactions to proceed, fusion reactions are also called *thermonuclear reactions*. The most easily fusionable elements are hydrogen and helium, the lightest elements.

B. *Thermonuclear weapons*, or *hydrogen bombs*, utilize fusion reactions. The principal reaction in these devices is $^2_1H + ^3_1H \rightarrow ^4_2He + ^1_0n$, which requires a

Lecture Outline – Chapter 21
Brown, LeMay, & Bursten, *Chemistry: The Central Science*, 6th Edition

temperature of at least 40,000,000 K and high pressure. The heat and pressure required to initiate the fusion reaction is provided by an atomic bomb built around the fusion fuel.

C. Useful power production using fusion has been proposed and ways of containing the reaction are being investigated. The most promising device is called a *tokamak*. It uses strong magnetic fields to contain the reaction because no structural materials are known that can contain it. Lasers to generate the heat required to initiate the reaction are also being investigated. The 1989 announcement of "cold fusion" has also caused a flurry of research directed toward these systems.

Discussion Question: If you are brave, discuss the relative merits and drawbacks of oil, fission, and fusion power generation. This will be a spirited discussion.

XI. Biological Effects of Radiation

A. Perhaps a more pressing issue than the explosive force of weapons and the dangers of power plants is the effect of radiation on humans and our environment. Within biological systems, radiation *ionizes* chemical compounds which, in turn, can react to form *free radicals*. Free radicals are chemical species that have one or more unpaired electrons. Free radicals, especially of carbon and oxygen compounds, can cause serious disruption of normal biological functions.

B. *Somatic damage* is that which affects the organism during its own lifetime. Somatic damage can cause illness, serious debilitation, or rapid death, depending on the radiation being absorbed, the dosage, the length of exposure, and the affected organs.

C. *Genetic damage* is that which has a genetic effect; it harms offspring through damage to the genes and chromosomes. Genetic damage is difficult to assess because it often takes several generations for abnormalities to become apparent.

D. The amount of radiation to which an individual is exposed is important in avoiding the consequences of radiation. *Radiation doses* are expressed in several ways.
1. The *becquerel* is the SI unit of radioactivity. A becquerel is one nuclear disintegration per second.
2. A more widely used unit is the *curie*, Ci. A curie is the number of nuclear disintegrations per second from a 1 g sample of radium. 1 Ci = 3.7 x 10^{10} disintegrations/s.
3. A *rad* is the amount of radiation that deposits 1 x 10^{-2} J of energy per kilogram of body tissue.
4. A rad of alpha particles causes more damage than a rad of beta particles. A dosage unit that takes the characteristics of different forms of radiation into account is the *rem*. The dosage in rems is given by the dosage in rads multiplied by a factor, the *RBE* or *relative biological effectiveness*, which denotes the amount of damage that can be done by that particular form of radiation. A dosage of 1 rem of gamma radiation then causes the same biological damage as 1 dosage of 1 rem of alpha radiation.

E. Of most concern as a society is the accidental release of radiation to the environment. The Chernobyl accident released enormous amounts of radiation, about 2,000,000 times more than was released at Three Mile Island. Although unlikely with proper management, nuclear accidents are possible in the future.

F. *Radon gas* has become a major health concern in the United States. Radon is produced naturally by decaying uranium-238; it is not a manmade pollutant.

Apparently with the push to better seal homes against winter heat loss, we have also sealed them so that radon gas that leaks into them cannot escape. Radon is considered a probable cause of lung cancer.

Discussion Question: Is any unintentional exposure to radiation truly harmless?

XII. Chemistry at Work: Radiation Therapy

Radiation is also used extensively in *medical therapy*. Radiation destroys cells and so can kill cancers and tumors. Unfortunately, radiation also kills healthy cells, leading to health complications from the therapy itself. Fortunately, malignant tumors are more susceptible to radiation than healthy cells. Because radiation therapy is nondiscriminating, though, it is often used only if treatments such as *chemotherapy*, drug therapy, have failed.

Lecture Outline – Chapter 21
Brown, LeMay, & Bursten, *Chemistry: The Central Science*, 6th Edition

SAMPLE QUIZ QUESTIONS

1. Which of the following is(are) "magic number" nuclei?
   (a) ^4_2He; (b) ^7_3Li; (c) $^{13}_6C$; (d) $^{208}_{82}Pb$

2. Write balanced nuclear equations for each of the following:

   (a) Potassium-40 undergoes radioactive decay by emitting a beta particle;
   (b) Polonium-210 decays forming lead-206;
   (c) Plutonium-239 (^{239}Pu) is bombarded with alpha particles producing a new element plus a neutron;
   (d) Uranium-235 undergoes neutron-initiated fission to form barium-144 and krypton-90;
   (e) Oxygen-17 undergoes alpha decay;
   (f) Carbon-10 undergoes positron emission;
   (g) Chromium-51 undergoes electron capture;
   (h) $^{43}_{20}Ca(\alpha,p)^{46}_{21}Sc$.

3. A nuclide whose neutron-to-proton ratio is too high can gain stability by: (a) beta emission; (b) gamma emission; (c) proton emission; (d) electron capture.

4. Radioactivity: (a) requires the emission of one or more particles; (b) is initiated by bombardment; (c) is only detectable for fission processes; (d) is the spontaneous alteration of nuclei to achieve stability; (e) is a process which reduces the *n/p* ratio in a nucleus.

5. Write the balanced nuclear equation for beta emission by ^{24}Mg:

   (a) $^{24}Mg \rightarrow {}^{24}Na + {}^{0}_{-1}e$;
   (b) $^{24}Mg \rightarrow {}^{24}Al + {}^{0}_{-1}e$;
   (c) $^{24}Mg \rightarrow {}^{24}Na + {}^{0}_{1}e$;
   (d) $^{24}Mg \rightarrow {}^{24}Al + {}^{0}_{1}e$;
   (e) $^{24}Mg \rightarrow {}^{24}Na + {}^{0}_{0}e$.

6. Write the balanced nuclear equation for alpha emission by ^{255}Lr:

   (a) $^{255}Lr \rightarrow {}^{255}No + {}^{0}_{1}He$;
   (b) $^{255}Lr \rightarrow {}^{259}Md + {}^{4}_{2}He$;
   (c) $^{255}Lr \rightarrow {}^{254}No + {}^{1}_{1}He$;
   (d) $^{255}Lr \rightarrow {}^{254}Lr + {}^{1}_{0}He$;
   (e) $^{255}Lr \rightarrow {}^{251}Md + {}^{4}_{2}He$.

7. Fill in the blank with the missing item in the reaction $^{27}_{13}Al + {}^{4}_{2}He \rightarrow$ _____ $+ {}^{1}_{0}n$:
   (a) $^{30}_{13}Al$; (b) $^{27}_{15}P$; (c) $^{30}_{16}S$; (d) $^{31}_{15}P$; (e) $^{30}_{15}P$.

8. Complete the nuclear equation $^{10}_5B + {}^{1}_1H \rightarrow {}^{4}_2He +$ _____:
   (a) $^{11}_6C$; (b) $^{15}_8O$; (c) $^{7}_4Be$; (d) $^{14}_7N$; (e) $^{7}_3Li$.

9. What nucleus is formed when $^{196}_{82}Pb$ undergoes two successive electron captures?
   (a) $^{196}_{81}Tl$; (b) $^{196}_{83}Bi$; (c) $^{196}_{84}Po$; (d) $^{196}_{80}Hg$; (e) $^{194}_{82}Pb$

10. What nucleus is formed when $^{69}_{33}As$ undergoes positron emission?
    (a) $^{68}_{33}As$; (b) $^{69}_{31}Ga$; (c) $^{69}_{34}Se$; (d) $^{70}_{35}Br$; (e) $^{69}_{32}Ge$

11. Complete the reaction $^{238}_{92}U + ^{1}_{0}n \rightarrow ^{235}_{92}U +$ _____:
    (a) $^{1}_{0}n$; (b) $4^{1}_{0}n$; (c) $3^{1}_{0}n$; (d) $^{0}_{-1}e + ^{0}_{1}e$; (e) $^{4}_{2}He$.

12. A radioactive disintegration series: (a) is a characteristic, fixed series of nuclear reactions; (b) is a series of ten disintegrations; (c) is the only way most radioisotopes can achieve stability; (d) is a variable pathway of reaction used by many unstable nuclei; (e) is a set of simultaneous reactions of an unstable nucleus.

13. Artificial radioactivity: (a) is initiated by bombarding an unstable nucleus; (b) is not a common occurrence; (c) uses bombardment to change a stable nucleus into an unstable one; (d) requires the use of neutrons to initiate the reaction; (e) is limited to bombardment by one kind of particle.

14. Francium-223 has a half-life of 22 minutes. How long will it take for 75% of the original material to disappear? (a) 33 min; (b) 44 min; (c) 11 min; (d) 16.5 min

15. For a nuclear reaction, what percentage of the sample will remain after six half-lives? (a) 12.5 %; (b) 6.25%; (c) 3.125%; (d) 1.5625%; (e) 0.78125%

16. 520 days after a pure sample of polonium-210 is prepared it is found to contain lead-206 in the amount of 8.6 g for each 10.0 g of ^{210}Po originally in the sample. What is the half-life of polonium-210?

17. Carbon-14 has a half-life of 5700 yr. Explain why radiocarbon dating cannot be used to date objects over 20,000 years old.

18. Describe the operation of either a cyclotron or a Geiger counter.

19. Platinum-192 decays to osmium-188 by alpha emission. The mass of platinum-192 is 191.9614 amu, that of osmium-188 is 187.9560 amu, and that of an alpha particle is 4.00150 amu. Calculate the energy change that accompanies this reaction.

20. Calculate the nuclear binding energy of $^{7}_{3}Li$ if this nucleus has a mass of 7.01435 amu (the mass of a proton is 1.00728 amu while that of a neutron is 1.00867 amu).

21. Mass defect: (a) is a phenomenon which is related to the amount of energy holding the nucleus together; (b) is the mass missing from an atom which cannot be accounted for; (c) is found to be independent of the energy of a nuclear reaction; (d) occurs in some radioactive isotopes; (e) does not occur in stable nuclei.

22. A nuclear power plant: (a) is inexpensive to build, compared to fossil fuel power plants; (b) cannot have a nuclear explosion; (c) contains a critical mass of fissionable material; (d) has no serious pollution problems; (e) is not a potential hazard to those who work in them.

23. Nuclear fission: (a) is characterized by gamma emission; (b) is characterized by absorbing neutrons; (c) generates fewer nuclei than it started with; (d) has yet to be used outside the laboratory; (e) generates more nuclei than it began with.

24. The hydrogen bomb: (a) produces a more destructive blast than the atomic bomb; (b) produces only low-level radiation after its initial blast; (c) has little in common with the atomic bomb; (d) is a fission device; (e) has been used in war.

25. Nuclear fusion: (a) is characterized by gamma emission; (b) is characterized by absorbing neutrons; (c) generates fewer nuclei than it started with; (d) has yet to be used outside the laboratory; (e) generates more nuclei than it began with.

26. The unit that measures the biological damage caused by radiation is: (a) the rem; (b) the rad; (c) the RBE; (d) the curie.

27. The Curie is a measurement of: (a) the amount of energy absorbed per gram of tissue; (b) the biological damage of 1 roentgen of radiation; (c) the amount of radiation which produces 2.1×10^{-9} coulombs of charge; (d) the number of nuclear disintegrations per second; (e) the amount of energy absorbed per kilogram of tissue.

28. The roentgen is a measurement of: (a) the amount of energy absorbed per gram of tissue; (b) the biological damage of 1 roentgen of radiation; (c) the amount of radiation which produces 2.1×10^{-9} coulombs of charge; (d) the number of nuclear disintegrations per second; (e) the amount of energy absorbed per kilogram of tissue.

29. In medical diagnostics, radioisotopes: (a) are injected and an X-ray photo is taken; (b) are limited to the study of blood disorders; (c) produced by the body are photographed with a scintillation camera; (d) are given in doses high enough to concentrate in all areas of the body; (e) are chosen by whether they selectively concentrate in the area of the body being studied.

# Chapter 22 — Chemistry of Hydrogen, Oxygen, Nitrogen, and Carbon

OVERVIEW: Chapter 22 begins a selective survey of the chemistry of the elements. The text weaves much descriptive chemistry into the first twenty-one chapters; the remaining chapters are primarily descriptive. The chapter begins with a brief review of periodic trends. Rather than attempt to fully cover them again, short explanations are given and references are made to the earlier sections of the book in which they were first developed.

Similarly, the principal types of chemical reactions are reviewed and, again, references are given to the appropriate sections earlier in the text.

The first element to be covered is hydrogen. The historical framework of its discovery is given, followed by a discussion of its isotopes, its properties, its preparation, and its uses. A survey of its binary compounds is then given.

Oxygen is described next, in much the same way as hydrogen was discussed. Ozone, superoxides, and peroxides are important to the overall chemistry of oxygen and so are included here. A brief description is given of the oxygen cycle in nature.

A similar coverage for nitrogen follows. The acids, oxides, and hydrides of nitrogen are important and so are covered. A natural cycle for nitrogen is also included.

The chapter concludes with a brief survey of inorganic carbon. Organic compounds are covered in Chapter 26. The allotropy of carbon is important, as are the oxides, carbides, and cyanide. All are described here, including the natural occurrence and processes of carbonates.

LECTURE OUTLINE

I. **Descriptive Chemistry**
    A. The systematic examination of the elements and the compounds they form is called *descriptive chemistry*.
    B. Descriptive chemistry is often organized by periodic groups. Here, we focus on particularly important elements.
    C. Periodic trends, such as in electronegativity, electron configuration, and atomic size, form a valuable framework from which to identify and understand variations in chemical behavior.

II. **Periodic Trends**
    A. *Periodic trends* are properties that correlate to an element's position in the periodic table. As review, some of the more important trends follow.
        1. The differentiation between *metals* and *nonmetals* is critical to identifying variations in chemical behavior.
        2. *Electronegativity* is a measure of an element's ability to compete for electrons in a bond. Electronegativity increases as you travel up a group and to the right in a period on the periodic table; fluorine has the highest value, while cesium and francium have the lowest. In general, then, nonmetals have higher electronegativities than do metals.
        3. Compounds formed between strongly metallic and strongly nonmetallic elements tend to be ionic, which are rigid and show high melting points.

Compounds formed between nonmetals are molecular substances, which are often gases or liquids at room temperature.
- B. Among the nonmetals, the first member of each group often differs in several important ways from subsequent members.
  1. These differences are due, in large part, to the smaller size and higher electronegativity of the first member.
  2. The first member of each group is restricted to forming a maximum of four bonds because it has only the 2s and the three 2p orbitals available for bonding. Lower members of each group have the use of *d* orbitals in addition to the *s* and *p* orbitals.
  3. The first member of each group has a greater ability to form π-bonds than subsequent members. This is primarily due to the smaller size of the first member, which allows for more effective π overlap of orbitals during bond formation.

III. Chemical Reactions

- A. *Combustion reactions* are those reactions in which a fuel reacts with molecular oxygen, O_2. In these reactions, the principal element in the fuel, usually C, is oxidized. In *complete combustion*, the element being oxidized is taken to its highest oxidation state. Therefore, the most common type of combustion reaction can be represented by the reaction of methane, $CH_4(g) + 2O_2(g) \rightarrow CO_2(g) + 2H_2O(l)$, in which the carbon in methane is oxidized and the molecular oxygen is reduced. Other elements can be oxidized in combustion. For example, the combustion of methylamine is $4CH_3NH_2(g) + 9O_2(g) \rightarrow 4CO_2(g) + 10H_2O(l) + 2N_2(g)$. Water, CO_2, and N_2 are common products of combustion, illustrating their high stabilities.
- B. *Metathesis* or *double-displacement reactions* are those in which two reactants that are ionic compounds react to switch the anion-cation pairings. Arrhenius acid-base reactions and precipitation reactions are more specific types of metathesis reactions.
- C. *Single-displacement reactions* are those in which an ionic compound reacts with an element, such that the elemental reactant replaces another element in the ionic compound. For example, consider the reaction, $Zn(s) + FeSO_4(aq) \rightarrow ZnSO_4(aq) + Fe(s)$.
- D. *Brønsted acid-base reactions* are also referred to as *proton-transfer reactions*. It must be recalled that the stronger a Brønsted base, the weaker its conjugate acid. For example, H_2, OH^-, NH_3, CH_4, and C_2H_4 show no tendency to lose a proton (very weak acids). Consequently, H^-, O^{2-}, NH_2^-, N^{3-}, CH_3^-, C^{4-}, and C_2^{2-}, gain a proton quite readily (very strong bases).
- E. *Redox reactions* are those in which one substance is oxidized (electrons are removed), while another substance is reduced (electrons are added). It is also possible for the element oxidized and the element reduced to both reside in the same reactant molecule.

IV. Hydrogen

- A. Hydrogen is the most abundant element in the universe, although its abundance is much lower on earth. Hydrogen has one proton in its nucleus. However, there are three isotopes of hydrogen.
  1. *Protium*, ^1_1H, is the isotope that has no neutrons. Consequently, its mass number is 1. It is the most abundant of the isotopes, comprising 99.9844% of the naturally occurring mixture.

2. *Deuterium*, ^2_1H, or D, is the isotope that has one neutron. Consequently, its mass number is 2. Deuterium comprises only 0.0156% of the naturally occurring mixture. It is not radioactive.
3. *Tritium*, ^3_1H, or T, is the isotope that has two neutrons. Consequently, its mass number is 3. It is rare in nature and is more commonly produced in nuclear reactors; its half-life is 12.3 years, converting to helium-3.
4. Deuterium and tritium are used as "labels", replacing protium in chemical compounds, to help determine the mechanisms of reactions.

B. Hydrogen is the only element that is not a member of a family in the periodic table. Hydrogen forms both a cation, H+, and an anion, H-. Hydrogen exists as a colorless, diatomic gas at room temperature. Its bond strength, 436 kJ/mol, is high for a single bond. Hydrogen forms strong covalent bonds to other elements.

C. Hydrogen can be prepared by reduction of acids. For example, consider Ca + 2H+ → Ca^{2+} + H$_2$. In industrial quantity, hydrogen is obtained from petroleum refining and from natural gas. It can also be obtained by treating coke with steam, C + H$_2$O → CO + H$_2$, the product mixture being called *water gas*.

D. Hydrogen primarily is used in the manufacture of other chemicals, most notably NH$_3$. It is used in lesser amounts to produce methanol and to hydrogenate double bonds. Hydrogenation reactions convert double bonds to single bonds, with addition of hydrogen, as in H$_2$C=CH$_2$ + H$_2$ → H$_3$C–CH$_3$.

E. Hydrogen forms binary compounds, referred to collectively as *hydrides*, of three types.
1. *Ionic hydrides* are formed between hydrogen and the alkali metals and heavier alkaline earth metals. In ionic hydrides, the hydrogen exists as H- because hydrogen is much more electronegative than the metal. Ionic hydrides are reactive, so they are generally stored in an air-free and moisture-free environment.
2. *Metallic hydrides* are formed between hydrogen and transition metals. The metallic properties of the metal is preserved in the hydride. In many metallic hydrides, the composition is variable, occurring within a range and including non-whole number ratios of hydrogen to metal. Such *nonstoichiometric* hydrides are sometimes called *interstitial hydrides*. They are often considered to be solutions of hydrogen atoms in the metal.
3. *Molecular hydrides* are formed between hydrogen and nonmetals or semimetals. They are discrete molecular units which normally exist as liquids or gases.

V. Chemistry at Work: Using Metal Hydrides

Metallic hydrides have been investigated for use in purifying H$_2$ from other gases and as a way to store H$_2$. It has been shown, however, that absorption of hydrogen structurally weakens metals.

Discussion Question: Why is hydrogen so abundant in the universe, but so comparatively uncommon on earth?

VI. Oxygen

A. Oxygen is the most abundant element in the earth's crust and in the human body. Oxygen occurs in two allotropes, molecular oxygen or dioxygen, O$_2$, and ozone, O$_3$. Unless noted otherwise, O$_2$ is the substance referred to as oxygen. Oxygen forms very stable bonds with many other elements. Consequently, there are many compounds of oxygen.

Lecture Outline – Chapter 22
Brown, LeMay, & Bursten, *Chemistry: The Central Science, 6th Edition*

- B. Oxygen can be prepared in many ways. Commercially, it is obtained by fractional distillation of air. In the laboratory, small amounts are obtained by decomposing oxygen-containing compounds, such as $KClO_3$.
- C. Oxygen is one of the most widely used elements, ranking third in 1988 in the United States, based on mass. It is an important oxidizing agent. Over half of the O_2 made in the United States is used in the manufacture of steel.
- D. Ozone, O_3, has a sharp, irritating odor and is toxic. The molecule contains a delocalized π bond. Ozone is a very strong oxidizing agent. It is produced by electrical discharge through dioxygen but decomposes readily to O_2, except at low temperatures. Ozone has limited commercial uses, being used mostly in the synthesis of organic compounds and pharmaceuticals. Naturally occurring ozone in the upper atmosphere is crucial to preventing ultraviolet radiation from reaching the earth's surface.
- E. *Oxides* are oxygen-containing binary compounds where oxygen is in the -2 oxidation state.
    1. *Acidic anhydrides* or *acidic oxides* are oxides that react with water to form acidic solutions, generally as oxyacids. Acidic oxides contain a nonmetal combined with the oxygen. A few nonmetal oxides—N_2O, NO, and CO—are not acidic oxides.
    2. *Basic anhydrides* or *basic oxides* are oxides that react with water to form basic solutions, generally as hydroxides. These oxides contain a metal combined with oxygen.
    3. *Amphoteric oxides* can exhibit either acidic or basic character. When a metal forms more than one oxide, the basic character decreases as the oxidation state of the metal increases.
- F. *Peroxides* contain the –O–O– unit with each oxygen in a -1 oxidation state. *Superoxides* contain an –O–O– unit with each oxygen in a -1/2 oxidation state. The most active alkali metals—Cs, Rb, and K—react with O_2 to give superoxides, while their slightly less active neighbors—Na, Ca, Sr, and Ba—give peroxides. Hydrogen peroxide is the most commercially important of these compounds, acting as either an oxidizing agent or a reducing agent. The –O–O– linkage occurs in other types of compounds, such as $S_2O_8^{2-}$, and particularly in the biochemistry of complex organisms.
- G. The *oxygen cycle* is a term applied to oxygen's repetitive collection, use, and return to the environment. Dioxygen is consumed by animals and released by plants during respiration. Many geological events use or replenish O_2. Dioxygen is also involved in photochemical events in the upper atmosphere. The energy that drives the oxygen cycle is solar.

Discussion Question: Why might hydrogen peroxide burn your skin?

VII. Nitrogen

- A. Nitrogen comprises almost 80% of the atmosphere as N_2. However, nitrogen is not abundant in the earth's crust. The N_2 molecule is unreactive, with a bond dissociation energy of 941 kJ/mol. In compounds, nitrogen exhibits all oxidation states between -3 and +5, the positive states being seen only in combination with oxygen, fluorine, and chlorine.
- B. Elemental nitrogen is produced by fractional distillation of air. Because of low reactivity, N_2 is used as a gaseous blanket to exclude O_2 during food preparation and packaging, in the manufacture of chemicals, and in the preparation of electronics. Liquid nitrogen is a coolant. Most nitrogen is used to produce nitrogen-containing compounds. Conversion of N_2 to nitrogen-containing

compounds is called *nitrogen fixation*. The principal use of nitrogen is in the manufacture of fertilizers. It is also used in the production of a wide range of other chemicals, from explosives to plastics.

C. *Ammonia*, NH_3, is one of the most important nitrogen compounds, and is a base. Ammonia can be produced in the laboratory by the action of NaOH on NH_4Cl.

D. Commercially, ammonia is produced by the *Haber process*, in which N_2 and H_2 are combined under high temperature and high pressure. The Haber process is the classic example of the systematic study of chemical equilibrium, $N_2(g) + 3H_2(g) \rightarrow 2NH_3(g)$.

E. The second important compound formed between nitrogen and hydrogen is *hydrazine*, N_2H_4. Hydrazine contains a reactive –N–N– linkage and is highly toxic. It is prepared by the reaction of ammonia and hypochlorite, OCl^-. Hydrazine is a strong reducing agent.

F. Nitrogen forms three common oxides and two common oxyacids.
1. *Nitrous oxide*, N_2O, or laughing gas is a colorless gas that is used as an aerosol propellant and as an anesthetic. It can be produced by the careful heating of NH_4NO_3.
2. *Nitric oxide*, NO, is also a colorless gas, but it is slightly toxic. It is prepared by reduction of dilute nitric acid by a metal, such as copper or iron. Commercially, NO is made by the platinum catalyzed reaction of ammonia and dioxygen.
    a. The catalyzed production of NO is first step of the *Ostwald process*, by which ammonia is first converted to NO by platinum catalysis. The NO then is combined with O_2 to produce NO_2. The last step converts the NO_2 into nitric acid, HNO_3, by dissolving the NO_2 in water.
    b. The last step of the Ostwald process is a *disproportionation* reaction, in which nitrogen is both oxidized and reduced. The balanced equation, $3NO_2(g) + H_2O(l) \rightarrow 2H^+(aq) + 2NO_3^-(aq) + NO(g)$, shows that 2 molecules of NO_2 are oxidized to NO_3^-, while 1 molecule of NO_2 is reduced to NO. Disproportionation can also be viewed as an atom-transfer reaction among molecules of the same reactant.
3. *Nitrogen dioxide*, NO_2, is a yellow-brown, toxic, and corrosive gas. Nitrogen dioxide *dimerizes* at low temperature, yielding N_2O_4.
4. *Nitric acid*, HNO_3, is a colorless, corrosive liquid that becomes discolored over time as small amounts of NO_2 collect due to photodecomposition. It is a strong acid that is a strong oxidizing agent. Most HNO_3 made in the United States is used in the production of fertilizers; it is also used in the manufacture of drugs, plastics, and explosives.
5. *Nitrous acid*, HNO_2, is somewhat unstable, disproportionating into NO and HNO_3. It is commonly formed by the action of a cold strong acid on a nitrite salt. Nitrous acid is a weak acid.

G. Nitrogen has a cycle of use and regeneration in nature called the *nitrogen cycle*. The nitrogen cycle involves conversion of N_2 to nitrogen oxides in the atmosphere and consumption of N_2 by some plants, notably legumes. Nitrogen oxides are also consumed by plants. Animals that eat plants, excrete ammonia which breaks down into N_2 and is released into the atmosphere. It is, of course, much more detailed and the effects of man's activities are generally ignored; man does release large amounts of nitrogen-containing compounds, most notably ammonia, into the environment.

VIII. Chemistry at Work: Nitrites in Food

There has been considerable controversy over *nitrites in food*. Sodium nitrite, NaNO$_2$, is used as a preservative in cured meats such as bacon. The debate centers on the observation that nitrite can react with food, especially upon heating, to produce *nitrosamines*, which show some cancer-causing activity in laboratory animals.

IX. Carbon

A. Carbon is not abundant on earth, but is concentrated in living organisms; over half the carbon in the earth's crust occurs in carbonate compounds. The term *organic* was originally applied to the chemistry of carbon because it was believed that compounds containing carbon and hydrogen could only be made in living organisms. We now make many organic compounds in the laboratory, but the term has endured. We often refer to compounds of carbon which have no C–H bonds as *inorganic carbon*, which is the focus of this section.

B. Carbon exists in several elemental forms.
1. *Graphite* consists of sheets; trigonal planar carbon atoms within each sheet are bonded together in a vast array of fused benzene rings. Because π electrons extend above and below the plane of each sheet, the sheets slide across one another. Graphite is used as a lubricant. It is also the common pencil "lead." Graphite appears metallic and conducts electricity.
2. *Diamond* is a clear, hard substance. Tetrahedral carbon atoms are bonded to one another in a 3-dimensional array. In addition to their esthetic use, diamonds are used in cutting, grinding, and polishing tools.
3. *Carbon black* is an amorphous (noncrystalline) form of carbon, which is produced by heating hydrocarbons, compounds of carbon and hydrogen, in a very limited supply of oxygen. It is used as a black pigment in inks and in automobile tires.
4. *Charcoal* is formed by heating wood in the absence of air. Because the wood is highly porous and of irregular structure, so too is the charcoal, giving it an enormous surface area per gram. Charcoal is used as a convenient fuel and to absorb contaminants, both in solution and in air.
5. *Coke* is an impure form of carbon formed when coal is heated in the absence of air. Coke is used primarily as a reducing agent in the refining of metals.

C. There are two oxides of carbon and several compounds based on the carbonate ion, CO$_3^{2-}$.
1. *Carbon monoxide*, CO, is formed when carbon-based fuels are burned in a limited supply of oxygen, as in 2CH$_4$ + 3O$_2$ → 2CO + 4H$_2$O. Carbon monoxide is colorless and odorless, and is toxic due to its ability to bind hemoglobin. It is a major environmental pollutant, but has many commercial uses. Carbon monoxide is used as a fuel, producing heat as it is burned. It is also a good reducing agent, being used extensively in metals refining.
2. *Carbon dioxide*, CO$_2$, is produced when carbon-based fuels are burned in excess oxygen and when many carbonates are heated or treated with acid. It is also formed as a product in fermentation. Carbon dioxide is colorless and odorless. It is not toxic, although it does not support life and so can cause suffocation. An important form of CO$_2$ is its solid, *dry ice*. Dry ice is used as a refrigerant, maintaining a temperature of -78°C.
3. Carbon dioxide is used to manufacture *washing soda*, Na$_2$CO$_3$·10H$_2$O, and *baking soda*, NaHCO$_3$. Washing soda is used to precipitate metal ions in soap. Baking soda is used for a variety of things, including baking, odor removal, and as an abrasive.

D. The only oxyacid of carbon is *carbonic acid*, H_2CO_3, which results from dissolving CO_2 in water, $CO_2(g) + H_2O(l) \rightleftharpoons H_2CO_3(aq)$. Carbonic acid is a modest acid; stepwise dissociation yields HCO_3^- and CO_3^{2-}. Solutions of HCO_3^- are slightly basic. Carbonate minerals are very common, particularly calcite or common limestone, $CaCO_3$. Carbonate minerals are generally slightly soluble in pure water, but dissolve more readily in acidic waters. Calcium carbonate, chemically pure $CaCO_3$, decomposes to CaO and CO_2 when heated. Quicklime, as CaO is commonly known, is important in the mortar used in the building trades.

E. Binary compounds of carbon with metals, semimetals, and certain nonmetals are called *carbides*. There are three types of carbides, differentiated by the electronegativity of the element to which carbon is bonded.
   1. *Ionic carbides* or *acetylides*, C_2^{2-}, are formed with electropositive metals. The most important of these is calcium carbide, CaC_2, which is used to make acetylene.
   2. *Interstitial carbides* are formed with transition metals. In these, the carbon atoms occupy open spaces between metal atoms, much like in the interstitial hydrides. These substances are hard. Tungsten carbide, for example is used in cutting tools.
   3. *Covalent carbides* are formed with boron and silicon. Silicon carbide is used as an abrasive and in cutting tools.

F. There are several other important inorganic compounds of carbon.
   1. *Hydrogen cyanide*, HCN, is a very toxic gas that smells like bitter almonds. The conjugate base of hydrogen cyanide, the *cyanide ion*, CN^-, forms stable complexes with most transition metals. Acidified solutions of cyanide release HCN gas.
   2. *Carbon disulfide*, CS_2, is an important industrial solvent for waxes and other nonpolar substances. It is volatile and the vapors are toxic and extremely flammable.

Discussion Question: Which element, C, O, or H, seems to be *the most important* element and why?

X. Chemistry at Work: Carbon Fibers and Composites

Carbon fibers, particularly of graphite, have gained widespread use in structural materials. Such fibers are embedded in other materials; the resulting structures are called *composites*. Composite materials are made to possess greater strength than any of the individual components.

Lecture Outline – Chapter 22
Brown, LeMay, & Bursten, *Chemistry: The Central Science*, 6th Edition

SAMPLE QUIZ QUESTIONS

1. Circle the formula that best fits into each of the following descriptions:
   (a) usual oxidation state is -2   N   O   H
   (b) laboratory source of O_2 NH_4NO_3 $KClO_3$ CaO
   (c) the superoxide ion O^{2-} O_2^{2-} O_2^-
   (d) forms a nitride when burned in air   Li   Na   K
   (e) nitric oxide N_2O NO NO_2
   (f) forms metal carbonyl compounds CO_3^{2-} CO_2 CO

2. Water gas, an important industrial fuel, consists of: (a) H_2 and CO_2; (b) H_2 and CO; (c) H_2O and CO; (d) H_2 and O_2.

3. Which of the following elements is the most likely to form an interstitial hydride?

   (a) Ca; (b) Mn; (c) Ge; (d) Se

4. Different forms of the same element in the same physical state, such as dioxygen and ozone, are known as _____.

5. Write balanced chemical equations for the following reactions: (a) sodium oxide is dissolved in water; (b) calcium hydride is dissolved in water; (c) sodium peroxide is dissolved in water; (d) $KClO_3(s) \xrightarrow{heat}$; (e) $CaCO_3(s) + H^+(aq) \rightarrow$; (f) zinc metal reacts with sulfuric acid.

6. Give the chemical formula for: (a) hydrazine; (b) ozone; (c) calcium carbide.

7. List two general ways that the first member of a periodic family differs from subsequent members.

8. Describe, using chemical equations, the preparation of nitric acid, starting with N_2, H_2, O_2, and water.

9. Which of the following elements forms the most acidic oxide?

   (a) B; (b) Al; (c) Ga; (d) In

10. The oxides of the alkali metals are: (a) covalent oxides; (b) amphoteric oxides; (c) basic oxides; (d) acidic oxides.

11. The oxides of the halogens are: (a) covalent oxides; (b) amphoteric oxides; (c) basic oxides; (d) acidic oxides.

12. The most common positive oxidation states of nitrogen are: (a) +2 and +4; (b) +3 and +5; (c) +3 and +7; (d) +4 and +6.

13. Which of the following contains the peroxide ion?

    (a) BaO; (b) Na_2O_2; (c) CsO_2; (d) SiO_2

14. Which of the following contains the superoxide ion?

(a) BaO; (b) Na$_2$O$_2$; (c) CsO$_2$; (d) SiO$_2$

15. The oxidation of NO in air produces: (a) N$_2$O; (b) NO$_2$; (c) N$_2$O$_4$; (d) N$_2$O$_5$.

16. Washing soda is: (a) NaHCO$_3$; (b) Na$_2$CO$_3$·10H$_2$O; (c) NaNO$_2$; (d) CaCO$_3$.

17. The most abundant element in the universe is: (a) H; (b) O; (c) N; (d) C.

18. List one commercially important reaction of each of the following: (a) H$_2$; (b) NH$_3$; (c) CaC$_2$.

19. Which of the following elements is the most likely to form an interstitial carbide?

    (a) Ca; (b) Se; (c) Ge; (d) Mn

Lecture Outline – Chapter 23
Brown, LeMay, & Bursten, *Chemistry: The Central Science*, 6th Edition

Chapter 23 — Chemistry of Other Nonmetallic Elements

OVERVIEW: Chapter 23 continues the survey of nonmetals, focusing on the noble gases and the nonmetallic members of groups 3A through 7A that were not covered in Chapter 22. The type of coverage here is very similar to that in Chapter 22.

The noble-gas elements are covered first. The few compounds of them are explained in terms of the conditions required to overcome the native stability of the elements.

The halogens are described, especially the periodic nature of their properties. The chemistry of the halogens is discussed as it relates to the oxides, oxyacids, hydrides, and interhalogen compounds.

The discussions of groups 6A, 5A, and 4A are similarly arranged. For each, the properties are given, as well as the typical methods of preparation and examples of their chemistries.

The final section of the chapter discusses boron, the only nonmetal in group 3A. Boric acid, borax, and borohydride ion are covered as representing the important chemistry of boron.

LECTURE OUTLINE

I. **The Noble-Gas Elements**

   A. The group 8A elements are all gases under normal conditions. All are found in the atmosphere except radon, which is radioactive and exists only briefly. Helium, neon, and argon have important commercial uses; krypton, xenon, and radon are only slightly used.

   B. Noble-gas compounds are rare because the elemental forms of the noble gases are so stable.
      1. Several fluorides of xenon are known, XeF_2, XeF_4, and XeF_6. There is only one confirmed fluoride of krypton, KrF_2.
      2. Xenon also makes several oxides and oxyfluorides, $XeOF_4$, XeO_2F_2, XeO_3, and XeO_4.

Discussion Question: What conditions of temperature and what properties of other reactants would be required to produce a compound of argon?

II. **The Halogens**

   A. The *halogens* comprise group 7A in the periodic table. Fluorine is found principally in minerals; the other halogens, chlorine, bromine, and iodine, are found in sea water. This indicates that halides tend to be water soluble, except those of fluorine. Astatine is a radioactive, synthetic element; discussions of the halogens usually ignore it.

   B. Each halogen has a high ionization energy, second in its period to the noble gas. Each also has the highest electronegativity in its period. Within the halogens, atomic radii increase down the group, and ionization energy, electronegativity, and oxidative activity decrease down the group as expected. Fluorine and chlorine are gases, bromine is a liquid, and iodine is a solid under normal conditions.

   C. All of the halogens are commercially important.
      1. Fluorine is used to produce *fluorocarbons*. Freon-12 is CF_2Cl_2, a *chlorofluorocarbon* which is an important refrigerant and aerosol

propellant. Other fluorocarbons include Teflon, which is polymeric tetrafluoroethylene.
2. Chlorine is the commercially most important halogen, with Cl_2 and HCl the principal forms in which it is produced. About half the inorganic chlorine production goes into the preparation of vinyl chloride, C_2H_3Cl, which is in turn used to make polyvinyl chloride, PVC, the solvent ethylene dichloride, CH_2Cl_2, and other organic compounds.
3. Chlorine also goes into the production of chlorine bleach, based on the ClO^- anion and present as NaClO.

D. Stable hydrogen halides exist for all the halogens. Aqueous solutions of all of them, except HF, are strongly acidic. The hydrogen halides can be made by direct combination of the elements, although the common technique for making HF and HCl involves reacting a salt of the halide with a strong, nonvolatile acid, such as sulfuric acid. The hydrogen halides form hydrohalic acids when dissolved in water. Hydrofluoric acid is used to etch glass.

E. *Interhalogen compounds* are binary compounds containing two different halogens which range stoichiometrically from XX' to XX'$_7$, where X and X' are the two halogens. The central atom in an interhalogen molecule is the less electronegative of the two elements. Other than ICl_3, the higher interhalogens, XX'$_3$, XX'$_5$, and XX'$_7$, contain fluorine as X' and chlorine, bromine, or iodine as X. These compounds are extremely reactive and readily attack glass. *Polyhalide ions* are polyatomic halogen ions. Examples include ICl_4^-, BrF_4^-, and I_3^-.

F. The halogens form *oxyacids* and *oxyanions*. Fluorine forms only one acid, HFO, but the other halogens form at least three, HXO_n, where n is 1, 3, or 4. Chlorine also forms $HClO_2$. Iodine makes two acids in which the iodine is in a +7 oxidation state, metaperiodic acid, HIO_4, and paraperiodic acid, H_5IO_6. All are unstable as pure compounds, decomposing readily and sometimes explosively when isolation is attempted. The oxyanions are generally more stable than the acids and many salts of the oxyanions are easily prepared. The acids and the oxyanions are also strong oxidizing agents. The acid strengths of the oxyacids increase with the increasing oxidation state of the central halogen atom.

III. A Closer Look: The Hydrolysis of Nonmetal Halides

A. Some halides of nonmetals are quite unreactive toward water because of kinetic factors, not thermodynamic factors.

B. Generally, a nonmetal halide is unreactive when the central atom is unable to expand its valence shell. Hence, NF_3, CCl_4, and SF_6 are unreactive, but $SiCl_4$ is reactive.

IV. The Group 6A Elements

A. The group 6A elements are all nonmetals, consequently, all form stable 2- anions and all are highly electronegative, have high ionization energies, and have high electron affinities within their individual periods. Here, we concentrate on sulfur, selenium, and tellurium.

B. Sulfur occurs in large underground deposits. The *Frasch process*, in which superheated water is forced into the deposit and compressed air is injected to force the molten sulfur out, is the principal method of obtaining sulfur. Sulfur is also obtained from minerals. Its presence in oil, coal, and metal ores makes sulfur an important environmental pollutant. Selenium and tellurium exist in rare minerals. Therefore, little industrial exploitation of them has occurred.

Lecture Outline – Chapter 23
Brown, LeMay, & Bursten, *Chemistry: The Central Science, 6th Edition*

- C. Sulfur is a yellow solid, cyclic S_8, at room temperature; upon heating, the S_8 molecules break and join into long chains. Further heating breaks the chains, producing short fragments. Selenium and tellurium exist in helical chains. Sulfur is biologically widespread, while selenium has a small biological role and tellurium has no known biological role. Selenium is a nonconductor in the dark but does conduct electricity in the light.
- D. Sulfur forms several important oxides, oxyacids, and oxyanions.
    1. *Sulfur dioxide*, SO_2, is formed by the action of acid on sulfite or by the combustion of sulfur in air. It is particularly toxic to lower life forms.
    2. *Sulfurous acid*, H_2SO_3, is a weak acid that forms upon dissolving SO_2 in water, although it may be more accurate to regard the solution as containing hydrated SO_2. The acid cannot be isolated, but salts of sulfite, SO_3^{2-}, and bisulfite, HSO_3^-, are easily isolated. Small amounts of Na_2SO_3 or $NaHSO_3$ are added to foods to inhibit bacterial growth, but both can be allergens.
    3. *Sulfuric acid*, H_2SO_4, is formed from its anhydride, SO_3. The SO_3 is formed catalytically from SO_2 from combustion of sulfur. Sulfuric acid is the leading chemical, in mass, produced in the United States. It is a strong acid, a good oxidizing agent, and a powerful dehydrating agent. The oxyanions of sulfuric acid, SO_4^{2-} and HSO_4^-, are also important, as is the related thiosulfate, $S_2O_3^{2-}$.
- E. Selenium and Tellurium also make oxides, oxyacids, and oxyanions.
    1. *Selenous acid*, H_2SeO_3, is produced by dissolving SeO_2 and is a weak acid.
    2. *Selenic acid*, H_2SeO_4, is produced by the hydrogen peroxide oxidation of selenous acid. It is a strong acid.
    3. *Tellurite* and *bitellurite* salts, such as Na_2TeO_3 and $NaHTeO_3$, can be isolated, but the acid, H_2TeO_3, does not exist because TeO_2 is insoluble in water.
    4. *Telluric acid*, H_6TeO_6, results from the oxidation of aqueous TeO_2. It is a weak acid.
- F. Sulfide, selenide, and telluride are the common monatomic anions of the elements, S^{2-}, Se^{2-}, and Te^{2-}. As sulfur is the most electronegative element of the group, ionic sulfides are more common than are selenides or tellurides. The most important compounds are the hydrides, H_2S, H_2Se, and H_2Te, although H_2S is the only commercially useful member. All are highly toxic and all are weakly acidic in aqueous solution.

V. The Group 5A Elements

- A. The group 5A elements generally occur in covalent molecules, although nitride, N^{3-}, does occur with highly electropositive metals. Common oxidation states range from -3 to +5. The variations by row in properties of these elements is striking; nitrogen and phosphorus are nonmetals, bismuth is a metal, and arsenic and antimony are semimetals.
- B. Phosphorus occurs principally in phosphate rock, which is mainly $Ca_3(PO_4)_2$. Elemental white phosphorus is produced by reduction of phosphate with coke and SiO_2. This allotrope consists of P_4 tetrahedra and spontaneously ignites on exposure to air. Heating in the absence of air produces red phosphorus, which is more stable and less toxic.
- C. Phosphorus makes halides of two important stoichiometries.

1. *Trihalides*, PX$_3$, are known for all of the halogens. The trichloride is the most important compound, being used in detergents, plastics, and insecticides.
2. *Pentahalides*, PX$_5$, are made by reacting the corresponding trihalide and the elemental halogen.
3. The phosphorus halides undergo hydrolysis on contact with water; most of them fume on exposure to water vapor.

D. *Oxy compounds*, those in which phosphorus is combined with oxygen, represent the most significant compounds of phosphorus. The two oxides, P$_4$O$_6$ and P$_4$O$_{10}$, illustrate the most common oxidation states of phosphorus, +3 and +5. Phosphorus(III) oxide, P$_4$O$_6$, is the anhydride of phosphorous acid, H$_3$PO$_3$, while phosphorous(V) oxide, P$_4$O$_{10}$, is the anhydride of phosphoric acid, H$_3$PO$_4$.
1. Phosphoric and phosphorous acids undergo *condensation reactions*, reactions in which two or more molecules combine to form a larger molecule by eliminating a small molecule such as water. Condensation of two molecules of H$_3$PO$_4$ yields H$_4$P$_2$O$_7$; continued condensation yields a polymeric chain with the empirical formula HPO$_3$. The acids are named orthophosphoric acid, pyrophosphoric acid, and metaphosphoric acid, respectively. All three acids contain phosphorus in the +5 oxidation state.
2. Phosphoric acid and phosphates find most use in detergents and fertilizers. The compound usually used in detergents is sodium tripolyphosphate, Na$_5$P$_3$O$_{10}$. Most phosphate rock is converted to fertilizers.
3. Phosphorus is critical to biological systems, occurring in DNA and RNA, the substances that control cell reproduction and transmittal of genetic information. It also occurs in adenosine triphosphate, ATP, which stores energy within cells; a phosphate can be released with the production of energy for the cell's use.

E. The remaining three members of the group, arsenic, antimony, and bismuth, are of lesser importance. All occur in nature in sulfide minerals and as minor components of other ores. As elements, arsenic and antimony are similar to phosphorus; bismuth is a reddish-white metal. Arsenic and antimony also resemble phosphorus chemically, while bismuth behaves as expected for a metal, with an extensive metallurgy.

Discussion Question: Why is bismuth a metal but antimony, one row above, is a semimetal that is more nonmetallic?

VI. The Group 4A Elements

A. Within group 4A, carbon and silicon are nonmetals, germanium is a semimetal, and tin and lead are metals. Their electronegativities, ionization energies, and electron affinities are moderate. The free ions tend to be 2+ cations, rather than 4+; +4 oxidation states are common in covalent compounds. Carbon exhibits *catenation*, which is the ability to form extended chains and rings using C-C bonds. The elements that surround carbon on the periodic table can form chains, but the maximum size is much smaller than for carbon, which shows no maximum number of atoms in a chain.

B. Silicon is the second most abundant element in the earth's crust, existing in nature as SiO$_2$ and in various minerals. The structure of elemental silicon is similar to diamond, with a 3-dimensional network of Si-Si bonds. Pure silicon is prepared by reacting the crude element with Cl$_2$, to form SiCl$_4$ which is distilled, reacting it

Lecture Outline – Chapter 23
Brown, LeMay, & Bursten, *Chemistry: The Central Science*, 6th Edition

with H$_2$ to reform the element, and then using zone refining. The process can be repeated until the resulting silicon is sufficiently pure.

C. Over 90% of the earth's crust is *silicates*. The basic structure involves each silicon atom being surrounded by four oxygen atoms. Various silicates are differentiated by how these tetrahedra are linked and, consequently, by the oxidation state of the silicon.

D. Aluminum ion is similar in size and is isoelectronic with Si^{4+}. Aluminum can therefore replace silicon in silicate structures, resulting in substances called *aluminosilicates*.
  1. In aluminosilicates, a small cation such as K$^+$ must be added due to the difference in charge. In both silicates and aluminosilicates, there is a sheet structure with cations between sheets. Muscovite and feldspar are typical of these *mica* aluminosilicate minerals.
  2. *Clay minerals* are hydrated aluminosilicates having small particle sizes. They are able to adsorb cations on their surfaces.

E. *Glass* is an amorphous solid composed mostly of silicon and oxygen; different substances can be added to impart various properties.
  1. *Soda-lime glass* contains CaO and Na$_2$O in addition to SiO$_2$ from sand.
  2. *Borosilicate glass*, which goes by the brand names of Pyrex® and Kimax®, results from adding B$_2$O$_3$. These glasses have a higher melting point and are more resistant to thermal shock than other glasses.
  3. *Photochromic glass* contains AgCl or AgBr. It darkens upon exposure to light by decomposing the salt to the elements; the salt reforms in the dark because the ions cannot move from one another in the solid.

F. *Silicones* consist of O–Si–O chains to which the remaining bonding positions on the silicon atoms are occupied by organic groups. Silicones tend to be oils or rubbery solids which are resistant to high temperatures and withstand attack by a wide variety of substances. Consequently, they are used in waterproofing and in forming seals.

Discussion Question: Why is it doubtful that life based on silicon instead of carbon exists?

VII. A Closer Look: The Asbestos Minerals

Asbestos minerals are those that contain fibrous silicates. The tetrahedra are linked in chains or in sheets. Although asbestos is an excellent thermal insulator, fibers of it can cause lung and digestive tract damage. The sites of damage may become cancerous over time.

VIII. Boron

A. Boron is a nonmetallic member of group 3A. A few of its compounds are important. The *borohydride ion*, or tetrahydroborate, BH$_4^-$, is a good reducing agent.

B. *Boric acid*, H$_3$BO$_3$ or B(OH)$_3$, is a very weak acid. It is used in eyewash solutions. Its anhydride, B$_2$O$_3$, is the only important oxide of boron.

C. *Borax*, Na$_2$B$_4$O$_7$·10H$_2$O, occurs in dry lake beds in California. It can be made from other borate minerals, as well. Borax is used in laundry products due to its weak basicity.

IX. Chemistry at Work: Optical Fibers

    A. Optical fibers are a relatively new application of glassmaking. Fibers transmit information by conducting light, just as metal wires conduct electricity.

    B. Impurities in the glass absorb light. (Have you ever seen the edge of a glass tabletop that looks green?) The key to good optical fibers, then, it to produce high-purity glass, with impurities below 1 ppb.

Lecture Outline – Chapter 23
Brown, LeMay, & Bursten, *Chemistry: The Central Science*, 6th Edition

SAMPLE QUIZ QUESTIONS

1. Circle the formula that best fits each of the following descriptions:
   a. most metallic   S   Se   Po
   b. best π-bonder   C   Si   Ge
   c. best oxidizing agent Cl_2 Br_2 I_2
   d. violet color in CCl_4 Cl_2 Br_2 I_2
   e. molecule is an 8-member ring   S   P   Si
   f. most abundant noble gas   Ar   Xe   Kr
   g. can oxidize Br^- Cl_2 I_2 HCl
   h. phosphoric acid H_3PO_4 HPO_3 H_3PO_3

2. Write balanced chemical equations for the following reactions:

   (a) $PBr_3 + H_2O \rightarrow$;
   (b) $SiO_2 + HF \rightarrow$;
   (c) $SiCl_4 + 2H_2 \rightarrow$.

3. Give the chemical formula for: (a) thiosulfate ion; (b) perchloric acid.

4. Explain why only a few noble-gas compounds are known.

5. Describe the most common sources and use of sulfur.

6. The standard potential for reduction of ClO_3^- to $HClO_2$ in acid solution is 1.21 V while that for the reduction of $HClO_2$ to HOCl is 1.64 V. Calculate the standard potential for the reduction of ClO_3^- to HOCl.

7. Predict the structure of the SeO_3^{2-} ion.

8. Draw the structure of the $Si_2O_7^{6-}$ ion.

9. What is the empirical formula for a sheet-type silicate anion?

10. How does a feldspar mineral differ from a silicate?

11. Write a balanced equation corresponding to a condensation reaction to form a chain of Si–O–Si linkages, beginning with H_4SiO_4.

12. Fill in the symbol of the element that: (a) exhibits the greatest tendency toward catenation _____; (b) is the most abundant element, after oxygen, in the Earth's crust _____; (c) forms a powerful reducing agent of the formula XH_4^- _____; (d) is produced by coke reduction of a calcium mineral _____.

13. Hypobromous acid is: (a) HOBr; (b) HOBrO; (c) $HOBrO_2$; (d) $HOBrO_3$.

14. Metaphosphoric acid is: (a) H_3PO_4; (b) HPO_3; (c) H_3PO_3; (d) $H_4P_2O_7$.

15. Which of the following is the weakest acid? (a) HOCl; (b) $HClO_2$; (c) $HClO_3$; (d) $HClO_4$

16. Which of the following statements does not describe sulfuric acid? (a) It is a good dehydrating agent; (b) It is a good reducing agent; (c) It is a strong acid; (d) It is commercially important.

17. Which of the following is an interhalogen compound?

    (a) NCl_3; (b) ClF_3; (c) $MgBr_2$; (d) $KBrO_3$

18. Which of the following is the weakest acid in water?

    (a) H_2SeO_4; (b) HI; (c) H_2S; (d) P_2O_5

19. Which of the following compounds is polar?

    (a) PCl_5; (b) SF_4; (c) XeF_2; (d) SF_6; (e) IF_7

20. Which of the following is the weakest acid? (a) HF; (b) HCl; (c) HBr; (d) HI

Lecture Outline – Chapter 24
Brown, LeMay, & Bursten, *Chemistry: The Central Science, 6th Edition*

Chapter 24 — Metals and Metallurgy

OVERVIEW: Chapter 24 is the first of two chapters dealing with metals; Chapter 25 looks at coordination complexes.

This chapter begins with a discussion of the distribution of metals in the earth's lithosphere. Examples of metals that occur in the free state, occur in ores, and occur in natural waters are given.

The topic of metallurgy is introduced, emphasizing the series of transformations required in going from the naturally occurring form to the purified metal. Each of the principal types of refining are then covered: pyrometallurgy, hydrometallurgy, and electrometallurgy.

Following the coverage of purifying metals, attention is turned to the peculiarities of metallic bonding. The electron-sea model and molecular-orbital theory are both discussed; insulators and semiconductors are also discussed in this context.

Alloys are covered next, emphasizing how the introduction of the alloying agents disrupt or enhance the bonding within the material. Illustrations of common alloys are given.

The chapter closes with a discussion of the chemistries of the transition metals. First, the physical properties of transition metals are described within the context of periodicity. The various types of magnetism are then covered because transition metals best illustrate paramagnetism and ferromagnetism. Finally, a short survey of the chemistry of chromium, iron, nickel, and copper is given.

LECTURE OUTLINE

I. **Occurrence and Distribution of Metals**

   A. Mineral deposits in the earth's crust that contain metals in economically exploitable quantities are called *ores*.

   B. *Minerals* are metal-containing inorganic compounds found in nature. They are generally impure and are referred to by their common geological names. Most metallic elements, except gold and the platinum-group metals, are found in minerals.

II. **A Closer Look: Metals as Strategic Materials**

   Many metals, as well as other substances, are referred to as *strategic materials*. These materials are those of which the United States does not have a dependable supply and are deemed critical to self-sufficiency.

Discussion Question: What are the reasons for maintaining a strategic materials reserve? Discuss the advantages versus the disadvantages.

III. **Metallurgy**

   A. *Metallurgy* is the art and science of extracting metals from their natural sources and preparing them for use. The process includes several steps.
      1. The ore of the metal must be mined or similarly collected from nature.
      2. The ore must be concentrated or prepared for further treatment.
      3. The ore must be reduced to yield the free metal.
      4. The metal obtained must be purified.

5. The metal must be alloyed, or mixed with other substances, if properties other than those of the pure metal are required.

B. Ores are pretreated in a number of ways. It is generally ground into small particles. The more concentrated metal is then separated from the *gangue*, the unwanted material in the ore. Concentration of iron is done with magnets, for example. Gold is separated from its gangue by washing the less dense gangue from a pool; the more dense gold does not wash away in the moving water as readily.

C. In the *flotation process*, the gangue is wetted while the ore is not. By bubbling oil through the mixture of the ore in water, the ore is attracted to the oil and floats to the top of the vessel where it is skimmed off; the gangue sinks.

IV. Pyrometallurgy

A. *Pyrometallurgy* is the use of heat to reduce ore to the free metal.

B. *Calcination* is the heating of an ore to cause its decomposition and the elimination of a volatile product. A hydrated substance would lose its water, for example.

C. *Roasting* is the heating of an ore to cause chemical reactions between the ore and the furnace atmosphere. Roasting may cause oxidation or reduction and may be accompanied by calcination.

D. *Smelting* is a process in which the products of chemical reactions separate into layers. The most common layers formed in this way are molten metal and *slag*.

E. *Refining* is the treatment of the metal to increase its purity or to better define the composition of a mixture of metals.

F. The *reduction of iron* is the most important pyrometallurgical operation. The reduction of various iron oxides takes place in a *blast furnace*; the crude iron produced is called *pig iron*. Coke reacts with O_2 and H_2O in the furnace to produce CO and H_2; the reduction is performed by the latter gases. Reduction of other elements present in the furnace also occurs. Molten iron and slag are periodically drained from the furnace.

G. *Steel* is an iron alloy. Pig iron is loaded into a *converter*, where impurities in the iron are eliminated by oxidation by O_2. When the impurity levels are acceptable, the converter is dumped into a ladle as alloying metals are added. The steel is then solidified in molds.

V. Hydrometallurgy

A. *Hydrometallurgy* utilizes aqueous reactions to purify a metal.

B. *Leaching* is the most important hydrometallurgical technique, in which the desired mineral is selectively dissolved. Commonly, an acid, a base, or a salt solution is required. In some applications, a coordination complex of the desired metal is formed. After leaching, the metal can be precipitated and purified.

C. Aluminum is second to iron in the amount of metal produced. Its principal ore is bauxite, $Al_2O_3 \cdot xH_2O$. Aluminum is purified hydrometallurgically by the *Bayer process*. In this process, the bauxite is ground and treated with aqueous NaOH at high temperature and pressure. The bauxite dissolves to form $Al(H_2O)_2(OH)_4^-$. Silica and iron(III), which are present as impurities, form a red "mud," which is filtered out. Dilution of the basic solution causes precipitation of hydrated aluminum hydroxide. The solid thus formed is then subjected to electrolysis to reduce the aluminum.

Lecture Outline – Chapter 24
Brown, LeMay, & Bursten, *Chemistry: The Central Science*, 6th Edition

VI. Electrometallurgy

A. *Electrometallurgy* uses electrolysis to reduce metals from compounds or to purify metal samples. Such systems can be aqueous solutions of the metal or molten salts of the metal.

B. Sodium is obtained by electrometallurgy. Using a *Downs cell*, molten NaCl, with $CaCl_2$ added, is electrolyzed to $Na(l)$ and $Cl_2(g)$. The products are removed continuously to prevent recombination or reaction with the environment.

C. Aluminum hydroxide, calcined to form Al_2O_3, is reduced to metallic aluminum electrometallurgically by the *Hall process*. The molten Al_2O_3 is combined with Na_3AlF_6 and electrolyzed using graphite rod electrodes. Carbon is oxidized to CO_2, while Al^{3+} is reduced to Al.

D. Pyrometallurgically obtained copper is insufficiently pure for many applications. It is further purified by Electrorefining. In this method, a crude copper slab serves as the anode and a sheet of purified copper is the cathode, both immersed in aqueous $CuSO_4$. Impurities are readily oxidized at the anode, but only the copper reduces at the cathode.

VII. A Closer Look: Charles M. Hall

A. By the mid-1800s, aluminum was recognized for its strength and low weight. It was, however, a very difficult metal to purify and so was costly ($545 in 1852).

B. Hall, a student, began working on the problem of purifying aluminum cheaply in his backyard shed in 1885.

C. His solution, which was almost simultaneously found in France by Héroult, was electrolysis using cryolite as an ionic molten solvent.

D. Today, most aluminum is made by the Hall process.

VIII. Metallic Bonding

A. The various metals all hold some physical properties in common, but differ greatly in others. The basis of these observed similarities and differences lies in the bonding employed in metals.
   1. Metals typically exhibit good *thermal* and *electrical conductivity*.
   2. Metals typically exhibit *malleability*, the ability to be hammered into thin sheets.
   3. Metals typically exhibit *ductility*, the ability to be drawn into thin wires.
   4. Metals typically exhibit *close packing structures*, much like marbles stack when placed in a box.

B. The *electron-sea model* is a simple descriptive approach to explaining metallic bonding. The metal atoms, stacked in a close packed structure, are viewed as cations floating in a "sea" of electrons. The electrostatic attraction between the cations and the electrons allow for fairly easy and uniform electron motion in the sample, accounting for the electrical conductivity of metals. The mobility of the electrons also allows them to transfer kinetic energy within the metal, accounting for the observed thermal conductivity. The electron-sea model, however, does little to explain quantitative comparisons of metals, such as melting points.

C. We can apply the molecular-orbital model to the bonding in metals. The valence atomic orbitals of the atoms overlap with those of their nearest neighbors, giving rise to many (twice the number of atomic orbitals) delocalized molecular orbitals. The energies of the molecular orbitals are very closely spaced in energy, so closely that we can think of a continuous *band* of allowed states. An electron in motion is in an excited state. In metals, the band is not completely filled, leaving unoccupied orbitals for the moving electrons to enter. The energy required to promote an electron into a previously unoccupied orbital is very small, accounting for conductivity.

Discussion Question: How does the electron-sea model or band theory explain other periodic properties of the transition metals?

IX. A Closer Look: Insulators and Semiconductors

For insulators and semiconductors, consider energy bands to also be formed. In metals, a single band is partially filled. In insulators and semiconductors, there are two bands; the one lower in energy is completely filled and is called the *valence band*, while the one higher in energy is empty and is called the *conduction band*. In order for conductivity to occur, electrons must be promoted, just as with metals. However, the *energy gap* or the amount of energy that must be supplied in order to promote an electron is greater in insulators and semiconductors than in metals. If the gap is fairly small, the material is a semiconductor; if the gap is large, the material is an insulator.

X. Alloys

A. An *alloy* is a mixture that has the characteristic properties of a metal.
   1. A *solution alloy* is a homogeneous mixture in which the components are uniformly distributed.
   2. A *substitutional alloy* is a mixture in which an atom of one element assumes the place of an atom of the major component in the structure.
   3. An *interstitial alloy* is a mixture in which an atom of one element assumes a position between atoms of the major component.
   4. *Steel* is an interstitial alloy, based on iron, in which carbon atoms occupy positions between iron atoms. Steel is categorized as *mild steel*, *medium steel*, or *high-carbon steel*, depending on the amount of carbon present. *Alloy steels* include other metals as well as carbon.
   5. A *heterogeneous alloy* is one in which the components are not uniformly distributed.

B. Intermetallic compounds are homogeneous alloys that have definite compositions and, therefore, definite properties. Examples include Cr_3Pt to harden razor blades and Ni_3Al in aircraft engines.

XI. Transition Metals

A. The transition metals have characteristic physical properties.
   1. Properties must be differentiated between those of isolated atoms and those of aggregate samples. Melting point, for instance, depends on the bonding among several atoms. Atomic radius and enthalpy of vaporization, on the other hand, are properties of a single atom. Properties of samples of transition metals tend to reach a maximum near the center of each period; properties of isolated atoms show relatively smooth variations across each period.
   2. The elements of the second transition row have properties and behaviors very similar to the elements of the third transition row. This is due to the occupation of the 4*f* subshell. As 4*f* electrons are added, the number of protons increases and, because *f* electrons do not screen well, the effective nuclear charge increases steadily. The increase in effective nuclear charge causes a contraction, referred to as *lanthanide contraction*, that offsets the size increase expected on jumping to a lower row.

- B. The transition metals have electron configurations and oxidation states that reflect the addition of electrons to a *d* sublevel.
  1. Transition metals commonly exhibit more than one oxidation state.
     a. Oxidation states of +2 and +3 are more commonly seen in the first transition series than in the second and third. Going down a group, higher oxidation states become increasingly more stable relative to the +2 and +3 states.
     b. From Sc through Mn, the maximum oxidation state increases from +3 to +7. Moving further to the right, the maximum oxidation state decreases.
  2. Many of their compounds are colored (are not white).
  3. The transition metals and their compound often exhibit interesting magnetic properties.
- C. Transition metals exhibit magnetism. There are several kinds of magnetism.
  1. When electrons are paired in an orbital, the interaction they have with an external magnetic field nearly cancels. The small amount of interaction that does not cancel is called *diamagnetism*. All substances, other than H^+, exhibit diamagnetism.
  2. Electrons that are unpaired interact with an applied magnetic field such that there is no cancellation. This type of interaction is called *paramagnetism*. Many transition metals and their compounds are paramagnetic; few nonmetallic compounds are paramagnetic.
  3. In *ferromagnetism*, unpaired electrons of several atoms align themselves with an applied field. Because this is a cooperative interaction involving many atoms, it is much stronger than paramagnetism alone. Common magnets exhibit ferromagnetism.

XII. Chemistry of Selected Transition Metals

- A. *Chromium* reacts with acid to form Cr^{2+} in the absence of O_2, but yields Cr^{3+} when oxygen is present. The +6 oxidation state of CrO_4^{2-} and $Cr_2O_7^{2-}$ is stable in aqueous solution, although both are oxidizing agents. Chromium is toxic in the concentrations used in laboratory work, but trace amounts are essential for proper cell function in humans.

- B. *Iron* exhibits +2 and +3 oxidation states, both of which are stable in aqueous solution; iron(II) is a frequent component of natural waters. Oxygen easily oxidizes Fe^{2+} to Fe^{3+}.

- C. *Copper* exhibits +1 and +2 oxidation states, although the +2 is more stable; Cu^+ disproportionates into Cu^0 and Cu^{2+}. Most salts of Cu^{2+} are water soluble.

Lecture Outline – Chapter 24
Brown, LeMay, & Bursten, *Chemistry: The Central Science*, 6th Edition

SAMPLE QUIZ QUESTIONS

1. Complete and balance the following equations:
   a. $PbCO_3(s) \xrightarrow{\Delta}$
   b. $WO_3(s) + H_2(g) \rightarrow$
   c. $ZnO(s) + C(s) \rightarrow$
   d. $ZnS(s) + O_2(g) \rightarrow$
   e. $PbO(s) + CO(g) \rightarrow$
   f. $CaO(l) + SiO_2(l) \rightarrow$
   g. $Au(CN)_2^-(aq) + Zn(s) \rightarrow$

2. Which of the following is least likely to be reduced by chemical means?

   (a) Mg^{2+}; (b) Na^+; (c) Pb^{2+}; (d) Ag^+

3. Iron is reduced in a blast furnace using: (a) carbon as a reducing agent; (b) water as a reducing agent; (c) limestone as a reducing agent; (d) an electrolytic method of reduction.

4. In a blast furnace, carbon, CO, and H_2 are all active ingredients. Describe how each of these reducing agents is added or formed in the furnace, using balanced chemical equations where appropriate.

5. What is the difference between a substitutional and interstitial alloy?

6. Describe, using equations, the Bayer process used to purify bauxite in the course of the hydrometallurgy of aluminum.

7. Which of the following metals has the highest melting point?

   (a) Ti; (b) Cr; (c) Co; (d) Au

8. Sodium is a highly malleable substance, whereas sodium chloride is not. Explain this difference in properties.

9. What is the purpose of utilizing the basic oxygen furnace in the production of steel?

10. At which electrode (anode or cathode) is the free metal produced in the electrolysis of metal compounds? Explain.

11. Describe the electrolytic refining of copper. What happens to more active metals like zinc that are impurities in the copper? What happens to the less active metals like gold?

12. Metallurgical applications represent the largest single use of liquid oxygen, formed by distillation of liquid air. Describe how the liquid oxygen is employed in the metallurgy of iron; write balanced chemical equations as part of your description.

13. What factors limit the use of aqueous solutions for electrorefining of metals? That is, what are the requirements that must be met for an aqueous-solution-based electrorefining process to be feasible?

14. Describe how a substitutional alloy differs from an interstitial alloy.

15. Write out the electron configurations for the following atoms or ions:

    (a) Co; (b) Co^{3+}; (c) Pt; (d) V^{3+}

16. Which of the following elements would you expect to exhibit the largest heat of atomization? (a) K; (b) Sc; (c) V; (d) Zn

17. Explain briefly what is meant by the term "lanthanide contraction." How does lanthanide contraction allow us to rationalize the strong chemical resemblance of Hf (element 72) and Zr (element 40)?

18. Explain, using one or more chemical equations, why a yellow solution of chromate becomes orange when acidified.

19. One chemical characteristic of the Cu$^+$ ion is its ready disproportionation in aqueous solution. Write a chemical equation that summarizes this disproportionation process.

20. Which of the following exhibits the highest positive oxidation state?

    (a) Sc; (b) Cr; (c) Mn; (d) Cu

21. Which of the following elements has common oxidation states of +2, +4, and +7?

    (a) Ti; (b) Mn; (c) Zn; (d) Cd

22. Which of the following oxides is the strongest oxidizing agent?

    (a) Sc$_2$O$_3$; (b) TiO$_2$; (c) CrO$_3$; (d) Fe$_2$O$_3$

23. Which of the following oxides is the most basic?

    (a) MnO; (b) Mn$_2$O$_3$; (c) MnO$_2$; (d) Mn$_2$O$_7$

24. Which of the following oxides is white? (a) TiO$_2$; (b) Cr$_2$O$_3$; (c) CoO; (d) CuO

25. How can the number of unpaired electrons per atom or ion be determined experimentally?

26. Which of the following substances is paramagnetic?

    (a) NaCl; (b) Zn; (c) TiO$_2$; (d) CoCl$_2$

27. What is ferromagnetism? List two elements that are ferromagnetic.

Lecture Outline – Chapter 25
Brown, LeMay, & Bursten, *Chemistry: The Central Science*, 6*th* Edition

Chapter 25 — Chemistry of Coordination Compounds

OVERVIEW: Chapter 25 completes the coverage of metals begun in Chapter 24. Here, the discussion centers on coordination complexes of the transition metals. The characteristics of complexes are described first, in order to develop a basis for the following topics.

As ligands are being described, bidentate and polydentate ligands are included, as is the chelate effect. With an appreciation for what groups commonly serve as ligands, the basic nomenclature of transition metal complexes is covered, building on the coverage of general inorganic nomenclature early in the text.

Isomerism is an important feature of complex chemistry. The various types of structural isomerism and stereoisomerism are covered in detail with several examples.

The rates of ligand substitution reaction, or ligand exchange reactions, are discussed briefly in order to build an understanding of stability as a kinetic property. Examples of reactions are given to emphasize the kinetic basis of stability. The pioneering work of Alfred Werner into the bonding, stability, and isomerism in transition metal complexes is described fully.

The observed colors and magnetic properties of metal complexes are also covered. The discussion here emphasizes the observations of these phenomena. The chapter concludes with an introduction to crystal-field theory, which explains nicely the observed properties. Emphasis is given to octahedral complexes, but square-planar complexes and tetrahedral complexes are also discussed.

LECTURE OUTLINE

I. **The Structure of Complexes**

   A. Metal atoms or ions act as Lewis acids. When a metal is surrounded by several anions or small molecules acting as Lewis bases, donating electron-pairs to the metal, the assembly thus formed is called a *complex, coordination complex,* or *complex ion*.

   B. The anions or small molecules that surround the metal in a complex are referred to as *ligands*. Molecules acting as ligands are generally polar; all ligands must possess an unbonded pair of electrons to donate to the metal.

   C. The central metal in a complex and its ligands constitute the *coordination sphere*. In writing chemical formulas, the contents of the coordination sphere are enclosed in square brackets, listing the metal first followed by the ligands, as in [Rh(NH$_3$)$_5$Cl](ClO$_4$)$_2$, where five ammonia molecules and a chloride ion are *coordinated* to the rhodium ion.

   D. As in other chemical formulas, the net charge of a complex is the sum of the charges of the metal ion and the ligands.
      1. A ligand's *donor atom* is that atom which directly attaches to the metal. Nitrogen, oxygen, and halide are the most common donor atoms.
      2. The number of donor atoms in a complex is the metal's *coordination number*. Note that this is not necessarily the number of ligands; it is the number of direct connections to the metal.
      3. Various metals form complexes with characteristic *geometries*. The most commonly seen geometries in complexes are the octahedron, the tetrahedron, and the square plane.

a. *Octahedral* complexes have a coordination number of 6. Small ligands and large metal ions tend to produce complexes with high coordination numbers, usually 6; large ligands and small metal ions tend to produce complexes with lower coordination numbers, usually 4.
b. *Tetrahedral* complexes have a coordination number of 4. Tetrahedral geometries are favored for 4-coordination in most cases, especially when large ligands or small metals are involved.
c. *Square planar* complexes also have a coordination number of 4. Square planar geometries are favored only for complexes in which the metal has a s^0d^8 valence electron configuration.

Discussion Question: Can students suggest other species that could effectively serve as ligands, both monodentate and polydentate?

II. Chelates

A. The common ligands discussed are *monodentate ligands*, ligands with a single donor atom. A monodentate ligand occupies a single coordination site, contributing 1 to the coordination number.

B. Ligands that utilize more than one donor atom are called *polydentate ligands* or *chelating agents*. Ligands that utilize two donor atoms are more specifically called *bidentate ligands*. Two of the more common chelates are ethylenediamine, H_2H-CH_2-CH_2-NH_2, and ethylenediaminetetraacetate, $(OOCCH_2)_2N$-CH_2-CH_2-$N(CH_2COO)_2^{4-}$.

C. It is observed that chelates form more stable complexes than do related monodentate ligands, a phenomenon referred to as the *chelate effect*. Because of the strong bonding of chelates to metals, chelates are often used to remove metals from solutions or to render it inactive. Thus, chelates are also referred to as *sequestering agents*.

D. Metals are critical to the survival of living organisms. Many of the compounds formed by metals in biological systems are complexes with chelates. Several examples are commonly cited. *Porphyrins* are complexes derived from a metal and a *porphine* molecule.
   1. *Heme* is a structure of iron and a porphine. Four of these heme groups are present in each molecule of *hemoglobin*, the substance responsible for the transportation of oxygen in the blood of mammals.
   2. *Chlorophyll* is a structure of magnesium and a porphine, which is responsible for photosynthesis, the process in plants that, among other things, produces energy for the plant from sunlight.

Discussion Question: What consumer products are students familiar with that contain chelating agents? What are the specific functions of the chelating agents in these products?

III. Closer Look: The Stability of Chelates

The stability of chelates is often due, in a thermodynamic sense, to the observation that chelation is usually accompanied by an *increase* in the *entropy* of the system, while complex formation using related monodentate ligands often has a negative entropy change.

IV. Chemistry and Life: The Battle for Iron in Living Systems

A. Most living organisms need iron, but few can easily assimilate it into their systems. Iron deficiency in animals is called *anemia*, while in plants it's called *chlorosis*.

B. Microorganisms absorb iron by excreting a substance called *siderophore*. Siderophore forms a soluble complex with iron called *ferrichrome*. The cell then absorbs the ferrichrome and extracts the iron from it.

C. Humans absorb iron with a protein called *transferrin* in the intestines. The complex formed between transferrin and the iron passes through the intestinal wall and the iron is transported throughout the body.

D. Current medical findings indicate that having an excess of iron in the body can be as damaging to health as having a deficiency.

V. Nomenclature

There are rules of nomenclature for complexes that are an extension of the rules for other inorganic substances.

1. In naming salts, the name of the cation is given, then the name of the anion.
2. Within a complex, the ligands are named before the metal. The ligands are listed in alphabetical order, with prefixes to state the number of each ligand present. The prefixes are not used to alphabetize.
3. Anionic ligands have names that end in -o. Neutral ligands simply use the name of the molecule, with few exceptions, most notably "aqua" for water, and "ammine" for ammonia.
4. A Greek prefix (*di-*, *tri-*, *tetra-*, *penta-*, etc.) is used to give the number of each ligand present, although no prefix is used for "1." When the name of the ligand contains a syllable that can be confused with a prefix for the ligand as a whole, slightly different prefixes (*bis-*, *tris-*, *tetrakis-*, *pentakis-*, etc.) are used.
5. If the complex is an anion, its name ends in *-ate*. If the element has a Latin atomic symbol, the Latin name of the metal is used as the base. For example, $Au(CN)_4^-$ is tetracyanoaurate(III).
6. The oxidation state of the metal is given in Roman numerals following the name of the metal. See #5 above. Note that there are no spaces in the name of a complex.

VI. Isomerism

A. *Isomers* are compounds with the same composition, but different arrangements of atoms. In *structural isomerism*, the bonding in the two isomers is different.

1. In *linkage isomerism*, a ligand contains two or more atoms that can serve as the donor atom, and a different isomer is formed when a different donor atom is involved. For example, NO_2^- can bond to a metal with the N or with one of the O atoms; when nitrogen is the donor, the ligand is called *nitrito*; when an oxygen is the donor, the ligand is called *nitro*.
2. *Coordination-sphere isomerism* occurs differ in the ligands that are directly bonded to the metal. For example, the compound $Co(NH_3)_5Cl_3$ can be assembled with different combinations of ammonias and chlorides in the coordination sphere: $[Co(NH_3)_5Cl]Cl_2$, $[Co(NH_3)_4Cl_2]Cl \cdot NH_3$, $[Co(NH_3)_3Cl_3] \cdot 2NH_3$.

B. *Stereoisomerism* is the most important type of isomerism; isomers differ in the spatial arrangement of ligands, but not in the bonds formed.

1. Consider Pt(NH$_3$)$_2$Cl$_2$, which is square planar. The pairs of ammonias and chlorides can be arranged in two ways; the chlorides can be adjacent to each other or they can be on opposite sides of the metal from each other. The same conclusion is reached by looking at the ammonias. The two possible configurations are *geometric isomers* of each other. The case where two identical ligands are adjacent to each other is the *cis* isomer; the case where they are across the metal from each other is the *trans* isomer. Similar arrangements occur in octahedral complexes, but not in tetrahedral complexes where each ligand is adjacent to all three of the others.
2. *Optical isomers* are nonsuperimposable mirror images of one another. Optical isomers are related to each other just as your right hand is related to your left hand. When optical isomers exist for a complex, the substance is said to be *optically active*.
   a. Substances that are optically active, or the metals of such complexes, are also said to be *chiral*. Experimentally, optical isomers are identified by the way in which each interacts with polarized light.
   b. A complex that rotates plane-polarized light to the right (clockwise) is said to be *dextrorotatory*.
   c. A complex that rotates plane-polarized light to the left (counterclockwise) is said to be *levorotatory*.
   d. In many laboratory preparations of chiral complexes, the synthesis produces equal amounts of both isomers, resulting in a product mixture. Such a mixture is referred to as a *racemic mixture*.

VII. A Closer Look: Alfred Werner and the Structures of Coordination Compounds

The premiere work in describing the bonding and isomerism in transition metal complexes was performed by Alfred Werner around 1900. Werner identified coordination and isomerism using cobalt complexes of chloride and ammonia. Complexes in which the donor atoms are nitrogen, oxygen, or halide are often referred to as Werner complexes in honor of Werner.

VIII. Ligand Exchange Rates

Ligand exchange or *substitution* reactions are common for complexes, particularly in solution. These reactions are often accomplished by dissolving a complex in a solution containing some new ligand material. When the new ligands replace those previously attached to the metal, a substitution reaction has occurred.
1. Complexes that readily undergo ligand exchange reactions are said to be *labile*.
2. Complexes that only slowly undergo ligand exchange reactions, if at all, are said to be *inert*.

IX. Color and Magnetism

A. Transition metal complexes are often colored. The color of a complex depends on the identity of the metal, its oxidation state, and the ligands surrounding it.
   1. Complexes in which the metal has a d^0 or a d^{10} electron configuration tends to be *colorless* in solution or white in the solid state.
   2. Colored compounds absorb light; the colors we see are those *not* absorbed by the compound. Using an artist's color wheel, the colors seen in a compound are the *complementary colors* to those absorbed by the compound. White or colorless compounds are those that absorb no visible light.

3. The amount of light absorbed by a sample as a function of wavelength is called its *absorption spectrum*. The wavelengths at which absorption occurs, as well as the intensity of the absorption, are characteristic of each compound and can be used to identify compounds and to determine their concentrations.

B. All substances, other than H^+, possess diamagnetism. Many transition metals and their complexes exhibit *paramagnetism*, in which unpaired electrons interact with an external magnetic field.

X. A Closer Look: Gemstones

Colored gemstones appear colored due to trace amounts of transition metals in their structures. For example, ruby is Al_2O_3 in which a fraction of the aluminum ions are replaced by chromium.

XI. Crystal-Field Theory

A. Many of the properties of transition metal complexes, such as their color and magnetic interactions, relate to the *d* electrons of the metals. A model of bonding in complexes, *crystal-field theory* or *CFT*, explains these phenomena. CFT assumes the interaction between a metal and its ligands is electrostatic. The interactions become more pronounced as the positive charge of the metal and the negative charge of the ligands increase. In an octahedron, the ligands are seen as approaching the metal along the x, y, and z axes. As such, the interaction between a set of ligands and the metal *d*-orbitals is greater for the d_{z^2} and $d_{x^2-y^2}$ than for the d_{xy}, d_{xz}, and d_{yz} orbitals. As a result, the d_{z^2} and $d_{x^2-y^2}$ orbitals increase in energy relative to the d_{xy}, d_{xz}, and d_{yz} orbitals. The two *degenerate* sets of orbitals describe the energies of the *d* orbitals in the complex.

B. The energy difference between the two sets of orbitals is called the *field splitting*, symbolized Δ. The magnitude of the field splitting, and therefore the color of the complex, depends on the metal and on the ligands.
  1. The electronic structure of a ligand determines its effect on the value of Δ. Ligands are commonly ranked in the order of their influence on Δ, the list being referred to as the *spectrochemical series*. Ligands considered "low" on the spectrochemical series, called weak-field ligands, include the electron-rich halide ions. Ligands considered "high" on the spectrochemical series, called strong-field ligands, include CO and CN^-, which contain triple bonds.
  2. The metal also influences the magnitude of Δ. The value of Δ increases as 1) the charge of the metal increases and 2) as the row number of the metal ion increases.

C. In *octahedral complexes*, electron configurations of the metal ions may be stated more precisely than in the isolated atoms. The placement of the *d* electrons is divided between the two sublevels. The first three *d* electrons are placed in the lower set of orbitals.
  1. The fourth electron can be placed in the higher set or can be spin-paired in the lower set, the deciding factor being whether the field splitting is greater than or less than the *spin-pairing energy*, the energy required to place two electrons into the same orbital.
  2. When the electrons are placed in the higher orbital set in preference to pairing them in the lower set, the situation is referred to as *high-spin*.
  3. When the electrons are paired in the lower orbital set in preference to placing them in the higher orbital set, the situation is referred to as *low-spin*.

D. *Square-planar complexes* can be seen as octahedral complexes in which the two ligands lying on the z axis are withdrawn from the complex. Much of the degeneracy of the *d* orbital energies is removed, with the energies of the $d_{x^2-y^2}$ and the d_{xy} rising, while the energies of the remaining *d* orbitals drop.

E. In *tetrahedral* complexes, the symmetry of the complex is such that the orbital energy diagram is the inverse of that for an octahedral complex. In addition, the magnitude of Δ for a tetrahedral complex is about $\frac{4}{9}$ as great as it is for an octahedral complex of the same metal and ligands. As a result, tetrahedral complexes are invariably high-spin.

SAMPLE QUIZ QUESTIONS

1. Name the following compounds: (a) $Na_3[CrF_6]$; (b) $[Co(NH_3)_4Cl_2]Cl$.

2. Write the chemical formulas for the following compounds:

    (a) diamminediaquadichloromanganese(II); (b) potassium trioxalatocobaltate(III)

3. What is the oxidation state of platinum in $K[Pt(NH_3)Cl_5]$?

4. Write the formula of the species formed when cobalt(III) is coordinated by three water molecules and three nitrate anions.

5. Draw all the possible isomers of $[Co(NH_3)_4Cl_2]^+$.

6. Draw the optical isomers of $[Cr(en)_2(NH_3)Cl]^{2+}$.

7. Describe how you would experimentally determine whether a compound is paramagnetic.

8. Which of the following compounds would you expect to absorb the highest energy light? $K_4[Fe(CN)_6]$ or $[Fe(H_2O)_6]SO_4$? Explain briefly.

9. Which of the following compounds as formulated by Werner do you expect to have the highest electrical conductivity in solution?

    (a) $PtCl_2 \cdot 2NH_3$; (b) $CoCl_3 \cdot 4NH_3$; (c) $CrCl_3 \cdot 5H_2O$; (d) $PtCl_4 \cdot 3NH_3$

10. The ion $[Co(NH_3)_6]^{3+}$ is a low-spin complex. Use crystal-field theory to describe the bonding in this complex.

11. Which of the following ions, if hydrated, would you expect to be practically colorless?

    (a) Ti^{2+}; (b) Ti^{4+}; (c) Co^{2+}; (d) Cr^{3+}

12. Which of the following electron configurations will give rise to a diamagnetic complex in a strong octahedral field? (a) d^2; (b) d^4; (c) d^5; (d) d^6

13. In an octahedral field, which of the following orbitals have the same energy?

    (a) d_{z^2} and $d_{x^2-y^2}$; (b) d_{xz} and d_{yz}; (c) d_{z^2} and d_{xy}; (d) $d_{x^2-y^2}$ and d_{xy}

14. In a tetrahedral field, which of the following orbitals are of lowest energy?

    (a) $d_{x^2-y^2}$ and d_{z^2}; (b) d_{xz} and d_{yz}; (c) d_{xy}, d_{yz}, and d_{xz}; (d) d_{xy} and $d_{x^2-y^2}$

15. Which of the following ions contain(s) a bidentate ligand?

    (a) $[Co(NH_3)_5SO_4]^+$; (b) $[Co(NH_3)_4CO_3]^+$; (c) $[Co(NH_3)_4(NO_2)_2]^+$; (d) $[Co(NH_3)_5NO_2]^{2+}$

16. The compound $[Co(NH_3)_5(H_2O)]Cl_3$ consists of: (a) 2 ions; (b) 3 ions; (c) 4 ions; (d) 5 ions.

17. Which of the following compounds exhibit geometric isomerism?

    (a) $[Pt(NH_3)_4]Cl_2$; (b) $[Pt(NH_3)_3Cl]Cl$; (c) $[Pt(NH_3)_2Cl_2]$; (d) $[Pt(NH_3)Cl_3]$

18. What is the coordination number of nickel in $[Ni(CN)_5]^{3-}$? (a) 3; (b) 5; (c) 8; (d) 10

Lecture Outline – Chapter 26
Brown, LeMay, & Bursten, *Chemistry: The Central Science*, 6*th* Edition

Chapter 26 — The Chemistry of Life: Organic and Biological Chemistry

OVERVIEW: Chapter 26 deals with organic chemistry and biochemistry. It is in no way a thorough treatment, that is left for future courses. The chapter begins by defining organic chemistry and by detailing the simplest of the organic compounds, the alkanes. Nomenclature, reactions, and stereoisomerism are emphasized.

Following alkanes, the alkenes and alkynes are covered. The characteristic reactions are discussed, notably addition reactions and Markovnikov's rule. Nomenclature is also touched upon as are the geometric isomers of alkenes.

A section is given on aromatic hydrocarbons. Rather than thoroughly developing aromaticity, benzene is used as the basis for the discussion.

With the basic hydrocarbons covered, a section now follows which describes the most common derivatives: alcohols, aldehydes, ketones, carboxylic acids, esters, amines, and amides. Included with each is a brief discussion on the synthesis and characteristic reactions of each functional group.

Most of the material on biochemistry describes the major types of biomolecules; the ways in which some of these substances are used in organisms are described briefly.

First is a discussion of the energy requirements of living systems. That the origin of all available energy is the sun is also covered. In this discussion, photosynthesis is used to illustrate the concepts.

The types of biomolecules are then described, separated by type. Proteins are covered first, with an introductory treatment of the amino acids. Here, chirality and condensation reactions are reviewed. Protein structure is also covered, with mention of globular and fibrous proteins, and the double helix.

Carbohydrates are covered next. These substances, like others in this chapter, are described in terms of their building blocks, the sugars. The differentiation of α and β sugars is given, and used in discussing the starch and glycogen.

The chapter closes with a discussion of nucleic acids, those substances that store the genetic code and control replication and chemical synthesis. The constituent parts of each nucleotide are described, as is the nature of the long-chain acid. The section ends with a description of the cleavage of the double helix and the synthesis of a new complementary strand.

LECTURE OUTLINE

I. **Organic Chemistry**

   A. *Organic chemistry* is the study of compounds in which carbon is bonded to hydrogen or elements replacing hydrogen. It was originally thought that only living organisms can synthesizes these compounds; we know now that carbon-based compounds can be made in the laboratory, but the term organic survives.

   B. The first organic compound synthesized in the laboratory was *urea*, H_2NCONH_2. It had long been known as a constituent of the urine of many mammals.

Lecture Outline – Chapter 26
Brown, LeMay, & Bursten, *Chemistry: The Central Science, 6th Edition*

    C.   The *hydrocarbons* are the simplest organic compounds. In these, the only elements present are carbon and hydrogen. Organic compounds containing other elements are seen as derivatives of the hydrocarbons.

II. Alkanes

A. Hydrocarbons contain stable carbon-carbon bonds, which can string together to form chains or rings. They tend to be nonpolar compounds.
   1. *Saturated hydrocarbons* are those in which all the carbon-carbon bonds are single bonds. Because carbon forms four bonds in virtually all stable species, each carbon in the midst of a chain is also bonded to two hydrogens, $-CH_2-$, while a carbon at the end of a chain is bonded to three hydrogens, $-CH_3$.
   2. *Unsaturated hydrocarbons* contain one or more double or triple bonds. Consequently, the carbons are bonded to fewer hydrogens than in the saturated hydrocarbon with the same number of carbons.
   3. *Alkanes* are hydrocarbons in which all carbon-carbon bonds are *single bonds*; alkanes are saturated hydrocarbons.
   4. *Alkenes* are unsaturated hydrocarbons in which at least one carbon-carbon bond is a *double bond*.
   5. *Alkynes* are unsaturated hydrocarbons in which at least one carbon-carbon bond is a *triple bond*.
   6. *Aromatic hydrocarbons* are unsaturated hydrocarbons in which the carbon atoms form a ring, with σ and π bonds between carbons.

B. The alkane series is the group of hydrocarbons in which each succeeding member has a CH_2 group added to the formula of the preceding member. Such a series is called a *homologous series*. The general formula for all alkanes, except ring structures, is C_nH_{2n+2}, where *n* is the number of carbon atoms. Properties such as melting point and boiling point increase as CH_2 groups are added.

C. VSEPR tells us that the structures of alkanes, in which carbon atoms are sp^3 hybridized with 109.5° bond angles, consist of the carbon chains being "zig-zags". We refer to *linear* or *straight-chain* geometries for the molecules when there is only one carbon chain in a molecule.

D. Alkanes above C_3 exhibit *structural isomers*, variations in the bonding. The straight-chain isomer of a given alkane has all carbon atoms attached to form one continuous chain. Other isomers consist of shorter "main" chains, with the remaining carbons attached to it. For example, pentane, C_5H_{12}, has three structural isomers: CH_3-CH_2-CH_2-CH_2-CH_3, CH_3-$CH(CH_3)$-CH_2-CH_3, and CH_3-$C(CH_3)_2$-CH_3.

E. Alkanes are named by a set of internationally agreed to rules.
   1. Each compound is named for the longest continuous chain of carbon atoms present.
   2. In general, a side group formed by removing a hydrogen atom from an alkane is called an *alkyl* group.
   3. The location of an alkyl group along a carbon chain is indicated by numbering the carbons along the chain, such that the alkyl group attaches closer to the "1" end than the other end.
   4. If there is more than one substituent group of a certain type, the number of groups is indicated by a numerical prefix.

F. *Cycloalkanes* are alkanes in which some or all of the carbon atoms form a ring. The general formula for cycloalkanes is C_nH_{2n}.

G. Alkanes undergo a variety of reactions, although they are relatively unreactive. At room temperature, they do not react with acids, bases, or strong oxidizing agents. Under appropriate conditions, they do react.
1. Alkanes generally undergo *combustion* in air to form CO_2 and H_2O.
2. Alkanes undergo *substitution* reactions with F_2, Cl_2, and Br_2 in which the halogen atoms replace one or more hydrogen atoms. The reaction with fluorine is vigorous; the others require heat or light to initiate them.
3. Light initiates chlorine substitution reactions by dissociating Cl_2 into chlorine atoms. Odd-electron atoms, like Cl, are called *free radicals*. Reactions that proceed due to the formation of radicals are referred to as *radical chain processes*.

III. Chemistry at Work: Gasoline

Gasoline is a volatile mixture of alkanes and aromatic hydrocarbons, ranging in size from C_5 to C_{12}.

A. The first step in processing petroleum is *refining*, which separates the petroleum mixture into fractions based on boiling points; it is a distilling process. The gasoline produced by refining is called *straight-run gasoline*.
1. The gasoline produced by refining is mostly straight-chain hydrocarbons that burns poorly, tending to "knock." The *octane rating* of a fuel describes the fuel's resistance to knocking when burned in an engine. Straight-run gasoline has an octane number of about 50. Branched hydrocarbons have significantly higher octane numbers.
2. In order to increase the content of branched hydrocarbons in gasoline, and hence its octane rating, the straight-run gasoline is subjected to *cracking*, which converts straight-chain hydrocarbons into branched ones. The use of additives also helps reduce knocking.

IV. Unsaturated Hydrocarbons

A. *Alkenes* are hydrocarbons that have at least one double bond. The inclusion of double bonds alters a hydrocarbon's structure and reactivity.
1. Alkenes are named in a way similar to alkanes; the *-ane* ending is changed to *-ene* and the position of the double bond is indicated by the carbon number at which the double bond begins. For example, CH_3-CH_2-CH=CH-CH_3 is 2-pentene, while CH_2=CH-CH=CH-CH_3 is 1,3-pentadiene.
2. Alkenes occur in *geometric isomers* based on the distribution of groups at the double bond. When viewed from above the plane of the double bond, if the longest carbon chain enters and leaves the double bond on the same side, it is a *cis* isomer; if the longest carbon chain enters and leaves the double bond on opposite sides, it is a *trans* isomer.

B. *Alkynes* are hydrocarbons that have at least one triple bond. Alkynes are named in a similar fashion to the alkenes; the ending used for alkynes is *-yne*. For example, CH_3-CH_2-C≡CH is 1-butyne.

C. Alkenes and alkynes undergo *addition reactions*. Addition reactions involve reaction of the π bond to 1) break the π bond and 2) add atoms or alkyl groups to the atoms that had possessed the π bond.
1. In *hydrogenation*, H_2 attacks the double bond of an alkene and adds a H atom to each of the two carbons of the double bond while converting the double bond to a single bond. If the reactant hydrocarbon is an alkyne, two moles of H_2 can react, forming first the related alkene and then the related alkane.

2. The addition of *hydrogen halides* presents two possibilities of which carbon atom receives the hydrogen and which receives the halogen atom. *Markovnikov's rule* states that the hydrogen atom will go to the carbon atom that *already has* more hydrogens attached to it.
3. Other substances, notably halogens, X_2, and water, H_2O, can add to double bonds.

D. Aromatic Hydrocarbons
1. *Benzene*, C_6H_6, illustrates the class of hydrocarbons called *aromatic hydrocarbons*. The delocalization of the π-electrons in benzene causes the molecule to be much less reactive than expected.
2. Aromatic hydrocarbons are named using a partially systematic scheme. Because all six carbons in benzene are equivalent, it is not necessary to state a location for one substituent. When there are two substituents, two schemes are used. In one, the carbons are numbered 1-6, and names such as 1,3-dimethylbenzene result, in keeping with the nomenclature of the alkanes. In the other scheme, prefixes are given to the possible combinations. By example, 1,2-dimethylbenzene is called *ortho*-dimethylbenzene, 1,3-dimethylbenzene is called *meta*-dimethylbenzene, 1,4-dimethylbenzene is called *para*-dimethylbenzene.
3. Aromatic hydrocarbons do not undergo addition reactions under normal conditions. However, they do undergo *substitution reactions*, in which hydrogen or some other group is removed, and the attacking agent replaces it on the ring.
   a. The isomers, *ortho*-, *meta*-, and *para*-, become important in the products of substitution reactions. Different substituting reagents lead to different principal isomers in the products. For example, attack by HNO_3 leads to substitution of two -NO_2 groups for hydrogens, yielding only the *meta* isomer.
   b. In *Friedel-Crafts reactions*, alkyl groups can substitute onto a benzene ring. The reagent that attacks the benzene is an alkyl halide. In all substitution reactions of benzene, the reactions are catalyzed by a positively charged species.

V. A Closer Look: Aromatic Stabilization

In aromatic compounds, like benzene, the bonding is stronger than would be expected for three single C–C bonds and three C=C double bonds. In fact, the heat of formation of benzene indicates that about 140 kJ/mol is attributable to the effect of aromaticity.

VI. Chemistry at Work: The Accidental Discovery of Teflon

A. A *polymer* is a high molar mass material formed from simple molecules called *monomers*. Polymers may be synthetic or naturally-occurring. Natural polymers include proteins, carbohydrates, and nucleic acids. Synthetic polymers include nylon and polystyrene.

B. Teflon was discovered completely by accident, after a "full" gas cylinder of tetrafluoroethylene registered "empty" for no known reason. The chemist at Du Pont, Roy Plunkett, cut the tank open and found the gas had polymerized to what we now call Teflon.

C. There are two principal reaction types leading to polymerization.
1. In *addition polymerization*, the monomer molecules contain π bonds that "open", leaving each of the two carbon atoms of the bond with an electron; carbon atoms on different molecules then share their unpaired electrons, forming new single bonds between molecules. The reaction

$nF_2C=CF_2 \rightarrow$ —[CF_2-CF_2]$_n$— describes the synthesis of Teflon from tetrafluoroethylene.

2. In *condensation polymerization*, monomer molecules combine while splitting out a small molecule. For example, the reaction of a dialcohol with a diprotic carboxylic acid yields polyesters and water.

VII. Hydrocarbon Derivatives

A. Reactions of hydrocarbons often yield organic compounds, called derivatives, with characteristic structural features and characteristic properties. *Alcohols* result from at least one hydrogen being replaced by a hydroxyl group, –OH. The names of alcohols are found by removing the final *-e* of the hydrocarbon's name and adding the suffix *-ol*.

1. A *primary alcohol* is one in which the carbon to which the –OH is bonded is attached to only one other carbon.
2. A *secondary alcohol* is one in which the carbon to which the –OH is bonded is attached to two other carbons.
3. A *tertiary alcohol* is one in which the carbon to which the –OH is bonded is attached to three other carbons.
4. Alcohols are synthesized in a variety of ways. Methanol, CH_3OH, can be made by reacting CO with H_2. Ethanol, CH_3CH_2OH, is produced in *fermentation*. Other alcohols are made by addition reactions of water to alkenes and alkynes.
5. *Polyhydroxyl alcohols* are alcohols that contain more than one -OH group. Ethylene glycol or 1,2-ethanediol is used in antifreeze. Others are commercially important.
6. *Phenol* is the simplest aromatic alcohol, having one H replaced by an –OH.

B. *Ethers* have two hydrocarbon groups bonded together through an oxygen. For example, CH_3CH_2–O–CH_2CH_3 is diethyl ether, which is used as a solvent and as an anesthetic. Ethers are formed by *condensation reactions*, in which two alcohol molecules combine and eliminate a molecule of water.

C. Aldehydes and ketones both contain a $-\overset{\overset{\displaystyle O}{\|}}{C}-$ group.

1. The $-\overset{\overset{\displaystyle O}{\|}}{C}-$ group is called the *carbonyl* group. It consists of the carbon double bonded to the oxygen; the carbon also makes two other bonds. If both of these other carbon bonds are to alkyl groups, the compound is a *ketone*. If at least one of these other carbon bonds is to hydrogen, the compound is an *aldehyde*.
2. The general structure of an aldehyde is $R-\overset{\overset{\displaystyle O}{\|}}{C}-H$, where R is an alkyl group.
3. The general structure of a ketone is $R-\overset{\overset{\displaystyle O}{\|}}{C}-R'$, where both R and R' are alkyl groups. The two groups need not be the same.
4. The synthesis of aldehydes and ketones is accomplished by careful oxidation of alcohols. An aldehyde is produced from a primary alcohol; a ketone is produced from a secondary alcohol.

Lecture Outline – Chapter 26
Brown, LeMay, & Bursten, *Chemistry: The Central Science*, 6th Edition

5. Aldehydes are generally more reactive than ketones. Ketones are used chiefly as solvents.

D. *Carboxylic acids* have the general structure, $R-\overset{\overset{O}{\|}}{C}-O-H$, in which R is an alkyl group. A very common example is acetic acid, CH_3COOH. Carboxylic acids are generally weak acids and are commonly found in nature.
   1. Carboxylic acids can be synthesized by the vigorous oxidation of a primary alcohol. Acetic acid can be made by a reaction between CO and CH_3OH known as a *carbonylation reaction*.
   2. Carboxylic acids are used in the manufacture of polymers for paints, films, and fibers.

E. *Esters* have the general structure $R-\overset{\overset{O}{\|}}{C}-O-R'$, where R and R' are both alkyl groups.
   1. Esters are synthesized by condensation reactions between carboxylic acids and alcohols. They are named by first using the name of the group from the alcohol, R', and then the name of the group from the acid; the suffix *-oate* is attached. Esters generally have pleasant odors, often associated with fruit and flowers.
   2. Esters can be *hydrolyzed*, split into their acid and alcohol components. Hydrolysis of an ester by base is called *saponification*.

F. Amines and amides are nitrogen-containing organic compounds.
   1. *Amines* have the general formula NR_3, although the three R groups need not be the same. Amines can be considered as derivatives of ammonia, NH_3.
   2. *Amides* have the general structure $R-\overset{\overset{O}{\|}}{C}-NR'_2$, in which R and the two R' need not be the same. Amides are formed by the condensation of a carboxylic acid and an amine.

VIII. Introduction to Biochemistry

A. All organisms require energy to live and flourish. Energy, in the form of enthalpy, is needed by an organism to synthesize compounds essential to existence. Because organisms are highly ordered systems, internal entropy must be decreased. Gibbs free energy, $\Delta G = \Delta H - T\Delta S$, describes the total energy requirement of an organism. The ultimate source of this energy is the sun.

B. *Photosynthesis* is the principal way in which solar energy is converted into forms useful to organisms. In photosynthesis, sunlight drives the reaction of CO_2 and H_2O to form carbohydrates and O_2. *Chlorophyll* is the pigment in leaves that absorbs photons of sunlight. The energy chlorophyll possesses by absorbing photons is converted to chemical energy needed to drive photosynthesis.

C. Much of the material in organisms is in the form of *biopolymers*, biological molecules of high molar mass. Biopolymers can be divided into three groups.
   1. *Proteins* are major sources of energy in the food of animals. They also comprise structural material, such as muscle.
   2. *Carbohydrates* are a source of energy in the food of animals. Complex carbohydrates, such as cellulose, serve as structural material in plants.
   3. *Nucleic acids* store and transmit genetic information and control cell reproduction and development.

Discussion Question: Can students think of any biologically available forms of energy that do not ultimately come from the sun?

IX. Proteins

A. *Proteins* are macromolecules present in all living cells. Proteins form skin, nails, cartilage, and muscle. They also catalyze reactions and serve to regulate cell processes. All proteins, regardless of function, are comprised of α-*amino acids*.
   1. The α-amino acids are carboxylic acids that also have an amine group on the carbon adjacent to the carbonyl carbon. The amino acids differ from one another in the identity of a substituent group, R, on the α-carbon. There are 20 common amino acids found in most proteins.
   2. Our bodies can synthesize 10 of the amino acids as needed. However, the other 10 amino acids, called the *essential amino acids*, must be ingested as food.
   3. Most (19) of the amino acids are *chiral*; they have nonsuperimposable mirror images. Nonsuperimposable mirror images are called *enantiomers*. Those isomers that rotate polarized light to the right are prefixed *R* or D- and those isomers that rotate polarized light to the left are prefixed *S* or L-.

B. A *polypeptide* is a structure formed when several amino acids are linked together by *peptide bonds*. *Proteins* are polypeptides with molecular weights over about 6000 amu.
   1. A peptide bond or linkage is formed when the amino group of one amino acid undergoes a condensation reaction with the acid group of another amino acid, with the elimination of water. The general form of the linkage is then R–C(O)NH–R', where the two R groups are generally different.
   2. A notation has been adopted for indicating the *sequence* of amino acids in polypeptides. Abbreviations for the amino acids are used and the amino acids are listed in the order they are linked, starting at the amine end and proceeding to the acid end. For example, gly-ala represents a dipeptide in which glycine and alanine are linked with the glycine maintaining its NH_2 group and alanine maintaining its COOH group.
   3. *Simple proteins* are those proteins composed *entirely* of amino acids.
   4. *Conjugated proteins* are those proteins that are comprised of simple proteins bonded to other kinds of biomolecules, such as carbohydrates or metal complexes.

C. The structure of a protein is described by many features.
   1. The *primary structure* of a protein is the sequence of amino acids within the protein.
   2. The *secondary structure* of a protein describes how the chains of amino acids bend and fold to form a gross shape for *sections* of the protein.
   3. The *tertiary structure* of a protein describes the *overall* shape of the protein.
      a. *Globular proteins* are those that fold into compact, roughly spherical shapes. Globular proteins are generally water-soluble and so are mobile within cells.
      b. *Fibrous proteins* are those that form long chains or fibers. Fibrous proteins are generally insoluble in aqueous solution.

Discussion Question: Beyond the general listings in the text, what functions do proteins play in the body?

D. *Enzymes* are biological catalysts that generally are globular proteins with molecular weights of about 10,000 to 1,000,000 amu. Most enzymes catalyze only specific reactions involving specific reactants. An enzyme may be a single

protein chain or it may consist of several chains held loosely together. They are sensitive to temperature and function less effectively at other temperatures, due to changes in the secondary and tertiary structure. They also tend to be sensitive to the pH of the medium of the reaction.
1. The specific location on the enzyme where the reaction occurs is called the *active site*.
2. The substance that undergoes reaction by attaching to the active site is called the *substrate*.
3. In some reactions, participants other than the enzyme and the substrate may be required for the reaction to proceed. These other substances are called *cofactors* or *coenzymes*.
   a. Carboxypeptidase catalyzes the hydrolysis of peptide bonds, releasing the amino acid at the acid end of the chain. The name of the enzyme is derived from the name of the substrate or the reaction of the substrate, followed by the suffix *-ase*.
   b. Protease catalyses the hydrolysis of peptide linkages, but not just the terminal peptide bond of the chain.
4. The efficiency with which an enzyme catalyses a reaction is expressed in its *turnover number*. The turnover number is the number of reaction events that occur at the active site of a single enzyme molecule in one second.

E. The mechanisms of enzyme-catalyzed reactions have been the subject of active research. The first step of the reaction involves a fast equilibrium between the enzyme and the substrate, $E + S \rightleftharpoons ES$, where *ES* is called the *enzyme-substrate complex*. The enzyme-substrate complex then decomposes to the product and free enzyme.

F. An old but still utilized picture of enzyme-catalyzed reactions is the *lock-and-key model*. The substrate (the key) is seen as fitting nicely into a depression (the lock, the active site) on the enzyme. The reaction occurs while the key is in the lock and then is withdrawn. This view also explains the observation that many enzyme-catalyzed reactions are *specific*, many enzymes only act on particular substrates.

Discussion Question: Discuss enzymes from the viewpoint of non-biological reaction kinetics and catalysts covered previously.

X. Carbohydrates

A. *Carbohydrates* are naturally occurring substances in plants and animals that generally have the empirical formula $C_x(H_2O)_y$. Carbohydrates are polyhydroxyl aldehydes and ketones. Small carbohydrates are called *sugars*.
1. *Glucose*, $C_6H_{12}O_6$, is the most abundant carbohydrate. It is a polyhydroxyl aldehyde that exists as a linear molecule or, by reacting with itself, as a six-membered ring. The ring structure is more common in nature.
2. *Fructose* can cyclize to form a five-membered or a six-membered ring.

B. Glucose and fructose are *monosaccharides*; they cannot be broken down into simpler sugars by hydrolysis. When two monosaccharides bond together by a condensation reaction, a *disaccharide* results. *Sucrose*, or table sugar, and *lactose*, or milk sugar, are two common examples. All monosaccharides and disaccharides are sweet, but vary greatly in the degree of sweetness we perceive upon tasting.

C. *Polysaccharides* are comprised of several monosaccharides bonded together as they are in disaccharides. Polysaccharides have many functions in organisms; three of them are particularly common.
   1. *Starch* is a mixture of polysaccharides found in plants. Starches store energy in plants and are especially prevalent in corn, potatoes, wheat, and rice. Humans gain the energy value of the starches when we eat them.
   2. *Glycogen* serves the same function as starches but is produced by animals. It is concentrated in muscle and the liver. In muscle, it serves as an intermediate storage location for glucose and helps maintain a constant blood-glucose level.
   3. *Cellulose* is the principal structural material of plants. It is similar to starch, but the glucose units are bonded somewhat differently; starch can be digested by humans, while cellulose cannot.

Discussion Question: Why are high-starch diets prone to causing weight gain?

XI. Nucleic Acids

A. *Nucleic acids* carry the body's genetic information. *Deoxyribonucleic acids* (DNA) are huge molecules with molecular weights of 6 million to 16 million amu. DNA is found in the nucleus of the cell. It stores the cell's genetic information and controls the synthesis of proteins.

B. *Ribonucleic acids* (RNA) are smaller than DNA molecules ranging in molecular weight from 20,000 to 40,000 amu. RNA is found in the cytoplasm of the cell. It carries information stored by DNA into the cytoplasm where it is used in protein synthesis.

C. The monomers of nucleic acids are called *nucleotides*. A nucleotide consists of three parts.
   1. It contains a *phosphoric acid* molecule, H_3PO_4. Some hydrogens may be lost in order to form bonds to other groups.
   2. It contains a *five-carbon cyclic sugar*. In DNA, the sugar is deoxyribose; in RNA, it is ribose. Ribose has one more -OH group than does deoxyribose.
   3. It contains a *nitrogen-containing organic base*, of which there are only five used, adenine, guanine, cytosine, thymine, and uracil.

D. *Polynucleotides* consist of two or more nucleotides bonded together by condensation of the phosphoric acid of one nucleotide with the sugar of another. Nucleic acids are long-chain polynucleotides. In DNA, two polynucleotide chains are wrapped around each other in a *double helix*. The two *complementary strands* are held together by *hydrogen bonding* between bases across the center of the double helix. In this hydrogen bonding, thymine and adenine interact, while cytosine and guanine interact.

E. It is believed that, during replication, the strands of DNA unravel; each strand synthesizes a new complement, yielding two identical DNA molecules.

Discussion Question: Name several genetic disorders and ask how DNA or RNA might become flawed to cause such disorders.

XII. Strategies in Chemistry: What Now?

A. So far, so good. Congratulations for sticking it out!
B. The key to whether you take additional chemistry courses is two-fold:
   1. Do your career plans, even tentative ones, call for more chemistry?

2. Did this exposure to chemistry challenge you and did you enjoy learning it, even if it was difficult. (Note: The person who constructed this outline went to graduate school because understanding chemical kinetics became a grudge match. Coincidentally, his doctorate is in transition metal kinetics.)

C. Whatever your future holds, remember the lessons of investigation and the satisfaction you received from this first course in chemistry.

Lecture Outline – Chapter 26
Brown, LeMay, & Bursten, *Chemistry: The Central Science*, *6th Edition*

SAMPLE QUIZ QUESTIONS

1. Name the following compounds:

   a. $CH_2=CH-CH_2-CH_2-CH_3$

   b. $CH_3-CH-CH_3$
        |
        OH

   c. $CH_3-CH_2-CH-CH_3$
          |
          CH_3

2. Hydrocarbons: (a) of fewer than 6 carbon atoms are liquids at room temperature; (b) tend to be colored materials; (c) tend to be fairly polar molecules; (d) tend to dissolve easily in water; (e) of more than 18 carbon atoms are solids at room temperature.

3. The number of isomers of C_5H_{12} is: (a) 2; (b) 3; (c) 4; (d) 5.

4. Propane is: (a) C_2H_6; (b) C_3H_8; (c) C_3H_6; (d) C_4H_{10}.

5. Which of the following molecules could exhibit optical activity?

   (a) 3-methylheptane; (b) 2-heptene; (c) 2-methylheptane; (d) 3-methylpentane

6. Natural gas is composed primarily of: (a) CH_4; (b) C_2H_6; (c) C_6H_6; (d) C_8H_{18}.

7. The general formula for open-chain alkanes is: (a) C_nH_{2n+4}; (b) C_nH_{2n+2}; (c) C_nH_{2n}; (d) C_nH_{2n-2}; (e) C_nH_{2n-4}.

8. Ethane has: (a) a π-bond; (b) 120° bond angles; (c) sp^3-hybridized carbon atoms; (d) none of these.

9. Which of the following is an open-chain alkane?

   (a) C_5H_{10}; (b) C_7H_{12}; (c) C_9H_{22}; (d) C_3H_6; (e) C_6H_{14}

10. Name the alkane:

    CH_2-CH_3
    |
    $CH_3-CH-CH_2-CH-CH_3$
                    |
                    $CH_2-CH_2-CH_3$

    (a) 4-ethyl-2-propylpentane; (b) 4-methyl-2-propylhexane; (c) 3,5-dimethyloctane; (d) 2-ethyl-4-methylheptane; (e) 3,4-dimethyldecane.

11. Give the structural formula for 2-methyl-3-ethylpentane.

12. Draw the structural formulas for the two isomers of butane.

Lecture Outline – Chapter 26
Brown, LeMay, & Bursten, *Chemistry: The Central Science*, 6th Edition

13. Describe, using an example, a condensation reaction.

14. Complete the following equations:

   a. $CH_2=CH_2 + Br_2 \rightarrow$

   b. $CH_3OH + CH_3CH_2\overset{\underset{\parallel}{O}}{C}OH \rightarrow$

   c. $CH_3\underset{\underset{CH_3}{|}}{C}=CH_2 + H_2O \rightarrow$

   d. $2CH_3OH \xrightarrow{H_2SO_4}$

   e.  ⬡ $+ Br_2 \xrightarrow{FeBr_3}$

15. Explain why aromatic hydrocarbons do not readily undergo addition reactions as do alkenes and alkynes.

16. Indicate the type of functional group present in each of the following molecules:

   a. $CH_3\overset{\underset{\parallel}{O}}{C}OH$

   b. $CH_2=CHCH_3$

   c. $CH_3CH_2\overset{\underset{\parallel}{O}}{C}H$

   d. CH_3NH_2

   e. $CH_3\overset{\underset{\parallel}{O}}{C}CH_3$

   f. CH_3OH

   g. $CH_3\overset{\underset{\parallel}{O}}{C}-NH_2$

17. Aldehydes are easily oxidized to: (a) alcohols; (b) esters; (c) ketones; (d) acids.

18. An sp^2 hybrid orbital: (a) is one of three equivalent orbitals in a set; (b) results from mixing an *s* atomic orbital and a *p* atomic orbital; (c) gives bond angles of 109.5°; (d) is one of four equivalent orbitals in a set; (e) results from mixing an *s* atomic orbital and three *p* atomic orbitals.

19. A triple bond: (a) gives bond angles of 120°; (b) consists of two σ bonds and one π bond; (c) contains four electrons; (d) is characteristic of alkenes; (e) consists of one σ bond and two π bonds.

20. Name the alkene:

    CH₃—CH₂
         |
         C=CH₂
         |
         CH₂—CH₂—CH₂—CH₂—CH₃

    (a) 2-ethyl-1-heptene; (b) 3-methyl-4-pentylpropane; (c) 3-methyl-3-octene; (d) 2-pentyl-1-butene; (e) 6-ethyl-6-heptene.

21. Name the alkyne:

    CH₃—CH—CH₂—CH—CH—C≡C—CH₂—CH₃
        | | |
        CH₃   CH₃ CH₂–CH₃

    (a) 5-ethyl-6,8-dimethyl-3-nonane; (b) 5-ethyl-2,4-dimethyl-6-nonyne; (c) 5-hexyl-3-heptyne; (d) 1,3-diethyl-4,6-dimethyl-1-heptyne; (e) 5-ethyl-6,8-dimethyl-3-nonyne.

22. Write a balanced chemical equation for each of the following: (a) formation of methyl ethyl ketone from an alcohol; (b) formation of an olefin from 2-butanol; (c) formation of isopropyl benzene from benzene; (d) formation of a polymer via a condensation reaction of 1,4-dihydroxybenzene.

23. The functional group $R-\overset{\overset{O}{\|}}{C}-H$ is characteristic of: (a) alcohols; (b) ketones; (c) ethers; (d) esters; (e) aldehydes.

24. What do we mean by the terms primary, secondary, and tertiary structure of a protein?

25. Write the structural formula for the dipeptide gly-ala.

26. Write the general formula for α-amino acids and describe the formation of a peptide bond.

27. Describe the lock-and-key model for enzyme activity.

28. What products are formed by the hydrolysis of each of the following?

    (a) proteins; (b) fats; (c) carbohydrates

29. What structural difference exists between starch and cellulose that makes one but not the other digestible by humans?

30. What feature of vegetable oils causes them to have lower melting points than animal fats of similar molecular weights? How are vegetable fats "hardened?"

31. Which of the following is not a carbohydrate: (a) glucose; (b) cellulose; (c) glycine; (d) starch?

32. Hydrolysis of fats yields: (a) proteins and fatty acids; (b) glycerol and fatty acids; (c) glycol and fatty acids; (d) sugars and fatty acids.

33. The only naturally occurring a-amino acid that does not exhibit optical activity is:

    (a) glycine; (b) lysine; (c) alanine; (d) valine.

34. Which of the following molecules will show optical isomerism?

    (a) HO–C(H)(Cl)–Cl;

    (b) HO–C(H)(Cl)–Br;

    (c) HO–C(H)(CH$_3$)–H;

    (d) H–C(H)(CH$_3$)–Cl

35. Which of the following does not describe glucose?

    (a) hexose; (b) pyranose; (c) sugar; (d) disaccharide

36. Which of the following bases are paired in DNA: (a) guanine and thymine; (b) guanine and adenine; (c) cytosine and guanine; (d) ribose and guanine?

37. Which of the following types of bonds connects the monomeric units in a nucleic acid?

    (a) C–N; (b) C–O–C; (c) P–O–C; (d) hydrogen bonds

38. Describe the role of chlorophyll in photosynthesis.

39. Compare starch, glycogen, and cellulose.

40. Describe hydrogenation of vegetable oils.

CHEMISTRY CONCEPTS AND TECHNIQUES

Skirius and Bergman, Florida State

This videotape, designed to accompany Brown, LeMay, Bursten's CHEMISTRY: The Central Science 6/e, is an invaluable addition to the supplementary materials accompanying this best-selling text. Approximately 100 minutes long, it treats reactions that are fundamental for student understanding of the basics of chemistry and is designed to convey the excitement and wonder of chemistry.

The ten segments composing the tape are organized along major topical divisions. Each segment takes one such topic, for example thermochemistry, and illustrates it with several different lab set-ups. In the example of thermochemistry, both exothermic and endothermic reactions, each of these is illustrated by different experiments.

The video tape is designed so that the footage can effectively demonstrate the exact process of a given experiment. In some instances, additional tools and techniques, such as the scanning electron microscope or microscopy, are used to provide a viewpoint rarely available to the student.

1. DISPLACEMENT REACTIONS
    A. Alkali metals in water
    B. Alkaline earths in water
    C. Copper metal in an aqueous solution of silver nitrate
        (Single displacement reaction)
    D. Aqueous solutions of KI and Pb $(NO_3)_3$
        (Double displacement reaction)

2. COMBUSTION REACTIONS
    A. Combustion of a hydrogen balloon
    B. Combustion of an oxygen balloon
    C. Combustion of a hydrogen/oxygen balloon
    D. Non combustion of a nitrogen (or a noble gas) balloon
    E. Combustion of Lycopodium powder in a milk can
    F. Miner lamps (CaC_2 and water)

3. ELEMENTS, COMPOUNDS AND MIXTURES
    A. Demonstration of elements in their natural state
    B. Combinations of elements to produce compounds and mixtures
    C. Thin layer chromatography to separate compounds
    D. Electrolysis of water

4. PHYSICAL PROPERTIES
    A. Demonstrations of density
        1. Wood on water (mention only)
        2. Water on lead (mention only)
        3. Lead on mercury
        4. Varying densities of sugar solutions

             5. Feon filled balloons
             6. Suspension of particles in a density gradient
        B. Changes of state of compounds and elements

   5. THERMOCHEMISTRY
        A. Exothermic reactions:
             1. Combustion of magnesium strip
             2. Dissolutions of salts
             3. Recrystalization of saturated salt solutions
             4. Balloon detonations
        B. Endothermic reactions:
             1. Dissolutions of salts
             2. Mixing of alcohols in water

   6. ELECTROCHEMISTRY
        A. Battery
        B. Corrosion
        C. Electroplating
        D. Increased current with increased ion concentrations

   7. ACID - BASE CHEMISTRY
        A. Definition of an acid and a base
        B. The pH value of common household items
        C. Indicators

   8. GAS LAWS
        A. Ideal gas law: PV - nRT
        B. Correlation of vapor pressure and temperature
        C. Crushing a soda can using vapor presssure
        D. Solubility of gasses with changing pressure

   9. INTERACTION OF LIGHT WITH MATTER
        A. Absorption of light
        B. Transmission of light
        C. The electromagnetic spectrum
        D. Refraction, reflection and dispersion
        E. Refractive index
        F. Polarimetry

   10. OXIDATION STATE OF MATTER
        A. Vanadium tower
        B. Flash powder(s)
        C. Combustion of organic compound with permanganate

2. Ira Remsen's Investigation of Nitric Acid

In this classic demonstration, the experience Ira Remsen had with nitric acid can be reenacted. Ira Remsen (1846-1927) was an influential chemist in America. He founded the chemistry department at Johns Hopkins University and initiated the first center for chemical research in this country. In this demonstration, he describes his experience with nitric acid.

Procedure

It is suggested that you read his account as you perform the demonstration.

> While reading a textbook on chemistry, I came upon the statement "nitric acid acts upon copper". I was getting tired of reading such absurd stuff and I determined to see what this meant. Copper was more or less familiar to me, for copper cents were then in use. I had seen a bottle marked "nitric acid" on a table in the doctor's office where I was then "doing time"! I did not know its peculiarities but I was getting on and likely to learn. The spirit of adventure was upon me. Having nitric acid and copper, I had only to learn what the words "act upon" meant. Then, the statement, "nitric acid acts upon copper" would be something more than mere words.
>
> All was still. In the interest of knowledge I was even willing to sacrifice one of the few copper cents then in my possession. I put one of them on the table; opened the bottle marked "nitric acid"; poured some of the liquid on the copper; and prepared to make an observation.

1. Place the penny in a 500-mL flask.
2. Carefully add 5 mL of concentrated nitric acid.
3. Place several damp, folded, paper towels over the top of the flask to catch the gas evolved.

> But what was this wonderful thing which I beheld? The cent was already changed, and it was no small change either. A greenish blue liquid foamed and fumed over the cent and over the table. The air in the neighborhood of the performance became dark red. A great colored cloud arose. This was disagreeable and suffocating—how should I stop this? I tried to get rid of the objectionable mess by picking it up and throwing is out of the window, which I had meanwhile opened. I learned another fact—nitric acid not only acts upon copper but it acts upon fingers. The pain led to another unpremediated experiment. I drew my fingers across my trousers and another fact was discovered. Nitric acid also acts upon trousers.

4. Display to the class a large piece of cloth on which a dropper full of nitric acid had previously been squirted.
5. Add water to the nitric acid to stop the reaction.
6. Rinse the beaker and remove the penny (now much smaller!).

> Taking everything into consideration, that was the most impressive experiment, and, relatively, probably the most costly experiment I have ever performed. I tell of it even now with interest. It was a revelation to me. It resulted in a desire on my part to learn more about that remarkable kind of action. Plainly the only way to learn about it was to see its results, to experiment, to work in the laboratory.

Reaction

Nitric acid reacts with copper to produce the brown gas nitrogen dioxide.

$$Cu(s) + 4H^+(aq) + 2NO_3^-(aq) \longrightarrow Cu^{2+}(aq) + 2NO_2(g) + 2H_2O(\ell)$$

Teaching Tips

NOTES

1. This wonderful excerpt is taken from Getman, Frederick H. *J. Chem. Educ.* **1940**, 9–10.
2. This demonstration is an excellent way to begin a course in chemistry. Have students find out more about Ira Remsen.
3. This demonstration must be done in a well-ventilated area or in the hood. NO_2 is a toxic gas.
4. Considering when Ira Remsen did this "experiment", have students comment on his investigation.

QUESTIONS FOR STUDENTS

1. Find out more about Ira Remsen.
2. What do you observe when copper is placed in nitric acid?
3. How does a scientist find out new things?

3. Burning Water

A "new" 500-mL Erlenmeyer flask is taken from a box and filled with water from a faucet. The top of the water is ignited, and a flame is produced. The "water" continues to burn for several minutes.

Procedure

1. Prior to performing this demonstration, place a squirt of cigarette lighter fluid in the flask, swirl the flask to distribute the fluid evenly so that the flask appears to be empty, and replace the flask in the original carton.
2. When performing the demonstration, produce the flask from its "original" container as if it were a new flask.
3. Fill the flask to the top with water from a faucet.
4. For a more dramatic effect, add a pinch of salt from a new box or sodium bicarbonate from a new box (let students see you open the new container).
5. Light the top of the flask. The lighter fluid will have floated to the top of the flask, unobserved by the students.
6. Ask for observations (flame, smoky product, kerosene-like odor, etc.).

Reaction

1. The real reaction is that of the students as they try to figure out what is going on. With a tongue-in-cheek presentation, the teacher can produce all sorts of reactions. Some students will believe that the water, salt, or baking soda is burning. Others will rely on their own observations and be skeptical; they know that another ingredient must be involved.
2. The chemical reaction is the combustion of cigarette lighter fluid, a mixture of light hydrocarbons that are "lighter" (less dense) than water. Lighter fluid is naphtha, a mixture of hexanes.

$$C_6H_{14}(\ell) + 9\tfrac{1}{2}O_2(g) \longrightarrow 6CO_2(g) + 7H_2O(g)$$

Material

Fluid for cigarette lighters. Do not use charcoal lighter fluid.

Teaching Tips

NOTES

1. If too much fluid is used, students will be able to see an "oily" layer, and the effect is lost.
2. Practice to get just the right amount. With larger flasks, more time is required for the fluid to reach the top.
3. If students assume that the "burning" is a result of some reaction with the salt, or baking soda, write their proposed reaction on the board.
4. This demonstration is FUN! Develop your own story line to go along with this demonstration.
5. Naphtha is distilled from petroleum at 70-90 °C.

QUESTIONS FOR STUDENTS

1. This really special water appears to be burning. Write down all your observations.
2. Using your observations, suggest an explanation for "burning" water.
3. Suggest a way that, using only H_2O, a burning reaction can take place. (Electrolyze the water to hydrogen and oxygen and then react these two substances to reproduce water!)

8.21

Conductivity and Extent of Dissociation of Acids in Aqueous Solution

The addition of a few drops of universal indicator to 0.1M solutions of a variety of strong and weak acids produces solutions of different colors, allowing the acids to be arranged in order of acidity. When the same solutions are tested with a conductivity probe, the weaker acids produce a dim glow of the light bulb, whereas the strong acids cause the bulb to glow brightly (Procedure A). A comparison of the conductivities of various concentrations of hydrochloric acid with that of 2M acetic acid allows an estimation of the value of the ionization constant of acetic acid (Procedure B).

MATERIALS FOR PROCEDURE A

For preparation of indicator solutions, see pages 27–29.
For preparation of stock solutions of acids and bases, see pages 30–32.

100 mL 0.1M hydrochloric acid, HCl

100 mL 0.1M acetic acid, $HC_2H_3O_2$

100 mL 0.1M citric acid, $H_3C_6H_5O_7$ (To prepare 1 liter of solution, dissolve 21 g of $H_3C_6H_5O_7 \cdot H_2O$ in 600 mL of distilled water and dilute the resulting solution to 1.0 liter.)

100 mL 0.1M malonic acid, $H_2C_3H_2O_4$ (To prepare 1 liter of solution, dissolve 10.4 g of $H_2C_3H_2O_4$ in 600 mL of distilled water and dilute the resulting solution to 1.0 liter.)

100 mL 0.1M ascorbic acid, $HC_6H_7O_6$ (To prepare 1 liter of solution, dissolve 17 g of $HC_6H_7O_6$ in 600 mL of distilled water and dilute the resulting solution to 1.0 liter.)

100 mL 0.1M propanoic acid, $HC_3H_5O_2$ (To prepare 1 liter of solution, dissolve 7.4 g of $HC_3H_5O_2$ in 600 mL of distilled water and dilute the resulting solution to 1.0 liter.)

100 mL 0.1M glycine, $HC_2H_4O_2N$ (To prepare 1 liter of solution, dissolve 7.5 g of $HC_2H_4O_2N$ in 600 mL of distilled water and dilute the resulting solution to 1.0 liter.)

100 mL 0.1M alanine, $HC_3H_6O_2N$ (To prepare 1 liter of solution, dissolve 8.9 g of $HC_3H_6O_2N$ in 600 mL of distilled water and dilute the resulting solution to 1.0 liter.)

2 mL universal indicator solution, 1–10 pH range
8 250-mL beakers or 300-mL tall-form beakers
8 labels for beakers
white background (e.g., 20-cm × 1-m poster board)
dropper
8 stirring rods
light-bulb conductivity tester (optional)
250-mL wash bottle, filled with distilled water (optional)
400-mL beaker (optional)

MATERIALS FOR PROCEDURE B

For preparation of stock solutions of acids and bases, see pages 30–32.

80 mL 2M hydrochloric acid, HCl
40 mL 2M acetic acid, $HC_2H_3O_2$
140 mL distilled water
special conductivity tester (See Procedure B for assembly instructions.)
7 100-mL beakers
50-mL graduated cylinder
10-mL pipette
4 stirring rods
3 4-mL pipettes

PROCEDURE A

Preparation

Label the eight 250-mL beakers with one each of these labels: "hydrochloric acid," "acetic acid," "citric acid," "malonic acid," "ascorbic acid," "propanoic acid," "glycine," and "alanine." Pour 100 mL of the appropriate 0.1M acid solution into each beaker. Arrange the beakers in a row before the white background.

Presentation

Add 4 drops of universal indicator to each beaker and stir each mixture. Arrange the beakers by color in spectral order from red to yellow (see the table on the following page).

Solution	Color
hydrochloric acid	bright red
malonic acid	orange-red
citric acid	orange-red
ascorbic acid	orange
acetic acid	orange
propanoic acid	orange
glycine	yellow
alanine	yellow

Turn on the conductivity tester and dip its electrodes to the bottom of the alanine solution. The bulb will glow very dimly. Rinse the electrodes with distilled water from the wash bottle, collecting the rinse in the 400-mL beaker. Dip the electrodes to the bottom of the beaker of glycine solution. Compare the brightness of the bulb with its intensity in the alanine solution. Proceed in this fashion, testing each solution in the order in which they are arranged by color. The brightness of the bulb will increase sequentially from alanine to hydrochloric acid.

PROCEDURE B

Preparation

Construct a special conductivity tester as illustrated in Figure 1. This tester allows a single light bulb to be connected to either of two sets of identical electrodes. Cut the wooden form out of a 30-cm piece of 1 × 6 lumber. Attach a light-bulb socket to the center of the wooden plank. Attach two momentary, normally open, push-button switches to the board near the handle. Fashion four 2-cm × 10-cm electrodes from 22-gauge copper metal sheet. Screw the electrodes to the wood. Use 18-gauge lamp cord to make the electrical connections indicated in the schematic (see Figure 2). Connect one lead from the 110-volt plug to one terminal of the light-bulb socket and the other lead from the plug to one pole of one of the button switches. Connect this pole to one pole of the other button switch. Connect the free pole of button 1 to one of the electrodes in set 1.

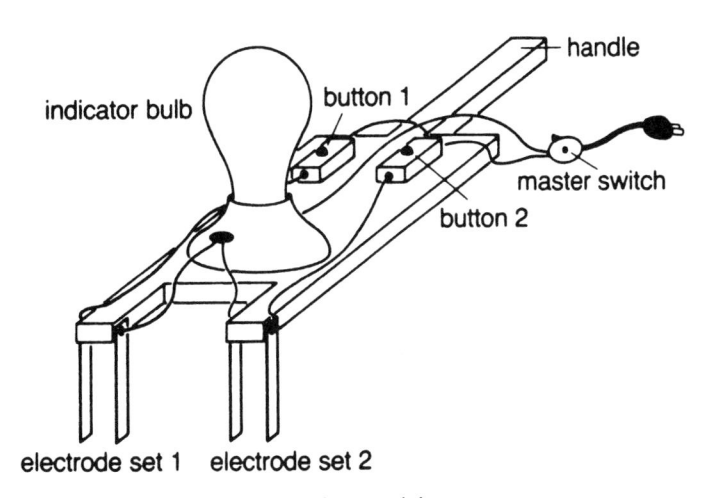

Figure 1. Conductivity tester.

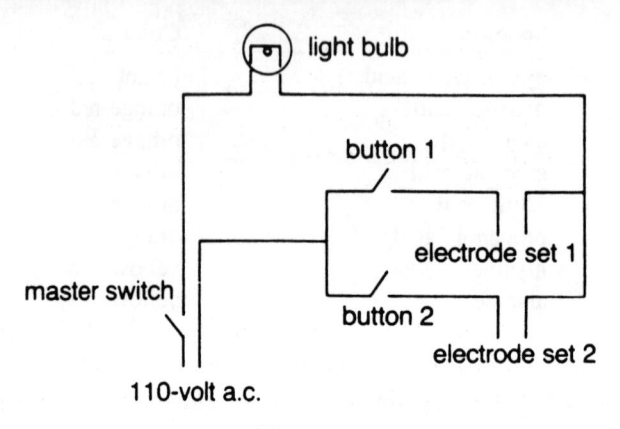

Figure 2. Wiring diagram for conductivity tester.

Likewise, attach the free pole of button 2 to one of the electrodes in set 2. Connect the two remaining electrodes to the free lamp terminal. Be certain that there are no bare electrical connections near the handle end of the tester. If necessary, cover bare connections with electrical tape.

Insert a master line switch in the cord going to the 110-volt plug. Mount a 150- or 100-watt light bulb in the socket.

Presentation

Pour 40 mL of 2M HCl into each of two 100-mL beakers. With the master switch off, insert one set of electrodes on the special conductivity tester in each beaker. Make sure both sets are immersed to the same depth. Turn on the master switch. Press button 1 and the bulb will light. Press button 2 and the bulb will glow just as brightly. Turn off the master switch.

Pour 40 mL of 2M acetic acid into another 100-mL beaker. Rinse set 2 of the electrodes with distilled water and immerse this set in the beaker of acetic acid. Turn on the master switch and alternately press buttons 1 and 2. When button 1 is pressed, the electrodes in the HCl beaker are connected, and the bulb glows brightly. However, when button 2 is pressed and the electrodes in the beaker of acetic acid are in the circuit, the glow from the bulb is much weaker. Turn off the master switch.

With a pipette, remove 10 mL of the 2M HCl solution from the beaker and transfer it to another 100-mL beaker. Add 30 mL of distilled water and stir the mixture. Rinse the electrodes that were in the 2M HCl with distilled water and insert them in the beaker of diluted (0.5M) HCl. Turn on the master switch and alternately press buttons 1 and 2 to compare the brightness of the bulb produced by the 0.5M HCl with the brightness produced by the 2M acetic acid. The 0.5M HCl produces a brighter glow than the 2M $HC_2H_3O_2$.

With a pipette, remove 4 mL of 0.5M HCl from the beaker, transfer it to another 100-mL beaker, add 36 mL of distilled water, and stir the mixture. Compare the brightness of the bulb when this solution of 0.05M HCl is used with that produced by the 2M $HC_2H_3O_2$. Still, the glow with HCl is brighter.

Transfer 4 mL of 0.05M HCl to another 100-mL beaker, add 36 mL of distilled water, and stir the mixture. Test this solution, 0.005M HCl, and the 2M $HC_2H_3O_2$ with the special conductivity tester. Now the glow of the bulb will be nearly the same with both solutions.

In another 100-mL beaker, dilute 4 mL of 0.005M HCl with 36 mL of distilled water to produce 0.0005M HCl. Compare the brightness of the bulb's glow produced by this solution with that produced by the 2M $HC_2H_3O_2$. The glow from the 2M $HC_2H_3O_2$ will be brighter than that of the 0.0005M HCl.

HAZARDS

Hydrochloric acid is harmful to the eyes and skin and irritating to mucous membranes. Its vapors are irritating to the eyes and respiratory system.

Glacial acetic acid can irritate the skin, and its vapors are irritating to the eyes and respiratory system.

Malonic acid is a strong irritant to the skin, eyes, and mucous membranes.

Severe electrical shock can result if the electrodes of the conductivity tester are touched while it is turned on and one of the buttons is pressed.

DISPOSAL

The waste solutions should be flushed down the drain with water.

DISCUSSION

The acidity of a solution depends on the concentration of the acid in the solution, and it is a measure of the extent of ionization of that acid. In Procedure A the acidities of solutions of various acids are compared. All acids are at the same concentration, namely 0.1M, but they have various values of pH, which is revealed by differences in the color of a universal indicator. Each of the solutions is also investigated with a light-bulb conductivity tester. This reveals that the more acidic solutions are more conductive. In Procedure B the conductivity of an acetic acid solution is compared with the conductivity of various concentrations of hydrochloric acid. By assuming that solutions which light the bulb to the same brightness contain the same concentration of ions, an estimate of the dissocation constant, K_a, for acetic acid can be made.

The universal indicator described in Procedure A exhibits the following colors in the acid pH range:

pH 1: deep red
pH 2: red
pH 3: orange-red
pH 4: red-orange
pH 5: orange
pH 6: yellow

Grouping the solutions by color arranges them by pH. The acids fall into one of four groups. The most acidic group contains only hydrochloric acid, whose solution is deep red. The next most acidic group, with an indicator color of orange-red, includes malonic and citric acids. Next comes the orange group: ascorbic, acetic, and propanoic acids. Least acidic are the acids that give a yellow color: glycine and alanine. When the acid solutions are tested with the conductivity tester, they fall into the same groups. The most acidic solution lights the bulb brightest, and the least acidic produces the faintest glow. Because conductivity depends on the concentration of ions in the solution, and each of the solutions has the same nominal concentration, 0.1M, each acid must dissociate or ionize to a different extent. Those that ionize the most are the most acidic; those that ionize the least are the least acidic. The results of the demonstration can be compared with the dissociation constants of the acids in the following table:

pK_a and pH Values for Several Aqueous Weak Acids

Acid	pK_a [1]	pH of 0.1M solution
Malonic	2.83	1.94
Citric	3.14	2.09
Ascorbic	4.10	2.56
Acetic	4.75	2.88
Propanoic	4.87	2.94
Glycine	9.78	5.39
Alanine	9.87	5.44

Solutions of acids with similar pK_a values have similar pH values and conductivities.

The conductivity of a weak acid solution can be used to estimate the value of its K_a. In Procedure B the conductivity of a 2M acetic acid solution is compared with that of solutions of hydrochloric acid with a range of concentrations. The comparison is made with a light-bulb conductivity tester. The 2M acetic acid lights the bulb to the same brightness as 0.005M hydrochloric acid and is assumed to contain the same concentration of ions. Because hydrochloric acid is a strong acid, it is completely ionized in solution, and the concentration of hydrogen and chloride ions in 0.005M HCl is 0.005M. Therefore, the concentration of hydrogen and acetate ions in 2M acetic acid is 0.005M, and the concentration of undissociated acetic acid molecules is 2M − 0.005M = 1.995M. Using these values, the value of K_a for acetic acid can be calculated:

$$K_a = \frac{[H^+][C_2H_3O_2^-]}{[HC_2H_3O_2]} = \frac{(0.005)(0.005)}{(1.995)} = 1.3 \times 10^{-5}$$

The calculated value for K_a, 1.3×10^{-5}, is within 30% of the literature value of 1.8×10^{-5} at 25°C [1]. Considering the somewhat subjective nature of assessing the brightness of the bulb, this result is certainly in reasonable agreement with the accepted value.

In this demonstration the conductivity of the solution is monitored with a light-bulb conductivity apparatus. This tester uses a light bulb whose power cord is interrupted by a pair of electrodes that are immersed in the solution being tested. In order for the bulb to light, current must flow between the electrodes. A 15-watt bulb requires about 0.13 amps of current to be fully lit, whereas a 150-watt bulb needs 1.3 amps. The more current needed for full brightness, the more ions must be present to carry this current. Therefore, a 150-watt bulb begins to diminish in brightness at a higher ion concentration than does a 15-watt bulb. Thus, a high-wattage bulb will show the decrease in concentration of ions in the solution over a wider range than a low-wattage bulb.

REFERENCE

1. R. C. Weast, Ed., *CRC Handbook of Chemistry and Physics*, 66th ed., CRC Press: Boca Raton, Florida (1985).

65. Name That Precipitate

Twelve large test tubes are arranged on the demonstration desk. As the solutions are mixed, two at a time, various precipitates are formed. These brown, red, white, blue, and yellow precipitates vividly show various ionic reactions.

Procedure

1. Arrange 12 test tubes in a test-tube holder.
2. Label the tubes as follows:

   1, HgCl$_2$
   2, Na$_2$CO$_3$
   3, NiCl$_2$
   4, Na$_2$CO$_3$
   5, Pb(NO$_3$)$_2$
   6, KI

   7, HgCl$_2$
   8, KI
   9, BaCl$_2$
   10, Na$_2$SO$_4$
   11, CuCl$_2$
   12, Na$_2$CO$_3$

3. Fill each test tube one-third full of its solution.
4. Pour the contents of test tube 1 into test tube 2. What do you observe?
5. Ask students to write an *ionic* equation for this reaction.
6. Ask students to name the precipitate.
7. Mix the test tubes as follows. In each case, have students describe the reaction and predict the formula for the precipitate.

   a. Test tube 3 into test tube 4.
   b. Test tube 5 into test tube 6.
   c. Test tube 7 into test tube 8.
   d. Test tube 9 into test tube 10.
   e. Test tube 11 into test tube 12.

Reactions

1. Mercury(II) chloride and sodium carbonate produce the BROWN precipitate, mercury carbonate.

   $$Hg^{2+}(aq) + 2Cl^-(aq) + 2Na^+(aq) + CO_3^-(aq) \longrightarrow HgCO_3(s) + 2Na^+(aq) + 2Cl^-(aq)$$

   (Ions, such as Cl and Na, that do not enter a reaction are often called *spectator* ions.)

2. Nickel chloride and sodium carbonate produce the pale BLUE precipitate, nickel carbonate.

   $$Ni^{2+}(aq) + 2Cl^-(aq) + 2Na^+(aq) + CO_3^-(aq) \longrightarrow NiCO_3(s) + 2Cl^-(aq) + 2Na^+(aq)$$

3. Lead nitrate and potassium iodide produce the YELLOW precipitate, lead iodide.

   $$Pb^{2+}(aq) + 2NO_3^-(aq) + 2K^+(aq) + 2I^-(aq) \longrightarrow PbI_2(s) + 2NO_3^-(aq) + 2K^+(aq)$$

4. Mercury(II) chloride and potassium iodide produce the ORANGE-RED precipitate, mercury iodide.

$$Hg^{2+}(aq) + 2Cl^-(aq) + 2K^+(aq) + 2I^-(aq) \longrightarrow HgI_2(s) + 2Cl^-(aq) + 2K^+(aq)$$

5. Barium chloride and sodium sulfate produce the WHITE precipitate, barium sulfate.

$$Ba^{2+}(aq) + 2Cl^-(aq) + 2Na^+(aq) + SO_4^{2-}(aq) \longrightarrow BaSO_4(s) + 2Cl^-(aq) + 2Na^+(aq)$$

6. Copper chloride and sodium carbonate produce the pale BLUE precipitate, copper carbonate.

$$Cu^{2+}(aq) + 2Cl^-(aq) + 2Na^+(aq) + CO_3^-(aq) \longrightarrow CuCO_3(s) + 2Cl^-(aq) + 2Na^+(aq)$$

Solutions

Make dilute solutions of each of the following by adding about 1 tsp of each to 500 mL of water.

1. Mercury(II) chloride, $HgCl_2$.
2. Sodium carbonate, Na_2CO_3.
3. Nickel chloride, $NiCl_2$.
4. Sodium carbonate, Na_2CO_3.
5. Lead nitrate, $Pb(NO_3)_2$.
6. Potassium iodide, KI.
7. Barium chloride, $BaCl_2$.
8. Sodium sulfate, Na_2SO_4.
9. Copper(II) chloride, $CuCl_2$.

Notice that some of the solutions [potassium iodide, sodium carbonate, and mercury(II) chloride] are used in more than one combination of solutions.

Teaching Tips

NOTES

1. Precipitates with characteristic colors and appearances result when certain ionic solutions are mixed. Recombination of certain positive and negative ions often results in the formation of a compound with low solubility. That compound will precipitate as a solid.
2. A precipitate will form in each case. The students will then "deduce" the formula of the precipitate. You should have solutions of NaCl, KCl, and KNO_3 available for them to see, as evidence that these compounds are soluble in water and would not appear as a precipitate.
3. You might try putting these solutions in dropper bottles and letting students mix drops of each on a glass plate.
4. Draw a large chart on the blackboard and have students enter data (colors, name, and formula of precipitate) for each reaction.
5. Positive ions are *cations*, and negative ions are *anions*. Notice that ionic reactions show only ions in solution.
6. See Appendix 4 for disposal of mercury, lead, and barium salts.
7. You might prefer, as the authors did, to use complete equations here, rather than net ionic equations.

QUESTIONS FOR STUDENTS

1. What is an *ionic* solution?
2. What causes a precipitate to form when ionic solutions are mixed?
3. What does this property have to do with solubility?
4. If copper chloride, $CuCl_2$, and sodium hydroxide are mixed, a blue precipitate is formed. What is this precipitate? Show the ionic reaction. [$Cu^{2+} + 2Cl^- + 2Na^+ + 2OH^- \longrightarrow Cu(OH)_2(s) + 2Cl^- + 2Na^+$. The precipitate must be $Cu(OH)_2$, because the only other possible new combination of ions would produce NaCl, table salt, which is soluble in water.]

20. A Hand-Held Reaction: Production of Ammonia Gas

Small amounts of two white solids are placed in the palm of the hand. When the solids are rubbed together, ammonia gas is produced and can be easily detected.

Procedure

1. If your hand is clean and dry, place an amount of ammonium chloride that would cover a dime in the palm.
2. Add an equal amount of calcium hydroxide on top of the ammonium chloride.
3. Rub the two solids together briskly and check for the odor of ammonia gas.

Reactions

1. The small amount of friction between the two solids is sufficient to begin the decomposition of ammonium chloride.

$$NH_4Cl(s) \longrightarrow NH_3(g) + HCl(g)$$

2. The calcium hydroxide reacts with the hydrogen chloride gas and prevents it from being detected.

$$Ca(OH)_2(s) + 2HCl(g) \longrightarrow CaCl_2(s) + 2H_2O(\ell)$$

Teaching Tips

NOTES

1. You should, of course, wash your hands thoroughly after this demonstration!
2. This demonstration is a good example of a reaction between two solids. Another good example is to mix solid potassium iodide and lead nitrate to form yellow lead iodide.
3. If you would like to confirm the presence of HCl (acid) and ammonia (base), *see* demonstration 21.

QUESTIONS FOR STUDENTS

1. Write the equation for the reaction to produce ammonia.
2. What does the calcium hydroxide do in this reaction?
3. Classify these reactions as exothermic or endothermic.

30. Exothermic Reaction: Sodium Sulfite and Bleach

Two solutions are mixed, and a significant increase in temperature results.

Procedure

1. Place 50 mL of laundry bleach in a 250-mL beaker. Record the temperature.
2. Add 50 mL of sodium sulfite solution.
3. Note the increase in temperature.

Reaction

This reaction is highly exothermic; the temperature should increase ~ 20 °C.

$$SO_3^{2-}(aq) + 2OCl^-(aq) \rightarrow SO_4^{2-}(aq) + 2Cl^-(aq) + heat$$

Solutions

1. The sodium sulfite solution is 0.5 M: 6.3 g of Na_2SO_3 per 100 mL of solution.
2. Laundry bleach is a 5.25% solution of sodium hypochlorite, NaOCl.

Teaching Tips

NOTES

1. Use a styrofoam cup for more accurate temperature change measurement.
2. Laundry bleach is a 5.25% solution of sodium hypochlorite, NaOCl.

QUESTIONS FOR STUDENTS

1. Write the chemical equation for this reaction.
2. Would the initial temperature of the solutions affect the final change in temperature? Try it!
3. Would changing the concentration of one of the solutions affect the final change in temperature? Try it!
4. Is this a *redox* reaction? If so, what was reduced and what was oxidized?
5. Could this demonstration be used to determine the *purity* of laundry bleach?

56. Flaming Cotton

A small amount of a substance is sprinkled over several balls of cotton. When 1 drop of water is added, the cotton balls burst into flame.

Procedure

1. Select about six to eight small cotton balls and spread them apart to increase their surface area.
2. Place the cotton on a fireproof or metal surface.
3. Place about 1 tsp of FRESH sodium peroxide on the cotton.
4. CAREFULLY add 1 drop of water from a dropper to the sodium peroxide.
5. Observe the exothermic reaction and production of flame that consumes the cotton.

Reactions

1. Sodium peroxide is a very good source of oxygen gas. When water is added, sodium peroxide reacts to give hydrogen peroxide.

$$Na_2O_2(s) + 2H_2O(\ell) \longrightarrow 2Na^+(aq) + 2OH^-(aq) + H_2O_2(aq)$$

2. The hydroxide ion, OH^-, acts as a catalyst to immediately decompose the hydrogen peroxide to form oxygen gas and water.

$$2H_2O_2(aq) \xrightarrow{OH^-} 2H_2O(\ell) + O_2(g)$$

3. The heat produced by the reaction, the high oxygen content, and the low ignition temperature of cotton produce the flame.

Teaching Tips

NOTES

1. The sodium peroxide must be fresh. Store the material in a tightly sealed jar.
2. Do not add too much water—only 1 or 2 drops is needed.
3. Most of the sodium peroxide produced in this country is used as a bleaching agent in the textile industry.
4. Sodium peroxide is formed when sodium metal burns in air.

QUESTIONS FOR STUDENTS

1. What is a *catalyst*?
2. What is the catalyst in this reaction? How is it produced?
3. Why does sodium peroxide often lose its "strength" when stored for a long period? (It reacts with water to form sodium hydroxide.)
4. Why does the cotton burn?

26. Endothermic Reaction: Ammonium Nitrate

A small amount of ammonium nitrate is added to 100 mL of water. As the solid dissolves, the temperature drops significantly.

Procedure

1. Place ~ 100 mL of water in a large beaker and record the temperature.
2. Quickly dump 10–15 g of ammonium nitrate into the water.
3. Note the change in temperature as the solid dissolves.

Reaction

The dissolving of ammonium nitrate in water is *endothermic*. The temperature should decrease 6 to 9 °C:

$$\text{heat} + NH_4NO_3(s) + H_2O(l) \rightarrow NH_4^+(aq) + NO_3^-(aq)$$

Teaching Tips

NOTES

1. Theoretically, 600–900 calories are absorbed per 100 mL water in this reaction.
2. You will get a more accurate reading by using styrofoam cups. The advantage of using a beaker is that it can be passed around the class for the students to observe.
3. Should you wish to calculate the heat change, 100 mL is a convenient volume.

QUESTIONS FOR STUDENTS

1. Can you think of any practical use for such a reaction (e.g., emergency cold packs)?
2. Why does the temperature drop?
3. Would the temperature drop more if twice as much solid is added? Try it!

33. Halogens Compete for Electrons

Water solutions change from colorless to yellow and brown and then to more surprising colors in another solvent when various combinations of halogen solutions are mixed.

Procedure

1. Prepare nine large test tubes, three each of 10 mL of solutions of KCl, KBr, and KI.
2. To the three test tubes of KCl add 10 mL of chlorine water to the first, bromine water to the second, and iodine water to the third. Set the test tubes aside.
3. Repeat procedure 2 with each test tube of KBr solution.
4. Repeat with each test tube of KI solution.
5. Shake each test tube and make observations.
6. Test the result of adding a little trichlorotrifluoroethane (TTE) to each of the stock halogens and halogen salt solutions. Stopper and shake the test tubes.
7. Note the appearance of molecular chlorine, bromine, and iodine in solution with TTE.
8. Add 10 mL of TTE to each of the nine test solutions from procedures 2, 3, and 4.
9. Account for the results you observe.

Reactions

1. The tendency for each halogen to gain and retain electrons is evident. Chlorine has the greatest tendency, then bromine, and then iodine. For example

$$Cl_2(aq) + 2e^- \longrightarrow 2Cl^-(aq)$$

2. Iodine has the greatest tendency to give up electrons (become oxidized).

$$2I^-(aq) \longrightarrow I_2(aq) + 2e^-$$

Solutions

1. The salt solution concentrations are 0.2 M: KCl, 7.5 g/500 mL; KBr, 12 g/500 mL; and KI, 17 g/500 mL.
2. Chlorine water: 100 mL of commercial bleach and 75 mL of 2 M HCl.
3. Bromine water: 150 mL of KBr solution; 0.1 g of potassium bromate, $KBrO_3$; and 25 mL of 2.0 M HCl.
4. Iodine water: 150 mL of water; a few grains of potassium iodate, KIO_3; 2 droppersful of KI solution; and 25 mL of 2.0 M HCl.
5. TTE, trichlorotrifluoroethane, is less toxic than the other nonpolar solvents.
6. KIO_3, potassium iodate, solution: 2.1 g/500 mL.
7. TTE, trichlorotrifluoroethane, is less toxic than the other nonpolar solvents.

Teaching Tips

NOTES

1. Stoppered, these test tubes make a nice display for a relatively long time.
2. Students can construct the reactivity series of the halogens and find electrode potentials for the reactions.
3. When finished with the test tubes containing TTE, pour the solutions together and predict the final result. The water can be removed from the top and the TTE and halogens evaporated in the hood on paper towels.
4. All the molecular halogens are toxic and should not be breathed or handled.

QUESTIONS FOR STUDENTS

1. Which halogen is the strongest oxidizing agent—takes electrons from other atoms? (Chlorine.)
2. Which halogen is most easily oxidized—gives up its electrons? (Iodine.)
3. Are the molecular halogens more soluble in water or TTE?
4. Predict what will happen if a solution of potassium iodide is added to a solution of bromine water.
5. What are the colors of the halogens in water solution and in TTE?

6.8

Preparation and Properties of Oxygen

A gas is prepared by the reaction of hydrogen peroxide with potassium permanganate and is collected by the downward displacement of water. This gas supports vigorous combustion. The generation of this gas produces a cloud.

MATERIALS FOR PROCEDURE A

100 mL 3% hydrogen peroxide, H_2O_2

60 mL 3M sulfuric acid, H_2SO_4 (To prepare 1.0 liter of solution, dissolve 170 mL of concentrated [18M] H_2SO_4 in 500 mL of water and dilute the resulting solution to 1.0 liter.)

3.0 g potassium permanganate, $KMnO_4$

right-angle glass bend, with outside diameter of 7 mm and length of each arm ca. 5 cm

2-holed #5 rubber stopper

long-stemmed glass funnel (or thistle tube)

250-mL Erlenmeyer flask

90-cm length plastic or rubber tubing to fit 7-mm glass tubing

pan or dish, preferably transparent, with capacity of 4–10 liters

2 500-mL Erlenmeyer flasks

small watch glass or glass plate to cover Erlenmeyer flask

wooden splint

matches

Bunsen burner

tongs

0.5 g steel wool, 000 grade, formed into tight ball

MATERIALS FOR PROCEDURE B

100 mL 3% hydrogen peroxide, H_2O_2

60 mL 3M sulfuric acid, H_2SO_4 (For preparation, see Materials for Procedure A.)

3.0 g potassium permanganate, $KMnO_4$

600-mL beaker (preferably tall form)

wooden splint
matches

MATERIALS FOR PROCEDURE C

4.0 g manganese dioxide, MnO_2
30 mL 30% hydrogen peroxide, H_2O_2
piece of facial tissue, ca. 6 cm square
10-cm length of thread
gloves, plastic or rubber
opaque bottle, ca. 500-mL capacity
2-holed rubber stopper to fit bottle

PROCEDURE A

Preparation

Assemble the apparatus as illustrated in the figure. Insert one end of the right-angle bend through the 2-holed rubber stopper. Insert the stem of the funnel through the other hole in the stopper so that the tip of the stem is within 5 mm of the bottom of the 250-mL Erlenmeyer flask when the stopper is seated in the mouth of the flask. Attach the 90-cm length of plastic or rubber tubing to the free end of the glass bend. Pour 100 mL of 3% hydrogen peroxide into the flask, and seat the stopper assembly in the mouth of the flask.

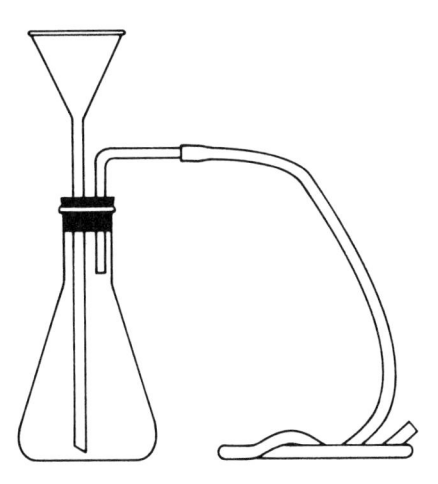

Dissolve 3.0 g of potassium permanganate in 60 mL of 3M sulfuric acid.
Pour water into the pan or dish to a depth of 10 cm. Fill two 500-mL Erlenmeyer flasks to the brim with water. Cover the mouth of each while inverting it in the pan of water.

Presentation

Pour 2–3 mL of the potassium permanganate solution into the funnel. Fizzing will appear where the permanganate mixes with the peroxide. Hold the open end of the plastic or rubber tubing under the water in the pan and observe the bubbles which emerge. Add more permanganate solution to the funnel to keep the bubbles forming at a moderate rate. After the bubbles have been forming for about 30 seconds, place the opening of the tubing under the mouth of one of the inverted Erlenmeyer flasks and fill this flask with gas by the downward displacement of its water. Place the small glass cover over the mouth of the flask and remove it from the bath, setting it upright on the table. Fill the other flask in the same manner.

Light a wooden splint and extinguish the flame so that only glowing embers remain. Plunge the glowing splint into one of the flasks of gas. Observe that the splint bursts into bright flames in the gas.

Light the Bunsen burner. Using tongs, hold the wad of steel wool in the flame of the burner until it glows. Plunge the hot steel wool into the other flask of gas. Observe that the steel wool bursts into flame in the gas.

PROCEDURE B

Preparation

Pour 100 mL 3% hydrogen peroxide into the 600-mL beaker. Dissolve 3.0 g potassium permanganate in 60 mL of 3M sulfuric acid.

Presentation

Pour about 5 mL of the potassium permanganate solution into the hydrogen peroxide in the 600-mL beaker. Observe that the mixture fizzes, indicating the production of a gas, and that the purple color of the permanganate solution disappears.

Light a wooden splint and extinguish the flame leaving a few glowing embers. Pour another 5 mL of the potassium permanganate solution into the beaker. Quickly insert the glowing splint into the gas in the beaker. The glowing embers will burst into flame. This may be repeated until all of the permanganate solution is consumed.

PROCEDURE C

Preparation

Wrap 4.0 g of MnO_2 in a small piece of facial tissue or similar paper. Tie the package with the 10-cm length of thread. Wearing gloves, pour 30 mL of 30% H_2O_2 into the opaque bottle having a capacity of about 500 mL. Carefully suspend the packet of MnO_2 inside the neck of the bottle and stopper the bottle so the stopper holds the packet by its thread.

Presentation

Do not have open flames nearby when presenting this demonstration. Remove the stopper from the bottle and step back. The packet of MnO_2 will drop into the H_2O_2 and cause a sudden decomposition of the peroxide, liberating oxygen gas. The rapid evolution of gas carries small droplets of liquid out of the bottle, producing a cloud over the mouth of the bottle. The bottle will become quite hot.

HAZARDS

Because sulfuric acid is both a strong acid and a powerful dehydrating agent, it must be handled with great care. Spills should be neutralized with an appropriate agent, such as sodium bicarbonate ($NaHCO_3$), and then wiped up. The dilution of concentrated sulfuric acid is a highly exothermic process and releases sufficient heat to cause burns.

Handle potassium permanganate with great care, because explosions can occur if it is brought into contact with organic or other readily oxidizable substances, either in solution or in the solid state.

A 3% solution of hydrogen peroxide is a topical antiseptic and cleaning agent. It decomposes rapidly with evolution of oxygen upon exposure to light, heat, or foreign materials.

Steel wool should be cut with scissors, not pulled apart, because pulling it with the fingers can cause skin lacerations.

Care must be taken when inserting burning materials into pure oxygen; the flames will increase dramatically. Therefore, the burning object should be grasped at a distance from the burning portion.

Because 30% hydrogen peroxide is a strong oxidizing agent, contact with skin and eyes must be avoided. In case of contact, immediately flush with water for at least 15 minutes; get immediate medical attention if the eyes are affected.

Avoid contact between 30% hydrogen peroxide and combustible materials. Avoid contamination from any source, because any contaminant, including dust, will cause rapid decomposition and the generation of large quantities of oxygen gas. Store 30% hydrogen peroxide in its original closed container, making sure that the container vent works properly.

Be careful not to tip the prepared bottle in Procedure C. The reaction can propel the stopper from the bottle with considerable force.

DISPOSAL

The oxygen-generation flask or bottle should be filled with water and the resulting solution flushed down the drain.

DISCUSSION

Oxygen is a colorless, odorless gas at room temperature and atmospheric pressure. It condenses to a pale blue liquid at $-182.962°C$ and 1 atm [1]. The preparation and properties of this liquid are investigated in the subsequent demonstration. Oxygen has a freezing point of −218.4°C [1]. The density of the gas at 0°C and 1 atm is 1.429 g/liter; the liquid at −183°C is 1.149 g/mL, and the solid at −252°C is 1.426 g/cm^3 [1].

The discovery of oxygen is generally attributed to Joseph Priestley. In 1774, Priestley produced a gas by the action of sunlight focussed by lens on a sample of mercury calx (HgO) prepared by heating mercury in air. He found that this gas had remarkable properties. It made a candle burn extremely brightly, and a mouse could survive in it about twice as long as in normal air [2].

Although Priestley realized that the gas he had produced differed from any other gas he had prepared, he was unable to recognize it as a new substance. He named it "dephlogisticated air" and considered it to be common air from which most of the phlogiston had been removed. Thus it was able to accept the phlogiston escaping from a burning candle more readily than normal air could. It was Lavoisier who realized that this gas was a unique substance which was a component of air. Lavoisier elucidated the role of this gas, which he named oxygen, in the formation of metal calces and in combustion.

The reaction which these early chemists used to produce oxygen gas is represented by the modern chemical equation.

$$2\ HgO(s) \longrightarrow O_2(g) + 2\ Hg(l)$$

This method is not used in this demonstration because of the toxicity of mercury and its compounds. Instead, oxygen is prepared from the decomposition of an acidic aqueous hydrogen peroxide solution by permanganate ion.

$$5\ H_2O_2(aq) + 2\ MnO_4^-(aq) + 6\ H^+(aq) \longrightarrow$$
$$5\ O_2(g) + 2\ Mn^{2+}(aq) + 8\ H_2O(l)$$

The production of oxygen by this method was developed in the early 19th century by Louis Thenard [3]. The equation indicates that the hydrogen peroxide is oxidized to oxygen by the purple permanganate ion, which is reduced to the nearly colorless manganese(II) ion.

REFERENCES

1. R. C. Weast, Ed., *CRC Handbook of Chemistry and Physics*, 59th ed., CRC Press: Boca Raton, Florida (1979).
2. A. F. Scott, *Sci. Am.* 250:126 (1984).
3. E. Farber, *The Evolution of Chemistry*, Ronald Press Co.: New York (1952).

5.4

Boyle's Law

The relationship between the pressure and volume of a fixed amount of gas at constant temperature is studied by monitoring the volume of the gas while varying the pressure, or vice versa.

MATERIALS FOR PROCEDURE A

100 mL mercury

130-cm length glass tubing, with outside diameter of 10 mm

glass-working torch

meter stick

ring stand, with clamp for glass tube

plastic tray, ca. 30 cm × 40 cm

white backdrop (e.g., piece of white poster board, 20 cm × 100 cm)

short-stemmed funnel, ca. 50 mm in diameter, nonmetallic

5-cm length rubber tubing to fit both glass tubing and stem of funnel

MATERIALS FOR PROCEDURE B

4 right-angle glass bends, with outside diameter of 7 mm, one arm 5 cm long, the other 20 cm

4 2-holed rubber stoppers to fit round-bottomed flasks

4 right-angle glass bends, with outside diameter of 7 mm and length of each arm ca. 5 cm

4 1-liter round-bottomed flasks

4 30-cm lengths copper wire, 16 gauge

3 15-cm lengths rubber tubing to fit 7-mm glass bends

4 cork rings to support flasks

2 40-cm lengths rubber tubing to fit 7-mm glass bends

mercury-filled U-tube manometer, with arms 2 m long

10 5-cm lengths copper wire, 16 gauge

PROCEDURE A

Preparation

Construct the apparatus as illustrated in Figure 1. Bend the glass tubing into a J having the dimensions in the figure. The distance between the two arms should be large enough to fit a meter stick easily between them. Seal the short end of the tube. Clamp the J-tube to the ring stand, and place the stand on a plastic tray to contain any spills. Place the white backdrop behind the tube. Attach the funnel to the open end of the tube with the rubber tubing.

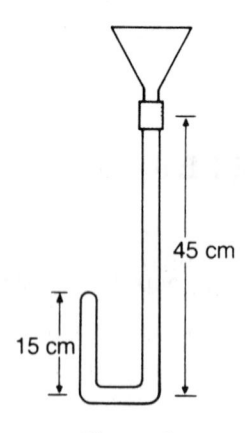

Figure 1.

Presentation

Pour enough mercury through the funnel to fill the curved portion at the bottom of the J-tube. As the mercury fills the bottom of the tube, it will trap the air in the sealed arm of the tube. Measure and record both the length of the column of air trapped in the short arm of the J-tube and the difference between the heights of the mercury in the two arms of the tube.

Pour a few more milliliters of mercury into the tube and repeat the two measurements. Continue to add mercury and record data until at least five sets of data have been recorded.

The data can be used to illustrate Boyle's law, as described in the Discussion section.

PROCEDURE B

Preparation

Assemble the apparatus shown in Figure 2 [1]. Insert the 20-cm end of the bends through each of the 2-holed rubber stoppers. Insert one of the short bends into each of the stoppers. Seat one of the stoppers in the mouth of each of the round-bottomed flasks. Fasten the stoppers to the flasks by wrapping wire over the top of each stopper

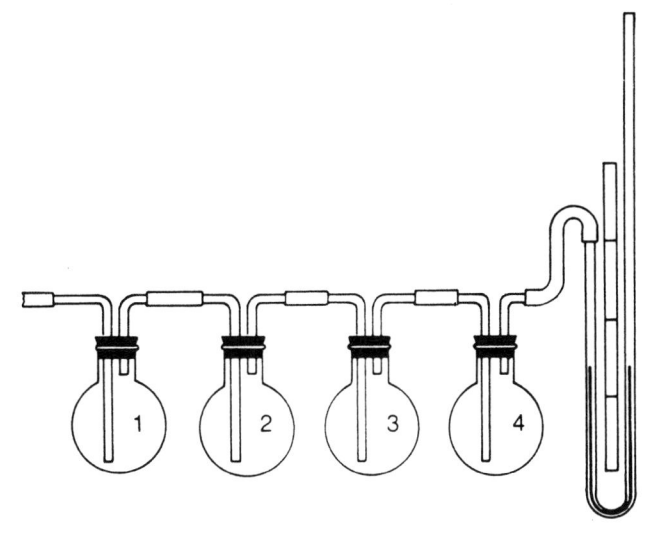

Figure 2.

and around the neck of the flask, in the manner a champagne cork is fastened. Using the short pieces of rubber tubing, connect the flasks together, the short glass bend of one to the long glass bend of the next. Support each flask on a cork ring. With longer pieces of rubber tubing, connect the free, *short* glass bend of an end flask to one arm of the manometer, and connect the free, *long* glass bend of the other end flask to a water tap. Secure all connections by tightening short pieces of wire around the rubber tubing wherever it joins the glass bends.

Presentation

Note that initially the gas in the apparatus has a volume of approximately 4 liters (the total volume of the flasks, neglecting the small volume of the tubing) at atmospheric pressure. Open the water tap and fill the first of the flasks with water. The volume of the gas within the apparatus is now 3 liters. Measure and record the difference in the levels of the mercury in the two arms of the manometer. Open the tap again, and fill the second flask. Now the volume of the gas is 2 liters. Record the difference in the mercury levels.

The third flask should not be filled with water. If the third flask were to be filled with water, the pressure of the gas would become quite high, too high for the manometer to read, causing the mercury to overflow.

The data collected in this procedure can be used to illustrate Boyle's law, as described in the Discussion section.

HAZARDS

Mercury is extremely toxic and should be handled with care to avoid prolonged or repeated exposure to the liquid or vapor. Continued exposure to the vapor may result in severe nervous disturbance, insomnia, and depression. Continued skin contact also can cause these effects as well as dermatitis and kidney damage. Mercury should be handled only in well-ventilated areas. Mercury spills should be cleaned up immediately by using a capillary attached to a trap and an aspirator. Small amounts of mercury in inaccessible places should be treated with zinc dust to form a nonvolatile amalgam.

DISPOSAL

Pour the mercury from the J-tube back into its container for reuse.

DISCUSSION

Robert Boyle (1627–1691) was fascinated by the behavior of gases, and devoted much of his life to the study of their properties. Among the properties he studied was the relationship between the pressure and the volume of a gas. He studied this relationship at pressures both above and below atmospheric pressure. To study gases at pressures higher than atmospheric, he used a J-shaped tube as used in Procedure A of this demonstration [2]. For studies at low pressures, he used a straight glass tube, sealed at one end and partially filled with mercury, with its open end immersed in a vessel of mercury. By raising the tube in the vessel of mercury, he could reduce the pressure of the gas in the tube and measure its volume. (This method is not used in this demonstration, because it requires a rather large amount of mercury.) Experiments with this equipment led him in 1662 to the law bearing his name, which states that the volume of a gas is inversely proportional to its pressure.

Expressed algebraically, Boyle's law can be represented by the equation

$$V = \frac{k}{P}$$

where V is the volume of the gas,
P is its pressure, and
k is a proportionality constant.

An equivalent form of the equation is

$$PV = k$$

Therefore, Boyle's law can be demonstrated by showing that the product of the volume of a sample of gas multiplied by its pressure remains constant as the pressure and volume change.

In the apparatus of Procedure A, the sample of gas is contained in the closed end of the J-tube, which is a cylindrical container. The volume of this gas is given by the equation for the volume of a cylinder,

$$V = \pi r^2 h$$

where r is the cross-sectional inside radius of the tube and
h is the height of the column of gas in the tube.

This equation indicates that the volume of the gas is proportional to the height of the column of gas. While the actual volume of the gas in the tube can be calculated, this is unnecessary to illustrate Boyle's law. Because the height of the gas column is proportional to the volume of the gas, Boyle's law can be demonstrated by showing that the product of this height multiplied by the pressure of the gas is constant.

The pressure of the gas in the J-tube can be determined by measuring the levels of the mercury in the two arms of the tube. The gas exerts pressure on the mercury, pushing some of it into the open arm of the tube. This pressure is exerted against the pressure of the atmosphere outside the apparatus and against the weight of the mercury that is elevated above the level of the mercury in the closed arm of the tube. Therefore, the pressure of the enclosed gas is the sum of atmospheric pressure plus the pressure produced by the weight of the raised mercury:

$$P_g = P_a + P_{Hg}$$

The mass of this mercury can be found by multiplying the density of mercury (13.6 g/mL [3]) by the volume of the mercury. The volume of the column of mercury, like the volume of the enclosed gas, is given by the equation for the volume of a cylinder. Therefore, it is proportional to the height of the mercury column, and the pressure of the gas produced by the mercury column is proportional to the difference between the heights of the mercury columns in the two arms of the J-tube.

The height of the column of gas and the difference in the levels of the mercury in the two arms of the J-tube can be used to illustrate Boyle's law. The pressure of the gas in the tube is the sum of the atmospheric pressure and the pressure produced by the weight of the column of mercury, which is proportional to the difference in the levels of the mercury in the two tubes. The pressure of the gas in the tube may be represented by a + d, where a is the atmospheric pressure, and d the difference between the two levels of mercury, both expressed in millimeters of mercury. The volume of the gas is proportional to the height of the gas column, h. To demonstrate Boyle's law, all that need be shown is that the product (a + d)h is a constant. This has been done with sample data presented in Table 1. In this example, atmospheric pressure was assumed to be 760 mm Hg, although for greater accuracy atmospheric pressure could have been measured.

Table 1. Sample Results of Procedure A

Difference in Hg levels (mm)	Height of gas column (mm)	Total pressure of gas[a] (mm Hg)	Product of pressure and height[b]
32	239	792	189,000
146	211	906	191,000
372	167	1132	189,000
512	149	1272	190,000
728	128	1488	190,000
947	114	1707	195,000

[a] Calculated by adding an atmospheric pressure of 760 mm Hg to the difference in the mercury levels.

[b] Calculated by multiplying the height of the gas column (in mm) by the total pressure of the gas (in mm Hg).

The data obtained in Procedure B are less precise than those from Procedure A, but their interpretation is similar. To illustrate Boyle's law with these data, one need show that the product of the volume multiplied by the pressure of the gas is constant. Data from a trial of Procedure B are presented in Table 2. Initially, the volume of the gas is 4 liters, neglecting the small volume of gas contained in the tubing, at which time its pressure is atmospheric pressure. As the gas is compressed into a volume of 3 liters, its pressure increases, supporting a column of mercury in the manometer. Then the pressure of the gas is the sum of the height of the column of mercury and atmospheric pressure. When the second flask is filled with water, the gas is compressed further, and its pressure increases. As the results in Table 2 show, the product of the volume multiplied by the pressure of the gas is virtually constant.

Table 2. Sample Results of Procedure B

Volume of gas (liters)	Difference in Hg levels (mm)	Total pressure of gas[a] (mm Hg)	Product of volume and pressure[b]
4.00	0	760	3040
3.00	258	1018	3050
2.00	775	1535	3070

[a] Calculated by adding an atmospheric pressure of 760 mm Hg to the difference in the mercury levels.

[b] Calculated by multiplying the volume of the gas (in liters) by the total pressure of the gas (in mm Hg).

REFERENCES

1. F. B. Dutton, *J. Chem. Educ.* 18:15 (1941).
2. J. R. Partington, *A Short History of Chemistry*, 3d ed., Macmillan and Co.: London (1957).
3. R. C. Weast, Ed., *Handbook of Chemistry and Physics*, 59th ed., CRC Press: Boca Raton, Florida (1978).

12. Charles' Law: The Relationship between Volume and Temperature of a Gas

Balloons are inflated to the same size and tied. One balloon is placed on the surface of a large beaker of ice water. Another balloon is placed on the surface of a beaker of hot water. At the decreased temperature, the balloon shrinks. At the increased temperature, the other balloon becomes larger.

Procedure

1. Blow up three balloons—not fully—to the same size and tie securely.
2. Leave one balloon for reference. Place one balloon on the surface of a large beaker (or pan) of ice water.
3. Place the third balloon on the surface of a large beaker of hot water.
4. Observe the effect of temperature on the volume of the gas in the balloons.
5. Remove both balloons and allow them to come to room temperature.

Reactions

As the temperature decreases, the average kinetic energy of the gas inside the balloon decreases, which decreases the pressure inside the balloon. When the balloon is heated, the volume increases as the internal pressure increases.

Teaching Tips

NOTES

1. Try various hot and cold temperatures to get maximum effects.
2. Jacques Alexandre Charles made the first observation of gases at different temperatures in 1787.
3. Putting his experience to practical use, Charles was one of the first hot-air balloonists. In 1783, he made the second ascent in a hydrogen-filled balloon carrying passengers. When he landed, his balloon was destroyed by terrified, pitchfork-wielding peasants.
4. The change in gas volume is essentially the same for each Celsius degree of temperature change, approximately $1/273$ of the volume that gas would have at 0 °C.

QUESTIONS FOR STUDENTS

1. What is Charles' law?
2. Why do the balloons behave as they do in hot and cold water?
3. Why do the balloons return to "normal" at room temperature?

10. Determining the Molecular Weight of a Gas: *Flick Your Bic*

A large graduated cylinder is filled with water and placed, inverted, in a water-filled trough. A small piece of rubber tubing from a pocket lighter is placed inside the cylinder. When the release button on the lighter is pressed, butane is released, displacing the water in the cylinder. By measuring the volume of water displaced and the weight of the gas, the molecular mass of butane is calculated.

Procedure

1. Remove the striking mechanism (flint, wheel, and spring) from a NEW disposable pocket lighter.
2. Weigh the lighter on a balance. If you plan to use a small cylinder and displace a small volume of water, use an analytical balance. If you use a large (500 mL) cylinder, you can use a triple-beam balance. Record the weight.
3. Attach a small rubber tube to the gas nozzle of the lighter.
4. Fill the largest graduated cylinder you have with water, invert it, and place it inside a trough half-filled with water. Be sure that the cylinder contains no gas bubbles.
5. Place the end of the rubber tube under and inside the cylinder and press the release button on the lighter.
6. Collect enough gas to displace 300–400 mL of water, or the largest volume possible.
7. Carefully measure the water remaining in the cylinder to determine the gas volume.
8. Remove the tube from the lighter and reweigh the lighter.
9. From these data, calculate the molecular mass of butane. (Remember to subtract the vapor pressure of water—see box—and to record the temperature and pressure.)

Calculations

1. PV equals nRT, where n is grams per molar mass. Thus, PV equals (grams/molar mass) RT.
2. Molar mass equals (grams) RT/PV, where R is the gas constant, 62,400 mL Torr·mol^{-1}·K^{-1}; T is the temperature in Kelvins (273 K = 0 °C); P is the pressure in Torr corrected for the vapor pressure of water; and V is the volume of gas in milliliters.

Teaching Tips

NOTES

1. The molecular mass of butane is 58.
2. Now is a good time to discuss experimental error.
3. If the tube on the lighter leaks, your results will be off. Try placing the lighter directly beneath the water-filled cylinder and releasing the gas.
4. A hair dryer is handy to dry the lighter prior to final weighing.

QUESTIONS FOR STUDENTS

1. How can you explain the fact that butane is a liquid in the lighter, but a gas when it is collected?
2. If you only have a triple-beam balance, why is it necessary to collect a large volume of gas?

8. Diffusion of Gases

A plug of cotton dipped in HCl is inserted in the open end of a graduated cylinder. The time required for a color change to occur in a strip of pH paper at the other end of the cylinder is noted. The procedure is repeated with a cotton plug dipped in NH_4OH. From these observations, Graham's law of diffusion is checked.

Procedure

1. Clean and DRY two 100-mL graduated cylinders.
2. With a glass rod, insert a moist piece of blue litmus paper in one cylinder. Push the paper to the bottom of the cylinder and see that it sticks to the bottom.
3. Place the cylinder on its side, making sure that it is level.
4. Dip a small piece of cotton in concentrated HCl (CAREFUL!) and place it just inside the opening of the cylinder.
5. Immediately place a piece of plastic wrap tightly over the opening of the cylinder.
6. As soon as the cotton plug is inserted, have a student assistant begin timing. Record the time required for the HCl gas to travel the length of the cylinder and cause the blue litmus paper to turn red.
7. Repeat the demonstration with red litmus paper and a cotton plug dipped in ammonium hydroxide in the other graduated cylinder.

Calculations

1. The data collected from this demonstration will be used to check Graham's law of diffusion of gases: The rate of diffusion of a gas is inversely proportional to the square root of its molecular mass.
2. Determine the rate: Divide the distance the gas traveled (length of the cylinder, in centimeters) by time (seconds).
3. Determine the ratio of the experimental diffusion rate: Divide the rate of diffusion of hydrogen chloride gas by the rate of diffusion of ammonia gas.
4. Check your answer against the theoretical ratio:

$$\frac{R_{NH_3}}{R_{HCl}} = \sqrt{\frac{M_{HCl}}{M_{NH_3}}} = 1.46$$

Teaching Tips

NOTES

1. Graduated cylinders are used because they are readily available and provide a short distance for the gas to travel.
2. An alternate method is to use a long piece of glass tubing. Simultaneously insert a cotton plug soaked in HCl in one end of the tube and a cotton plug soaked in ammonium hydroxide in the other end. Note the appearance of a white ring of ammonium chloride in the tube. Measure the distance from each end of the tube to this ring, and use this data to calculate the rate. (The white ring is often difficult to detect!)

QUESTIONS FOR STUDENTS

1. Why did the ratio calculated from the data collected in this demonstration differ from the theoretical ratio? (This is a good place to discuss experimental error.)
2. What is the relationship between the mass of gas and its rate of diffusion?
3. Why is it necessary to use DRY cylinders for the demonstration?
4. How would the calculated ratio be effected if the cylinders were not level during the demonstration?

9. Surface Tension of Water: The Floating Needle

A needle floats on the surface of water in a beaker. One drop of soap solution is added, and the needle sinks to the bottom.

Procedure

1. Fill a large beaker almost full of water. Be sure that the beaker is clean.
2. Place a sewing needle on top of a small piece of tissue paper.
3. Place a large spatula or scoopula beneath the tissue paper and carefully lower it and the needle onto the water. The paper and needle will float. Soon, the paper will become soaked and sink, and the needle will be left floating on the surface.
4. After discussing the role of surface tension, add 1 drop of soap solution to the surface of the water.
5. Observe that the needle sinks in a few seconds.

Reaction

Surface tension of a liquid is the energy required to increase the surface area. This tension causes the surface to act as if it had a membrane, or "skin". This property is partly due to the fact that surface molecules are only pulled inward, whereas other molecules in the liquid are pulled equally in all directions.

The needle is denser than water, so if pushed under, the needle will sink. On the surface, however, energy would be required to increase the water surface enough for the needle to submerge, so it floats.

When detergent is added, the surface tension is drastically reduced and the needle sinks.

Teaching Tips

NOTES

1. Surface tension also allows some insects (water bugs, etc.) to walk on water, even though they are heavier than water; causes water to rise in a capillary tube; causes a drop of rain to be nearly spherical; and is one of the forces that causes water to rise in a plant.
2. The surface tension of water is 7.3×10^{-2} J/m^2.
3. You may have difficulty placing the needle directly on the surface of the water.

QUESTIONS FOR STUDENTS

1. Why does the heavy needle float?
2. What is *surface tension*?
3. What does surface tension have to do with volume and area?
4. What other effects can you think of that are due to surface tension?

9.6

Evaporation as an Endothermic Process

An aspirator reduces the pressure in a boiling flask containing acetone. The acetone then boils, and its temperature decreases (Procedure A). A "drinking duck" toy sips water from a glass until the level of the water in the glass falls below the reach of its beak (Procedure B).

MATERIALS FOR PROCEDURE A

See Materials for Demonstration 1.1 in Volume 1 of this series.

MATERIALS FOR PROCEDURE B

tap water, to fill drinking glass

drinking glass, ca. 10 cm deep (e.g., "old-fashioned" glass or jelly tumbler)

"drinking duck" toy (see figure in Procedure B) †

PROCEDURE A

See the Procedure of Demonstration 1.1 in Volume 1 of this series.

PROCEDURE B

Preparation and Presentation

Fill the drinking glass with tap water. Set the "drinking duck" next to the glass and tilt the body of the duck until its beak dips in the water (see figure). Hold the beak of the duck in the water until the duck's head becomes damp, then release the duck so its body returns to an upright position. After several seconds the liquid in the lower bulb of the duck's body will begin to rise in the central tube. The liquid will flow through the tube into the head of the duck. At the same time, the duck will gradually tip, until its beak dips in the water. When the body of the duck becomes horizontal, the liquid will flow from the duck's head, through the tube, and back into the lower bulb. When the liquid returns to the lower bulb, the duck returns to a vertical position. The duck will

† Such toys are available under a variety of names from toy stores and novelty shops.

repeat this "drinking" cycle until the level of the water in the glass falls below the reach of its beak.

HAZARDS

The drinking duck toy is made of very thin glass and breaks easily. Breaking the toy will release the volatile liquid it contains, which may be flammable.

DISPOSAL

The drinking duck toy can be retained for repeated use. It will continue to function indefinitely. If the toy breaks, the liquid contents should be allowed to evaporate in a well-ventilated area. The remaining materials can be discarded in a solid-waste receptacle.

DISCUSSION

This demonstration shows the spontaneous transfer of matter from the liquid to the vapor phase, evaporation, when the system pressure is less than the vapor pressure. It also shows that vaporization is an endothermic process. Procedure A shows that the rate of evaporation of acetone is increased when the pressure of the gas over the liquid is lowered: the acetone begins to boil. The temperature of the acetone also decreases, indicating that evaporation is an endothermic process. Procedure B uses the evaporation process to produce motion in the drinking duck toy.

The endothermic nature of the evaporation process is responsible for the operation of the drinking duck [1]. The duck consists of a central glass tube and two glass bulbs (see figure). The upper end of the glass tube opens directly into the upper bulb that forms the head of the duck. This upper bulb has a small protrusion on the side, representing the duck's beak, and the entire bulb is coated with an absorbant material that functions as a wick. The lower end of the glass tube extends almost to the bottom of the lower glass bulb. The lower bulb is a reservoir of volatile liquid (often dichloromethane colored with a dye). The duck is supported on a pivot near the center of the tube.

When the liquid is in the lower bulb, the mass of the bottom of the duck is greater than that of the top, and the duck assumes an upright position. When the head of the duck is dry, the vapor pressure of the liquid in the bottom is the same as it is in the head (ignoring the small hydrostatic pressure of the liquid in the tube), and the liquid stays in the lower bulb. When the head of the duck is moistened with water, the water evaporates from the wick, and the endothermic evaporation process cools the head. Because the head is cooler than the lower bulb, the vapor pressure of the liquid is lower in the head than in the lower bulb. The pressure differential between the two bulbs causes the liquid to rise through the tube into the head. As liquid flows into the head, the head becomes heavier, and the duck begins to tip at the pivot. When the duck tips far enough, the lower end of the tube rises out of the liquid, allowing the higher pressure vapor in the lower bulb to enter the upper bulb. As the vapor pressures equalize between the bulbs, the liquid returns to the lower bulb, and the duck returns to an upright position. If the duck is positioned in such a way that its beak dips in water when it tips, its head will remain damp, the evaporation of water will continue, the head will remain cooler than the lower bulb, and the tipping cycle will repeat.

REFERENCE

1. C. F. Bohren, *Clouds in a Glass of Beer*, John Wiley and Sons: New York (1987).

3.2

Polyurethane Foam

When two viscous liquids are mixed, a rigid foam is produced, whose volume is 20–30 times that of the original mixture [1].

MATERIALS FOR PROCEDURE A

40 ml Part A and 40 ml Part B of a two-component polyurethane foam system [2]

gloves, plastic or rubber

2 50-ml beakers or paper cups

paper towels to cover area ca. 0.5 m × 0.5 m

1-liter beaker, preferably tall-form

2 stirring rods, ca. 25 cm

food-coloring dye (optional)

MATERIALS FOR PROCEDURE B

40 ml Part A and 40 ml Part B of a two-component polyurethane foam system [2]

200-ml disposable cup

paper towels to cover area ca. 0.5 m × 0.5 m

gloves, plastic or rubber

2 50-ml beakers or paper cups

stirring rod

food-coloring dye (optional)

PROCEDURE A

Perform this demonstration in a well-ventilated area.

Wearing gloves, place approximately 40 ml of each component in separate small beakers. The exact amount is not critical as long as equal amounts are used. Place the paper towels under the large beaker. Pour the contents of both beakers into the large beaker. With a stirring rod, mix the contents thoroughly. When the foam begins to expand, stop stirring. The volume will increase 20–30 times that of the original mixture. If carefully supported by two glass rods, the rigid foam will form a column about twice the height of the beaker.

Since the freshly prepared foam usually contains unreacted isocyanate, it should not be handled until it has cured several hours in a well-ventilated area.

Foams of different colors can be produced by adding a few drops of a food-coloring dye to the lighter-colored component prior to mixing with the other component.

PROCEDURE B

Place the 200-ml disposable cup in the center of a mat of paper towels 0.5 m × 0.5 m. Wearing gloves, mix equal amounts of Parts A and B in the disposable cup. When the foam begins to expand, stop stirring. As the mixture foams, it rises out of the cup, overflows, and forms a bell-shaped solid that adheres to the paper towels.

Because the freshly prepared foam usually contains unreacted isocyanate, it should not be handled until it has cured several hours in a well-ventilated area.

As in Procedure A, a few drops of a food-coloring dye can be added to the lighter-colored component.

HAZARDS

Isocyanates are irritants to the skin, eyes, and respiratory system. Toluene 2,4-diisocyanate, which has been used in many polyurethane systems, is reported to cause skin irritation, allergic reactions, and bronchial asthma [3, 4]. The foam product should be allowed to cure several hours in a well-ventilated area before handling.

DISPOSAL

Since the individual components are soluble in acetone, they can be dissolved and the solutions flushed down the drain with water. The rigid foam, when cured, can be discarded in a waste container.

The fully cured polyurethane is soluble in solvents such as dimethylformamide. Before it is fully cured, the foam can be removed from the beaker by using a spatula and small amounts of acetone.

DISCUSSION

The polyurethane foam system used in this demonstration consists of two viscous liquids. Part A, light amber in color, contains a polyether polyol, a blowing agent, a silicone surfactant, and a catalyst. Part B, which is dark, contains a polyfunctional isocyanate.

A polyurethane foam is formed by producing a polyurethane (polycarbamate) polymer in the presence of a fluorocarbon blowing agent. The polymer is formed by the reaction of a polyester polyol, HO-R-OH (whose probable molecular weight range is 400–4000), with a polyfunctional isocyanate, OCN-R'-NCO [for example, methylene bis(4-phenylisocyanate)]. Amine compounds or metal salts are used as catalysts. The reaction can be represented as:

$$HO-R-OH + O=C=N-R'-N=C=O \longrightarrow \left[O-R-O-\underset{\underset{O}{\|}}{C}-\underset{\underset{H}{|}}{N}-R'-\underset{\underset{H}{|}}{N}-\underset{\underset{O}{\|}}{C} \right]$$

The polyfunctional character of the reactants results in a high degree of cross-linking in the product, forming a rigid foam.

REFERENCES

1. Dirreen, G. E.; Shakhashiri, B. Z. *J. Chem. Educ.* **1977**, *54*, 431.
2. We have used a product, Super Foam, marketed by Edmund Scientific Co., 101 E. Gloucester Pike, Barrington, New York 08007. Similar products may be available at paint, hardware, or lumber stores.
3. Hocking, M. B.; Canham, G. W. R. *J. Chem. Educ.* **1974**, *51*, A580.
4. Windholz, M., Ed. "The Merck Index," 9th ed.; Merck and Co.: Rahway, New Jersey, 1976; p 1225.

3.8

Polybutadiene (Jumping Rubber)

A violet suspension of an "alfin" catalyst is added to a bottle containing a solution of 1,3-butadiene in pentane. The bottle is corked and shaken for several seconds. The mixture sets to a gel, and within 2 minutes the contents erupt from the bottle [1].

MATERIALS

Jumping Rubber kit [2]. The kit includes:
  wax-sealed, screw-capped vial containing the alfin catalyst
  small sealed bottle containing 1,3-butadiene dissolved in dry pentane
  cork for bottle

gloves, plastic or rubber

bottle opener

stirring rod

tongs

PROCEDURE

Warning! The alfin catalyst is a fire hazard.
Perform this demonstration in a well-ventilated room and wear gloves.

Open the catalyst vial and stir its contents with the glass rod. Remove the crown seal from the bottle. Quickly add *all* the catalyst to the bottle. Immediately cork the bottle and shake. Do not point the corked bottle at anyone. The temperature increases slightly (to about 50°C), and the pressure increases (perhaps 2–3 atmospheres) until, within 2 minutes, the cork is forced from the mouth of the bottle and a polymer "snake" shoots out. The bottle is left almost dry, and the liquid is trapped in the polymer. Use a pair of tongs to handle the polymer. During the next hour or so, the trapped pentane will evaporate and the polymer will shrink.

HAZARDS

This dramatic demonstration should be performed only by individuals who understand the fire hazard and reactivity of organosodium reagents. We have performed this demonstration over three hundred times without a single accident. Under hot or extremely humid conditions, the catalyst could ignite in air or ignite the butadiene-pentane solution. A carbon dioxide fire extinguisher must be available.

Pentane is a highly flammable and volatile liquid (boiling point: 36°C), which can explode when exposed to heat, sparks, or flame. Avoid inhalation of pentane vapors since they are slightly toxic. In high concentrations pentane is a narcotic.

The compound 1,3-butadiene (boiling point: −5°C) can be irritating to skin and mucous membranes and is a narcotic in high concentrations.

DISPOSAL

Since pentane is trapped in the product, the polymer should be kept away from flames. After several hours the pentane will evaporate, and the shrunken polymer can be discarded.

The empty screw-capped vial should be rinsed carefully with water and discarded.

DISCUSSION

Alfin catalysts are a class of heterogeneous catalysts which cause rapid polymerization of butadiene, isoprene, and other monomers resulting in polymers with very high molecular weights (1–2 million or higher). The alfin catalyst used in this demonstration is a solid surface catalyst developed by A. A. Morton [3–5] and co-workers. It is produced by reacting amyl chloride ($C_5H_{11}Cl$) with sodium, which is then reacted with isopropyl alcohol [$(CH_3)_2CHOH$]. The resulting mixture contains amylsodium ($C_5H_{11}Na$), sodium isopropoxide [$(CH_3)_2CHONa$], and sodium chloride. Propylene ($CH_2=CHCH_3$) is added to produce allylsodium ($CH_2=CHCH_2Na$) from amylsodium. The following sequence shows the necessary stoichiometry [6]:

$$1.5\ C_5H_{11}Cl + 3\ Na \longrightarrow 1.5\ C_5H_{11}Na + 1.5\ NaCl$$

$$1.5\ C_5H_{11}Na + (CH_3)_2CHOH \longrightarrow 0.5\ C_5H_{11}Na + (CH_3)_2CHONa + C_5H_{12}$$

$$0.5\ C_5H_{11}Na + 0.5\ CH_2=CHCH_3 \longrightarrow 0.5\ CH_2=CHCH_2Na + 0.5\ C_5H_{12}$$

The sodium isopropoxide-allylsodium combination gives the highest polymer yield. The role of the sodium chloride is not clear, although it could be acting as a support for the catalyst.

The catalyst is the mixture of the three sodium compounds: allyl sodium, sodium isopropoxide, and sodium chloride. All are essential constituents. The name "alfin" was derived from the words "alcohol" and "olefin" because both are involved in the preparation of the reagent.

The alfin catalyst is believed [7, 8] to adsorb and orient the monomer prior to the chain-growth process. Surface effects presumably influence the stereochemistry [6], and the polymer consists mainly of trans-1,4 repeating units. Since one allyl group is incorporated into each polymer chain, the process resembles Ziegler-Natta catalysis [6]:

$$n\ CH_2=CH-CH=CH_2 \xrightarrow{\text{alfin catalyst}} \left[\begin{array}{c} CH_2 \\ \diagdown \\ \end{array}\!\!\!\!C=C\!\!\!\!\begin{array}{c} H \\ \diagup \\ \end{array}\right]_n$$

trans-1,4-polybutadiene

The predominance of 1,4-polymerization has been suggested as evidence for a free radical propagation mechanism. It has been proposed [6] that complexes such as

$$\begin{array}{cc} CH_3-CH-CH_3 & CH_3-CH-CH_3 \\ | & | \\ O^- & O^- \\ Na^+ \quad Na^+ & Na^+ \quad Na^+ \\ H_2C^{\delta-} \quad {}^-CH_2 & H_2C^- \quad {}^{\delta\cdot}CH_2 \\ \diagdown_{\delta+}\diagup & \diagdown_{\delta\cdot}\diagup \\ CH & CH \end{array} \quad \longleftrightarrow$$

are formed and that the adsorbed monomer displaces the allyl anion from the complex to form an ion pair, which then reacts to form a radical pair:

$$[CH_2\mathord{=\mkern-3mu=}CH\mathord{=\mkern-3mu=}CH\mathord{=\mkern-3mu=}CH_2Na]^{+\,-}[CH_2\mathord{=\mkern-3mu=}CH\mathord{=\mkern-3mu=}CH_2] \longrightarrow$$
$$\cdot CH_2-CH=CH-CH_2{}^-Na^+ \; + \; \cdot CH_2-CH=CH_2$$

The radical anion initiates polymerization, which continues until combination with an allyl radical occurs. This combination does not occur very readily because the allyl radical is bound to the catalyst surface [6], and hence very high molecular weights are obtained.

An anionic mechanism for polymerization with the alfin catalyst has been proposed [9]. According to this mechanism, monomer molecules are inserted into the chain:

[reaction scheme showing insertion of monomer via anionic mechanism with Na⁺Cl⁻, Na⁺⁻OR, and catalyst surface]

$$\begin{array}{c} CH_2-CH_2-CH=CH-CH_2R \\ | \\ CH \\ \| \\ CH \\ \diagup \\ CH_2{}^- \\ \\ Na^+ \\ \wr \\ \text{catalyst} \\ \text{surface} \end{array}$$

The references cited include further discussion of these and other hypotheses.

REFERENCES

1. Shakhashiri, B. Z.; Dirreen, G. E.; Williams, L. G. *J. Chem. Educ.* **1980**, *57*, 738.
2. The Jumping Rubber Kit can be obtained from Organometallics, Inc., Route 111, East Hampstead, New Hampshire 03826.
3. Morton, A. A.; Magat, E.; Letsinger, R. L. *J. Am. Chem. Soc.* **1947**, *69*, 950.
4. Morton, A. A.; Welcher, R. P.; Collins, F.; Penner, S. E.; Combs, R. D. *J. Am. Chem. Soc.* **1949**, *71*, 487.
5. Sorenson, W. R.; Campbell, T. W. "Preparative Methods of Polymer Chemistry," 2nd ed.; Interscience Publishers, John Wiley and Sons: New York, 1968; pp 305–7.
6. Stevens, M. P. "Polymer Chemistry: An Introduction"; Addison-Wesley Publishing Co.: Reading, Massachusetts, 1975; pp 179–80.
7. Morton, A. A.; Lanpher, E. J. *J. Polymer Sci.* **1960**, *44*, 233. (Includes bibliography.)
8. Ravve, A. "Organic Chemistry of Macromolecules"; Marcel Dekker: New York, 1967; p 125.
9. Reich, L.; Schindler, A. "Polymerization by Organometallic Compounds"; Interscience Publishers, John Wiley and Sons: New York, 1966; Ch. 5.

64. Supersaturation

A flask containing a liquid is shown to the class. The stopper is removed from the flask, and a single solid crystal is added. Immediately, the entire contents of the flask solidify and the flask becomes warmer.

Procedure

1. Prepare a flask as described under Solution.
2. Present the flask containing a liquid to the class.
3. Carefully remove the stopper from the flask and add *one* tiny crystal of solid sodium acetate trihydrate.
4. Stopper the flask. Observe as the liquid in the top part of the flask begins to solidify and the solidification increases throughout the flask.
5. Note that the flask has become warmer.

Reactions

The supersaturated solution of sodium acetate trihydrate holds more dissolved solute than would normally be in equilibrium with undissolved solute. When this equilibrium is suddenly disturbed by adding a small crystal of solute, by sudden cooling, or even by scratching the container, all of the excess solute precipitates from solution.

Solution

Fill a clean, dry round-bottomed flask with solid sodium acetate trihydrate ($CH_3COONa \cdot 3H_2O$). Slowly heat the flask on a hot plate until the material completely liquifies.

Heat for a few minutes but do not boil. Remove the flask from the heat and carefully rinse down the neck with a small amount of water from a wash bottle. Insert a stopper and allow the flask to cool at room temperature.

Teaching Tips

NOTES

1. You can use the same flask over and over, just heat after each use to liquify the crystals.
2. You can add small crystals of other substances, such as copper sulfate, to induce solidification.
3. The flask becomes warm because of heat of crystallization of sodium acetate.
4. You can also use sodium thiosulfate.
5. A supersaturated solution of acetanilide, when cooled, will form crystals when the temperature is lowered to 50 °C or when a small crystal is introduced.
6. Honey is a good example of a supersaturated solution. Some water evaporates in the beehive as honey is being made. When honey stands on the shelf for a long time, excess sugar may precipitate.

QUESTIONS FOR STUDENTS

1. Describe the reaction. Was heat evolved? Why?
2. What would happen if a crystal of another substance (copper sulfate, for instance) was added instead of sodium thiosulfate?
3. What function does the added crystal serve?

9.20

Osmotic Pressure of a Sugar Solution

A membrane bag is filled with sugar solution and attached to a glass tube. When the bag is immersed in water, the level of liquid rises in the glass tube.

MATERIALS

    2-liters distilled water

    ca. 400 g sucrose (table sugar), $C_{12}H_{22}O_{11}$

    ca. 10 mg iodine, I_2 (or other colored substance soluble in hexane)

    2 mL hexane, C_6H_{14}

    hot plate

    1-liter beaker

    25 cm of cellulose dialysis tubing or synthetic sausage casing (available at meat markets)

    2-m glass tube, with inside diameter of 8 mm

    glass-working torch

    1-holed rubber stopper, no. 4–6

    single-edged razor blade

    2-holed rubber stopper, no. 8–12 (hole sizes compatible with glass tubing)

    10 cm glass tube, with inside diameter of 8 mm

    20-cm length of rubber tubing to fit 10-cm glass tube

    pinch clamp

    ring stand, at least 50-cm tall

    clamp for ring stand

    30 cm string

    white poster board, strip 10 cm wide with a total length of 2 m

    thermometer, $-10°C$ to $+110°C$

    gloves, plastic or rubber

    test tube, 10 mm × 75 mm

    60-mL beaker

PROCEDURE

Preparation

Boil the dialysis tubing or sausage casing for several minutes in water. Allow the tubing to cool in the water.

Construct the apparatus illustrated in the figure. Flare one end of the 2-m glass tube to allow liquid to be poured easily into the tube. Insert the other end of the tube through the 1-holed stopper and position the stopper about 35 cm from the unflared end of the tube. Cut a groove around the side of the 2-holed rubber stopper. Insert the 10-cm glass tube through one of the holes of the stopper, and insert the 2-m glass tube through the other hole. Attach the 20-cm rubber tubing to the 10-cm glass tube in the rubber stopper and close the free end of the rubber tubing with a pinch clamp. Mount the 2-m glass tube to the ring stand by clamping it at the 1-holed stopper. Tie a knot in one end of the boiled dialysis tubing. Insert the 2-holed stopper in the open end of the dialysis tubing and secure it with a string tied snugly in the groove in the stopper. Fasten a 10-cm wide strip of white poster board along the height of the 2-m glass tube. This can be accomplished by punching small holes in the poster board and looping string through the holes and around the tube.

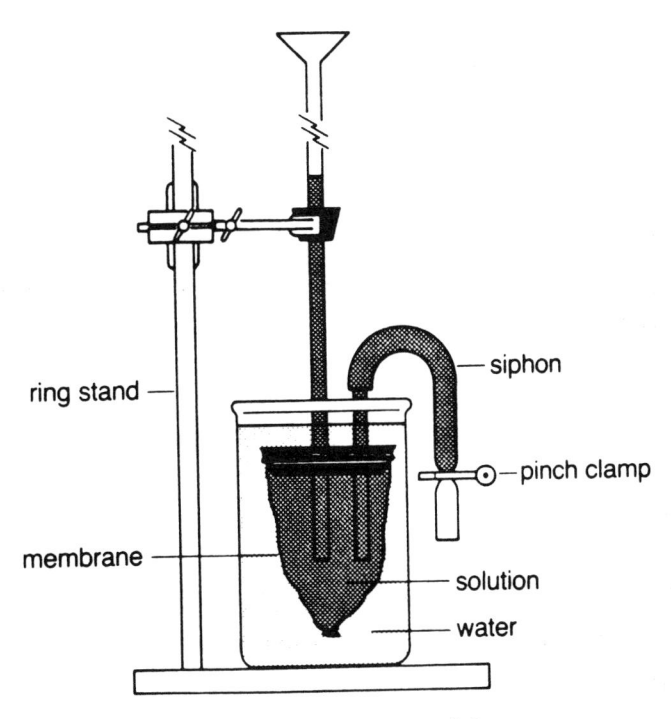

Osmotic pressure apparatus.

Place the 1-liter beaker on the stand and lower the apparatus so the cellulose "bag" is in the beaker. Fill the beaker with distilled water to cover the 2-holed stopper.

Prepare a sucrose solution by adding table sugar to 100 mL of boiling distilled water until the temperature of the boiling solution rises to 108°C. This will require about 400 g of sugar. Allow the sugar solution to cool. The concentration of this so-must be applied to the sucrose solution to stop the net flow of water through the membrane is the osmotic pressure of the solution.

No attempt is made to stop the flow of water through the membrane in this demonstration. Rather, the effect of this flow is observed. As the water flows into the bag formed by the membrane the volume of the solution is increased, causing its level in the attached tube to rise. The solution in the tube will rise until the downward pressure of the liquid is equal to the osmotic pressure of the solution.

The osmotic pressure of the original sucrose solution in the bag can be calculated using the following equations [3]:

$$\bar{V}^0 \pi = -RT \ln(a) \qquad \ln(a) = \frac{\Delta H^0_{vap}}{R}\left(\frac{1}{T'} - \frac{1}{T_0}\right)$$

where π is the osmotic pressure,

$\bar{V}^0$ is the molar volume of the solvent at the temperature, T, of the osmotic pressure experiment,

ΔH^0_{vap} is the enthalpy of vaporization of the solvent at its boiling point, T_0,

and T' is the boiling point of the solution at the same pressure.

These equations indicate that the osmotic pressure of the original solution is over 400 bar (1 atm = 1.01325 bar). The familiar equation for osmotic pressure, $\pi = cRT$, where c is the molarity of the solution, is an approximation that applies only to dilute solutions. The solution used in this demonstration is most certainly not dilute, and an application of this equation will give a result far from the actual value of the osmotic pressure.

The pressure of 400 bar corresponds to a height of a solution having a density of 1.39 g/mL [2] of 3000 meters, indicating the impracticality of observing the equilibrium height. The actual equilibrium height will be less, because the solution is diluted when water crosses the membrane into the solution. Dilution of the solution reduces the osmotic pressure and makes accurate calculations on this system difficult. Even if a tall tube were used to contain the rising column, the membrane would rupture once the pressure differential reaches a few bar.

The (equilibrium) osmotic effect can be understood as a balance between the tendency of water molecules to flow from a region of high concentration to one of lower concentration (the solution) and the effect of increasing pressure that tends to drive solvent molecules out of the solution phase.

REFERENCES

1. *International Critical Tables*, Vol. 3, 1st ed., McGraw-Hill: New York (1928).
2. R. C. Weast, Ed., *CRC Handbook of Chemistry and Physics*, 66th ed., CRC Press: Boca Raton, Florida (1985).
3. G. W. Castellan, *Physical Chemistry*, 3d ed., Addison-Wesley Publishing Co.: Reading, Massachusetts (1983).

lution is 10.1 molal [*1*], which corresponds to a concentration of 77.6% (w/w) sucrose [2].

Empty the 1-liter beaker of water and place the empty beaker under the apparatus. Open the pinch clamp on the rubber tubing and pour sugar solution into the 2-m glass tube until the solution fills the cellulose bag. Close the pinch clamp and pour more sugar solution into the 2-m tube until the level of the solution in the tube is about 15 cm above the 2-holed stopper. Tighten the strings if any of the solution leaks from the cellulose bag. Wearing gloves, dissolve a crystal of iodine in 2 mL of hexane in the test tube and pour the solution into the 2-m tube. Put the clamped end of the rubber tubing in the 600-mL beaker.

Presentation

Fill the 1-liter beaker with distilled water to cover the 2-holed stopper. The level of liquid in the 2-m tube, as indicated by the colored hexane floating on its surface, will begin to rise at an initial rate of about 3 cm/minute. After the solution column rises above the top of the siphon, the pinch clamp should be opened to allow any air remaining in the siphon to be expelled. The rate at which the column rises will decrease as the sugar solution inside the membrane is diluted by the water diffusing into it. After about an hour, the liquid will approach the top of the 2-m tube. To prevent it from overflowing, drain off some of the liquid into the 600-mL beaker by opening the pinch clamp. The rise of solution and draining of the 2-m tube can be repeated many times. Replenish the water in the 1-liter beaker as necessary to keep its level above the 2-holed stopper.

HAZARDS

Wear gloves while handling the iodine container. Iodine is a very strong oxidizing agent, and contact with the skin can result in severe burns. Because iodine vaporizes readily at room temperature to yield toxic fumes, adequate ventilation must be provided when it is handled.

Hexane is flammable and should be kept away from open flames.

DISPOSAL

Use a dropping pipette to remove the hexane solution of iodine from the apparatus and discard the solution in a container for waste organic solvents. Flush the remaining aqueous solutions down the drain with water. Discard the membrane in a standard waste receptacle.

DISCUSSION

In this demonstration, pure water and a concentrated sucrose solution are separated by a membrane that is permeable to water molecules but not to sucrose molecules. Water molecules flow through the membrane from a region of high water concentration (pure water) to one of low water concentration (the sucrose solution). The pressure that

9.41

Color of the Sunset: The Tyndall Effect

When a beam of white light passes through a sol, the beam appears blue when viewed from the side and orange-red when viewed on end.

MATERIALS FOR PROCEDURE A

water to fill aquarium

for each gallon of water:

either

20 mL saturated sodium thiosulfate, $Na_2S_2O_3$ (To prepare ca. 100 mL of stock solution, combine 50 mL of distilled water and 90 g of $Na_2S_2O_3 \cdot 5H_2O$. Warm the mixture until it is liquid and allow it to cool; decant the liquid, which is the saturated solution.)

5 mL 6M hydrochloric acid, HCl (To prepare 1 liter of stock solution, pour 500 mL of concentrated [12M] HCl into 300 mL of distilled water, and dilute the resulting solution to 1.0 liter.)

or

50 mL milk

small aquarium, 1–5 gallons

slide projector

mirror, ca. the size of a smaller side of the aquarium

stirring rod

MATERIALS FOR PROCEDURE B

overhead projector, with screen

400 mL 10% sodium thiosulfate, $Na_2S_2O_3$ (To prepare the solution, dissolve 68 g of $Na_2S_2O_3 \cdot 5H_2O$ in 375 mL of distilled water.)

10 mL 1M hydrochloric acid, HCl, in dropper bottle (To prepare 1 liter of stock solution, pour 83 mL of concentrated [12M] HCl into 600 mL of distilled water, and dilute the resulting solution to 1.0 liter.)

single-edged razor blade

poster board, slightly larger than the stage of the overhead projector

600-mL beaker

glass stirring rod

MATERIALS FOR PROCEDURE C

1 liter 1M sodium chloride, NaCl (To prepare 1 liter of solution, dissolve 58 g of NaCl in 600 mL of distilled water and dilute the resulting solution to 1.0 liter.)

enough unflavored gelatin mix to prepare 1 liter (140 g; 5 oz.)

2 1-liter transparent containers with flat sides

slide projector

PROCEDURE A [1]

Preparation

Arrange the aquarium, slide projector, and mirror as shown in the figure. The beam of the projector should pass through the aquarium side to side, so the audience can see the length of the beam as it passes through. The mirror should be placed at the side opposite from the projector and adjusted so the reflection of the beam shines toward the audience. Add water to the aquarium until it is nearly full.

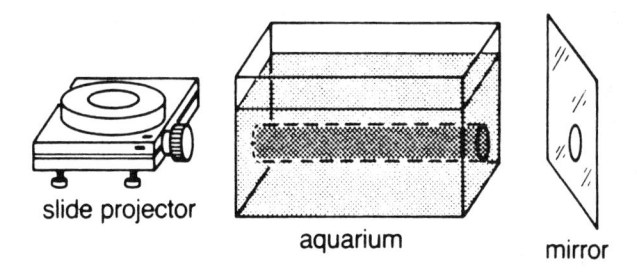

Presentation

Turn on the projector. The beam will be virtually invisible where it passes through the aquarium and bright white where it emerges and shines toward the viewers.

If the effect of the gradual formation of a sol is desired, follow the steps described in the next paragraph of Procedure A. If the effect of the immediate formation of a sol is desired, follow the steps described in the last paragraph of Procedure A. Both effects cannot be shown with a single aquarium.

To produce a sol gradually, for each gallon of water in the aquarium, add 20 mL of saturated sodium thiosulfate solution and 5 mL of 6M hydrochloric acid. Stir the contents of the aquarium. The mixture will become turbid in 1–2 minutes. The turbidity will gradually increase for about 10 minutes. As the turbidity increases, the beam will become visible where it passes through the mixture. The beam passing through the solution appears blue, and where it emerges it is orange. Furthermore, the blue beam will gradually widen.

To produce a sol quickly, for each gallon of water in the aquarium, add 25 mL of milk. Stir the contents of the aquarium. The beam will become visible where it passes through the mixture. The beam passing through the solution is blue, and where it emerges it is orange-red. Add another 25 mL of milk. The beam will still appear blue, but it will be wider than before.

PROCEDURE B [2]

Preparation

With the razor blade, cut a round hole slightly smaller than the diameter of a 600-mL beaker in the center of the poster board. Place the poster board on the stage of the overhead projector and center the 600-mL beaker over the hole.

Presentation

Pour 400 mL of 10% sodium thiosulfate solution into the beaker. Focus the overhead projector to produce a bright spot on the screen. Note that, when viewed from the side, the solution in the beaker is clear. Add 2–3 drops of 1M HCl to the solution in the beaker and stir the mixture. The solution in the beaker will become slightly turbid, and it appears brighter when viewed from the side of the beaker. At the same time, the projected spot on the screen will become less bright. Add several more drops of 1M HCl to the beaker and stir the mixture. The mixture will become more turbid, the beaker will appear brighter when viewed from the side, and the projected spot will become dimmer. Continue to add drops of 1M HCl to the beaker while stirring the mixture. Eventually, the projected spot will become too dim to see, and the mixture in the beaker will appear very bright at the bottom, and dim at the top.

PROCEDURE C [3]

Preparation

Fill one of the containers with 1M sodium chloride solution. Prepare 1 liter of gelatin according to the directions on the package and pour it into the second container before it sets. Allow the gelatin to set.

Presentation

Arrange the containers side by side and shine the beam of the slide projector through both containers simultaneously. Where the beam passes through the gelatin it is visible, but not where it passes through the NaCl solution.

HAZARDS

Hydrochloric acid can cause severe burns. The vapors are extremely irritating to the skin, eyes, and respiratory system.

DISPOSAL

The waste mixtures should be flushed down the drain with water.

DISCUSSION

It has been known for a long time that colloidal suspensions scatter light strongly, producing the Tyndall effect, named after the British physicist John Tyndall (1820–1893), who extensively studied this phenomenon. This phenomenon is shown here for sols, dispersions of a solid in a liquid phase, and for gels, dispersions of a liquid in a solid phase. The theory of light scattering from these suspensions is well developed [4]. Shorter wavelengths are scattered more, and thus, the scattered light is rich in the blue region of the visible spectrum. The transmitted light is correspondingly poor in this region, so it is relatively rich in the red. According to Rayleigh's limiting law for a system dilute in particles that are small relative to the wavelength of the light, the intensity of the scattered light is inversely proportional to the fourth power of its wavelength, that is, it is proportional to λ^{-4}. Because the wavelength of orange light is a factor of 2 longer than that of blue light, the scattering of blue is an order of magnitude greater than that of orange. For coarser particles, the dependence is weaker. Thus, blue light scatters more than orange light, because the wavelength of blue light is shorter than that of orange light. The color of a colloid need not result only from light scattering; gold sols containing relatively large particles absorb as well as scatter light, and are red instead of blue [5].

In Procedure A, the beam gradually widens (or appears wider after the second addition of milk) because the amount of scattering increases. At first the light is scattered only once, and the beam appears narrow. As the number of suspended particles in the colloid increases, the scattering increases. Some of the light is scattered more than once, and the beam appears wider.

Lasers should not be used, because even a small helium-neon laser has a power output 100 times higher than the threshold of eye damage.

Nature provides its own sunset demonstration daily [6]. When viewed at an angle to the light source (the sun), the atmosphere appears blue because it scatters the shorter (blue) wavelengths of light preferentially. When viewed directly toward the light source, the atmosphere appears orange because the longer wavelengths are not scattered as much. This is most obvious at sunset or sunrise because the direct light of the sun passes through the greatest thickness of atmosphere at these times. It is inadvisable to attempt to look directly at the light of the sun at any time other than at sunrise or sunset, because of potential damage to the retina of the eye.

Scattering is also responsible for the color of clouds [7]. When a thick cloud is viewed upward toward the sun, it appears dark, because most of the light heading toward the observer is scattered and some is absorbed. On the other hand, when the same cloud is viewed from above, for example, from an aircraft or a mountaintop, reflected light from the cloud makes it appear white.

REFERENCES

1. W. G. Lamb, *Sci. and Children* (Jan.): 101 (1984).
2. R. H. Goldsmith, *J. Chem. Educ.* 65:623 (1988).
3. D. D. Ebbing, *General Chemistry*, Houghton Mifflin Co.: Boston (1984).
4. H. R. Kruyt, Ed., *Colloid Science*, Vol. 1, *Irreversible Systems*, Elsevier: Amsterdam (1952).
5. B. Jirgensons and M. E. Straumanis, *A Short Textbook of Colloid Chemistry*, John Wiley and Sons: New York (1954).
6. J. Trefil, *Meditations at Sunset*, Charles Scribner's Sons: New York (1987).
7. C. F. Bohren, *Clouds in a Glass of Beer*, John Wiley and Sons: New York (1987).

48. The Old Nassau Clock Reaction

Three colorless solutions are mixed. In a few seconds, the solution turns bright orange, then suddenly turns dark blue.

Procedure

1. Label three 250-mL beakers A, B and C.
2. Place 50 mL of solutions A, B, and C into their respective beakers.
3. Mix the solutions IN THIS ORDER: Add A to B to C.
4. Hold the beaker in view of the class.

Reactions

$$IO_3^- + 3HSO_3^- \rightarrow I^- + 3SO_4^{2-} + 3H^+$$

$$Hg^{2+} + 2I^- \rightarrow HgI_2 \text{ (orange)}$$

$$6H^+ + IO_3^- + 5I^- \rightarrow 3I_2 + 3H_2O$$

$$I_2 + starch \rightarrow \text{(blue)}$$

Solutions

1. Solution A: Dissolve 4 g of soluble starch in 500 mL of boiling water. (Make a paste with a few milliliters of water first.) Cool, add 15 g of $NaHSO_3$ and dilute to 1 liter with distilled water.
2. Solution B: 3 g of $HgCl_2$ per liter distilled water.
3. Solution C: 15 g of KIO_3 per liter distilled water.

Teaching Tips

NOTES

1. To speed up the reaction, use less of solution B.
2. The reaction is called the *Old Nassau* reaction because it produces the colors of Princeton University (orange and black). Nassau Hall is one of the older buildings on the Princeton campus.

QUESTIONS FOR STUDENTS

1. Propose a mechanism to explain how this reaction can produce two distinct colors.
2. How can the reaction rate be increased?
3. Is it necessary to mix the solutions in a particular order? Try it!
4. What compound is formed when the solution turns orange?

50. A Traffic Light Reaction

A flask containing a pale yellow solution is gently swirled. The solution turns red. The flask is shaken, and the solution turns green.

Procedure

1. Place 50 mL of solution A in a 250-mL flask.
2. Add 5–10 mL of indicator solution.
3. Stopper the flask.
4. At the beginning of the demonstration, the solution should be light yellow.
5. Gently swirl the flask to produce the red color.
6. Give the flask a quick shake to produce the green color.

Reactions

1. The indicator is reduced by alkaline dextrose, and a yellow color is produced.
2. When the flask is swirled, oxygen is added, the indicator is oxidized, and the red color is produced.
3. Shaking the flask introduces even more oxygen and causes further oxidation of the indicator to the green color.
4. Upon standing, the dextrose reduces the indicator back to the yellow color.

Solutions

1. Solution A: 3 g of dextrose (glucose) and 5 g of NaOH in 250 mL of water.
2. The indigo carmine indicator is a 1.0% solution. (With practice, you can use a small amount of solid.)

Teaching Tips

NOTES

1. If the red color does not persist, adjust the number of drops of indicator.
2. This traffic light has an advantage over others—a magnetic stirrer is not required.
3. For a variation of this reaction, see Demonstration 59, The Blue Bottle Reaction.

QUESTIONS FOR STUDENTS

1. Propose a chemical equation for the reaction.
2. Is this a redox reaction? If so, what is oxidized and what is reduced?
3. What role does the indicator play?
4. What happens when the flask is swirled? Shaken?
5. Will this reaction *run down* if the stopper remains in the flask?

44. Catalytic Decomposition of Hydrogen Peroxide: Foam Production

A tremendous amount of foam shoots from a graduated cylinder when detergent and potassium iodide are added to hydrogen peroxide.

Procedure

1. Place a large graduated cylinder (500 mL) in a plastic tray or in a laboratory sink.
2. Pour ~ 50 mL of 30% hydrogen peroxide into the cylinder. (CAUTION!)
3. Add a squirt of dishwashing detergent and a drop of food coloring.
4. Add about one-fourth of a spoonful of solid KI.

Reactions

1. The rapidly catalyzed decomposition of hydrogen peroxide produces oxygen gas, which forms a foam with the liquid detergent:

$$2H_2O_2(aq) \rightarrow 2H_2O(l) + O_2(g)$$

2. The actual decomposition of H_2O_2 in the presence of iodide ion occurs in two steps. The first reaction is the rate-determining reaction.

$$H_2O_2(aq) + I^-(aq) \rightarrow H_2O(l) + OI^-(aq)$$

$$H_2O_2(aq) + OI^-(aq) \rightarrow H_2O(l) + O_2(g) + I^-(aq)$$

Solution

You must use 30% hydrogen peroxide for best results. Do not use the 3% hydrogen peroxide available from a drugstore.

Teaching Tips

NOTES

1. Be careful when using 30% hydrogen peroxide. Wear gloves and avoid contact with this solution.
2. You can also show the decomposition of hydrogen peroxide on an overhead projector. Place a small amount of 3% hydrogen peroxide in a petri dish on an overhead projector. Add a pinch of potassium iodide, or manganese dioxide, and note the evolution of oxygen gas bubbles.
3. The catalytic decomposition of hydrogen peroxide occurs when the 3% solution is placed on a wound. Catalase, an enzyme in the blood, catalyzes the reaction.

QUESTIONS FOR STUDENTS

1. How does a catalyst work?
2. What happened to the KI?
3. How can you account for the large amount of foam produced?
4. What evidence is there that iodine is produced? (The brown color of the foam.)

36. Equilibrium in the Gas Phase

A reddish-brown gas is prepared and placed in two small test tubes. When one tube is placed in boiling water, the color of the gas changes to a deep brown color. When the other tube is placed in an ice bath, the gas becomes almost colorless. When both tubes are allowed to reach room temperature, the gas in both again becomes reddish-brown.

Procedure

PREPARING THE GAS TUBES

1. Prepare a gas generator by attaching a rubber tube to a small side-arm flask.
2. IN A HOOD, add 10 mL of concentrated nitric acid to the flask.
3. Drop in a copper penny.
4. A deep-red gas, NO_2, will immediately form. Allow enough of the gas to form to displace the air in the flask and two test tubes. Colorless NO forms first. Colorless bubbles rise to the surface where they mix with O_2 in the air and immediately form NO_2.
5. Fill the two tubes with the brown gas. Stopper the tubes.
6. Stop the reaction in the flask by filling the flask with water. Notice the blue color of the solution and the small size of the penny.

DEMONSTRATING EQUILIBRIUM

1. Point out the color of the gas in the tube. Put the equilibrium reaction on the board.
2. Place one tube in a beaker of boiling water. Notice the change in color.
3. Place the other tube in an ice bath. Notice the formation of a colorless gas.
4. Remove both tubes and allow them to come to room temperature. Note the restoration of the brown color in both tubes.

Reactions

1. The equations for the production of the gas are as follows:

$$3Cu + 8H^+ + 2NO_3^- \rightarrow 3Cu^{2+} + 2NO + 4H_2O$$

$$2NO + O_2 \rightarrow 2NO_2$$

2. The equilibrium mixture in the tubes consists of NO_2 and N_2O_4. They react according to the equation:

$$2NO_2(g) \rightleftharpoons N_2O_4$$

(red)   (colorless)

3. When the equilibrium mixture is heated, the equilibrium shifts toward the formation of brown NO_2.
4. When the mixture is cooled, the equilibrium shifts toward the formation of more colorless N_2O_4.

Teaching Tips

NOTES

1. The preparation of the gas tubes must be done IN A HOOD: NO_2 is TOXIC. It has been estimated that five pennies could produce enough poisonous NO_2 to fill a laboratory!
2. Fat, short, clear plastic test tubes work best. They can be used for several weeks, if tightly stoppered.

QUESTIONS FOR STUDENTS

1. Write the chemical equation for the production of the gas.
2. Why is the solution formed in the reaction vessel blue?
3. Could the chemical change in the tubes be due to an increase in pressure, rather than an increase in temperature?
4. Devise an experiment to prove or disprove your hypothesis.
5. Is this an exothermic or endothermic reaction?

32. Effect of Temperature Change on Equilibrium: Cobalt Complex

An equilibrium system involving the dehydrated–hydrated cobalt complex is produced. When this system is heated, a color change from pink to blue indicates a shift of equilibrium to the right. When the solution is cooled, the color change from blue to pink indicates a shift to the left.

Procedure

1. Place 100 mL of $CoCl_2$ solution in a 250-mL beaker.
2. Add concentrated HCl until the solution changes from pink to blue.
3. Divide the solution into three smaller beakers and treat them as follows:

    a. Place one beaker on a hot plate.
    b. Place one beaker in an ice bath.
    c. Leave one beaker at room temperature as a standard.

4. After a few minutes, show that the heated sample has turned a darker blue and that the cooled sample has turned a light pink.

Reactions

1. This reaction involves the following equilibrium:

$$\text{heat} + [Co(H_2O)_6]^{2+}(aq) + 4Cl^-(aq) \rightleftharpoons [CoCl_4]^{2-}(aq) + 6H_2O$$

   (pink)   (blue)

2. Addition of heat causes a shift of equilibrium toward products, the blue solution.
3. Cooling causes a shift of equilibrim to the left, the pink hydrated complex.

Solutions

1. The $CoCl_2$ solution is 0.4 M: Dissolve 5.2 g per 100 mL of water.
2. The HCl solution is concentrated.

Teaching Tips

NOTES

1. As indicated in the equation, you may have to add quite a bit of HCl to get the formation of the blue complex.
2. The blue color is due to the tetrachlorocobalt(II) complex, and the pink color is due to the hexaaquacobalt(II) complex.

QUESTIONS FOR STUDENTS

1. Write an equation for the equilibrium system.
2. Why was it necessary to add HCl to establish equilibrium?
3. How does heating shift the equilibrium?
4. What do you think would happen to the equilibrium system if water is added? Try it!

8.4

Acid-Base Indicators Extracted from Plants

A purple liquid is extracted from red cabbage, and this liquid changes color when added to various substances found around the house (Procedure A). A colored liquid is extracted from plant material, and this extract has different colors at different pH values, as observed directly (Procedure B) or by overhead projection (Procedure C). "Indicator paper" is prepared by soaking a sheet of paper in the plant extract; the paper changes color when it is dampened with solutions of various pH values (Procedure D). A multicolored design is created when a sheet of paper treated with plant extract is "painted" with colorless liquids (Procedure E).

MATERIALS FOR PROCEDURE A

Provide household materials in their original containers.

ca. 100 g red cabbage (half of a small head is sufficient)

ca. 2 liters distilled water

125 mL vinegar

125 mL laundry ammonia

5 mL (1 teaspoon) baking soda

125 mL colorless carbonated beverage (e.g., lemon-lime flavor)

5 mL (1 teaspoon) laundry detergent

125 mL milk

knife to cut cabbage

electric blender or food processor

kitchen sieve or colander

1-liter beaker

7 (or more) 250-mL beakers

spoon or stirring rod

MATERIALS FOR PROCEDURE B

For preparation of stock solutions of acids and bases, see pages 30–32.

100 g plant material, such as:

   vegetables (e.g., red cabbage, beets, red onions, radishes, rhubarb) fresh or frozen fruit (e.g., blueberries, cherries, red grapes)

    flowers (e.g., day lilies, roses)

    plants (e.g., carrot greens, tomato plants, black tea)

100 mL 95% ethanol or distilled water

600 mL distilled water

200 mL 2M hydrochloric acid, HCl

200 mL 2M sodium hydroxide, NaOH

electric blender or food processor

kitchen sieve or colander

2 1-liter beakers

50-mL graduated cylinder

glass stirring rod

MATERIALS FOR PROCEDURE C

For preparation of stock solutions of acids and bases, see pages 30–32.

overhead projector

100 g plant material, such as:

    vegetables (e.g., red cabbage, beets, red onions, radishes, rhubarb)

    fresh or frozen fruit (e.g., blueberries, cherries, red grapes)

    flowers (e.g., day lilies, roses)

    plants (e.g., carrot greens, tomato plants, black tea)

100 mL 95% ethanol or distilled water

10 mL distilled water

20 mL 2M hydrochloric acid, HCl

20 mL 2M sodium hydroxide, NaOH

electric blender or food processor

kitchen sieve or colander

1-liter beaker

50-mL graduated cylinder

50-mL beaker

MATERIALS FOR PROCEDURE D

For preparation of stock solutions of acids and bases, see pages 30–32.

100 g plant material (See Materials for Procedure B.)

100 mL 95% ethanol or distilled water

10 mL each of buffer solutions with pH of 1, 3, 5, 7, 9, 11, and 13 (For preparation, see Procedure A of Demonstration 8.1.)

10 mL 2M hydrochloric acid, HCl

10 mL 2M sodium hydroxide, NaOH

electric blender or food processor

kitchen sieve or colander

1-liter beaker

sheet of white cotton cloth or filter paper, 50 cm × 50 cm

2 ring stands

100 cm adhesive tape (optional)

9 20-mL test tubes (optional)

meter stick (optional)

MATERIALS FOR PROCEDURE E

For preparation of stock solutions of acids and bases, see pages 30–32.

100 g red cabbage

100 mL distilled water

10 mL each of buffer solutions with pH of 1, 5, 7, 11, and 13 (For preparation, see Procedure A of Demonstration 8.1.)

10 mL 2M sodium hydroxide, NaOH

knife to cut cabbage

electric blender or food processor

kitchen sieve or colander

1-liter beaker

sheet of white cotton cloth or filter paper, 50 cm × 50 cm

2 ring stands

6 50-mL beakers

6 small paint brushes

PROCEDURE A

Preparation and Presentation

Prepare red cabbage extract as follows. Cut the red cabbage into 1-inch cubes. Place the cabbage cubes in the blender and add enough water to cover the cabbage. Blend the mixture until the cabbage has been chopped into uniformly tiny pieces. Using the sieve, strain the liquid from the mixture into the 1-liter beaker, pressing the cabbage to speed the straining. The strained liquid is the red cabbage extract.

Pour 125 mL of vinegar into one of the 250-mL beakers. Add about 5 mL of red cabbage extract and stir the mixture. Record the color of the mixture.

Pour 125 mL of laundry ammonia into another 250-mL beaker. Add about 5 mL of red cabbage extract and stir the mixture. Record the color of the mixture.

swirling the beaker while adding 2M NaOH drop by drop. Continue to shift the color back and forth by alternately adding acid and base.

PROCEDURE D

Preparation

At least one day before the demonstration is to be presented, prepare an extract as described in the first paragraph of Procedure B. Soak a cloth or large piece of filter paper in the extracted liquid and hang the cloth or paper until it is dry.

Suspend the dried cloth or paper like a curtain between two ring stands.

Presentation

Dribble 10 mL of each of the buffer solutions, in separate streaks, as well as 10 mL of 2M NaOH and 10 mL of 2M HCl, down the vertically mounted, extract-stained cloth or paper. (One way to accomplish this is by attaching nine upright test tubes with adhesive tape at intervals along with a 30-cm portion of a meter stick. Starting at one end of the row of test tubes, pour 10 mL of 2M HCl into the first tube, 10 mL of pH 1 buffer into the second, 10 mL of pH 3 buffer in the third, and so forth through 10 mL of pH 13 buffer solution in the eighth tube. Pour 10 mL of 2M NaOH into the ninth tube. Then tip the meter stick so that the solutions streak down the extract-stained material.)

PROCEDURE E

Preparation

At least one day before the demonstration is to be presented, prepare a red cabbage extract as described in the first paragraph of Procedure A. Then prepare the cloth or paper as described in the Preparation section of Procedure D.

Pour about 10 mL of each buffer solution of pH 1, 5, 7, 11, and 13 and 10 mL of 2M NaOH into separate 50-mL beakers.

Presentation

Dip a paint brush in one of the colorless buffer solutions or in the 2M NaOH, and "paint" with the brush on the vertically mounted, extract-stained cloth or paper. The color produced on the cloth when touched by the brush depends on the solution on the brush, as indicated below.

Solution	Color
pH 1	red
pH 5	violet
pH 7	blue
pH 11	blue-green
pH 13	yellow-green
2M NaOH	yellow

A multicolored picture can be created by painting with the brushes, dipping each in a different one of the colorless buffer solutions or 2M NaOH.

HAZARDS

Care should be taken to avoid spilling ethanol near the blender. Sparks in the electric motor of the blender can ignite ethanol vapor. Ethanol should not be used in a food processor because vapors are free to enter the motor housing from the bowl of the processor; sparks from the motor can ignite ethanol vapors.

Concentrated solutions of sodium hydroxide can cause severe burns to the eyes, skin, and mucous membranes.

Hydrochloric acid can irritate the skin. Its vapors are extremely irritating to the eyes and respiratory system.

DISPOSAL

The waste solutions should be flushed down the drain with water.

DISCUSSION

An extract of red cabbage is prepared in Procedure A and this extract changes color when it is mixed with different household substances. In Procedures B and C, a plant extract is prepared and its color is changed reversibly by the alternating addition of acid and base. In Procedure D, the plant extract is used to prepare test paper, and solutions of various pH values are dribbled onto the paper to show the effect of each pH on the color of the paper. Procedure E involves "painting" with colorless buffer solutions on a sheet of paper that has been treated with red cabbage extract.

For well over 300 years investigators have been studying the effects of the addition of acids and bases to colored plant extracts. The ability of certain substances to change the colors of plant extracts was one of the earliest defining characteristics of acids. In 1664 Robert Boyle published in *The Experimental History of Colours* that the extracts of certain plants such as red roses and brazil wood changed color reversibly when made alternately basic and acidic. These extracts could be used as acid-base indicators. In 1670–1671 DuClos was perhaps the first to use "turnesole," an early impure form of litmus. Over a hundred years later James Watt complained that red rose paper didn't keep its color for more than a few months and suggested the use of red cabbage indicator paper. In 1801 Vauquelin spoke of using litmus and violets, and Welter advocated using litmus and radishes; Lampadius reported success using litmus, curcuma, red cabbage, alkanet, rhubarb, violets, columbine, and red roses.

The table lists the variations in the colors of several plant extracts with changes in their pH values. Most plant extracts will undergo at least a faint color change if the pH is changed sufficiently. Not all such color changes are reversible, so not all extracts are, strictly speaking, pH indicators. However, the extracts listed in the table do undergo reversible changes, although the initial color change may not be completely reversible. This is most likely due to the irreversible destruction of some pigment in the extract when the pH is changed. Subsequent changes in pH will result in reversible color

Colors[a] of Selected Plant Extracts at Various pH Values

Plant extract	Solvent	Initial appearance of extract[b]	1	2	3	4	5	6	7	8	9	10	11	12	13
Beets	ethanol	dk rd	vt	rd-vt						rd-vt			vt	br	
Beets	water	dk rd	rd-vt				rd		rd						yl-gn
Blackberry juice	water	dk rd			rd						rd	br	vt	bl	gn
Blueberry juice	water	dk rd			rd						br		vt	bl	bl-gn
Carrot greens	ethanol	dk gn	yl-br							gn-yl					
Carrot greens	water	yl-br	lt yl-br												dk yl-br
Cherries	ethanol	rd	rd	pk										br	gn
Cherries	water	rd-br	rd-br												gn-br
Daisy top	ethanol	yl-gn	pa yl												dk yl
Daisy top	water	br	pa br												dk br
Day lily	ethanol	gn-br	pk		pa br		pa yl				yl-gn	gn			yl-gn
Day lily	water	br	pk			rd				dk br	rd-bl	pa yl		bl	dk yl
Grape juice	water	dk rd	rd		dk vt			pa bl							gn-bl
Red cabbage	ethanol	rd-br	rd		vt	rd-vt		vt	bl	gn		bl-gn		gn	yl
Red cabbage	water	vt	rd				pa pk			gn-bl			gn		yl-gn
Red onions	ethanol	dk vt	rd		pa pk		pa pk		pa yl	yl				gn	yl
Red onions	water	rd-vt	pk		pa pk				pa gn	yl				gn	yl
Radishes	ethanol	pk	pk		pk					cl	br		vt	gn	yl
Radishes	water	pk	pk							cl	br		vt	gn	yl
Rhubarb	ethanol	rd	rd				cl			pa vt			bl		dk br
Rhubarb	water	or	pk							pa yl		yl		gn	yl
Rose petals	ethanol	yl-gn	pk							br	gr	yl	gr		br
Rose petals	water	br	pa yl-br												br
Black tea	ethanol	br	pa yl									yl-br			dk br
Black tea	water	br	pa br												dk br
Tomato leaves	ethanol	gn	pa br		pa yl-gr					dk yl-gr		gn			yl gn
Tomato leaves	water	gn-br	pa yl-br												dk yl-br

[a] The following abbreviations have been used for the colors in this chart:

bl = blue   dk = dark   pa = pale   vt = violet
br = brown   gn = green   pk = pink   yl = yellow
cl = colorless   or = orange   rd = red

[b] The abbreviations in regular type in this column indicate that the extract initially has a clear appearance; the abbreviations in *italic* type, that the extract initially has a cloudy appearance.

changes, as long as pH extremes (below 1 and above 11) are avoided. The initial pH of the extracts can be discerned by comparing the initial color of the extract with its color as it is affected by the different pH values (see the table).

Most plant extracts contain a mixture of pigments [3]. Because of this, there is usually no sharp color change with pH, but instead a gradual fading from one color into another over a range of several pH units. However, some extracts do change sharply, some at more than one pH value (e.g., red cabbage extract). As illustrated in Procedure A, red cabbage can be used to determine the approximate pH of a number of common household substances (or their solutions).

Most of the pH-sensitive red, blue, and violet pigments in plants are water-soluble anthocyanins which are easily extracted from the plant. A typical anthocyanin is red in acid solution, purple in neutral, and blue in basic. The blue cornflower, burgundy dahlia, and red rose all contain the same anthocyanin, but they differ in the acidity of their sap. Many white flowers contain anthoxanthin, which turns yellow when treated with base. A green color can result in a plant extract made basic from the combined effect of blue anthocyanin and yellow anthoxanthin pigments. More than one anthocyanin can be present in a flower, and changes in the pH of its sap can produce subtle changes in shade of the flower [4].

red color
of rose

yellow color
of rose

The pH-sensitive pigments in most plants are soluble in ethanol as well as in water. Many of the other components of the sap are not readily soluble in ethanol, and therefore, ethanol is a more effective extracting solvent for the pigments than is water. The extracts produced with ethanol are generally clearer and have more vivid colors than the water extracts.

REFERENCES

1. S. Sharpe, *The Alchemist's Cookbook: 80 Demonstrations*, Shell Canada Centre for Science Teachers, McMaster University: Hamilton, Ontario (undated), p. 20.
2. M. Forster, *J. Chem. Educ.* 55:107 (1978).
3. E. Bishop, *Indicators*, Pergamon Press: Oxford (1972), Ch. 1 of Vol. 51 in International Series of Monographs in Analytical Chemistry.
4. G. S. Losey, "Biological Coloration," in *Encyclopaedia Britannica*, 15th ed. (1983), Vol. 4, p. 917.

8.20

Differences Between Acid Strength and Concentration

The pH of 0.1M hydrochloric acid is about the same as that of 0.1M sulfuric acid and lower than that of 0.1M acetic acid. All three acids are neutralized with sodium hydroxide solution. Neutralization of the hydrochloric acid requires only half as much sodium hydroxide as does neutralization of the sulfuric acid, although they have nearly the same pH. Neutralization of the hydrochloric acid requires the same amount of sodium hydroxide as does the neutralization of the acetic acid, although they have different pH values (Procedure A) [1, 2]. Similar differences between pH and neutralizing capacity are also demonstrated with bases (Procedure B).

MATERIALS FOR PROCEDURE A

For preparation of indicator solutions, see pages 27–29.
For preparation of stock solutions of acids and bases, see pages 30–32.

40 mL 0.1M hydrochloric acid, HCl

40 mL 0.1M sulfuric acid, H_2SO_4

40 mL 0.1M acetic acid, $HC_2H_3O_2$

2 mL phenolphthalein indicator solution

200 mL 0.1M sodium hydroxide, NaOH

3 250-mL beakers

3 labels for beakers

dropper

3 glass stirring rods

pH meter, standardized, or pH-indicating paper, pH 0–3 range

100-mL graduated cylinder

MATERIALS FOR PROCEDURE B

For preparation of indicator solutions, see pages 27–29.
For preparation of stock solutions of acids and bases, see pages 30–32.

40 mL 0.1M sodium hydroxide, NaOH

40 mL 0.1M barium hydroxide, $Ba(OH)_2$ (To prepare 1 liter of solution, dissolve

17 g of Ba(OH)$_2$ in 600 mL of distilled water, and dilute the resulting solution to 1.0 liter.)

40 mL 0.1M ammonia, NH$_3$ (To prepare 1 liter of stock solution, pour 7 mL of concentrated [15M] NH$_3$ into 600 mL of distilled water, and dilute the resulting solution to 1.0 liter.)

2 mL bromocresol green indicator solution

200 mL 0.1M hydrochloric acid, HCl

3 250-mL beakers

3 labels for beakers

dropper

3 glass stirring rods

pH meter, standardized, or pH-indicating paper, pH 11–13 range

100-mL graduated cylinder

PROCEDURE A

Preparation

Label the three 250-mL beakers with the names of the three acids: hydrochloric acid, acetic acid, and sulfuric acid. Pour 40 mL of the appropriate 0.1M acid into each beaker. Add 10 drops of phenolphthalein indicator solution to each beaker and stir the mixtures thoroughly.

Presentation

Use the pH meter or the pH test paper to measure the pH of each of the three acid solutions. The pH of the HCl and H$_2$SO$_4$ solutions will be near 1, and that of the HC$_2$H$_3$O$_2$ will be about 3. Record the pH values.

While stirring the HCl solution, *slowly* pour 0.1M NaOH from the 100-mL graduated cylinder until the phenolphthalein indicator changes from colorless to pink. The volume of NaOH solution required to do this will be close to 40 mL. Record the volume.

Repeat the procedure described in the preceding paragraph with the beaker of HC$_2$H$_3$O$_2$ solution and with the beaker of H$_2$SO$_4$ solution. The HC$_2$H$_3$O$_2$ will require about 40 mL of NaOH, and the H$_2$SO$_4$ will use about 80 mL.

PROCEDURE B

Preparation

Label the three 250-mL beakers with the names of the three bases: sodium hydroxide, ammonia, and barium hydroxide. Pour 40 mL of the appropriate 0.1M base into each beaker. Add 10 drops of bromocresol green indicator solution to each beaker and stir the mixtures thoroughly.

Presentation

Use the pH meter or the pH test paper to measure the pH of each of the three base solutions. The pH of the NaOH and Ba(OH)$_2$ solutions will be near 13, and that of the NH$_3$ will be about 11. Record the pH values.

While stirring the NaOH solution, *slowly* pour 0.1M HCl from the 100-mL graduated cylinder until the bromocresol green indicator changes from blue to green or yellow. The volume of HCl solution required to do this will be close to 40 mL. Record the volume.

Repeat the procedure described in the preceding paragraph with the beaker of NH$_3$ solution and with the beaker of Ba(OH)$_2$ solution. The NH$_3$ will require about 40 mL of HCl, while the Ba(OH)$_2$ will use about 80 mL.

HAZARDS

Soluble barium compounds are toxic and, if ingested, cause nausea, vomiting, stomach pains, and diarrhea.

DISPOSAL

The waste solutions should be flushed down the drain with water.

DISCUSSION

The notions of acid concentration (as expressed by molarity), acid equivalence (as determined by a titration), and acid strength (as expressed by pH) are sometimes confused. Procedure A of this demonstration illustrates that these are three different properties of acids. The concentration (molarity) of an acid is an operational expression of how the solution can be prepared. That is, it indicates how much of the acid must be used to prepare the solution. The equivalence of an acid solution expresses how much base is required to neutralize it. The equivalence depends on the concentration of the solution and on the number of acidic hydrogens in each molecule of acid. The pH of an acid solution depends not only on the concentration and equivalence of the solution of the acid, but is also related to the degree of ionization of the particular acid in the solution. In Procedure B, these three concepts—strength, concentration, and equivalence—are also applied to bases.

The differences between strength, concentration, and equivalence are illustrated for acids in Procedure A. The procedure uses solutions of hydrochloric acid, sulfuric acid, and acetic acid that have the same concentration, namely 0.1M. This means that each solution can be prepared by dissolving 0.1 mole of its acid in 1 liter of solution. Although the solutions of these three acids have the same concentration, they do not have the same pH. The pH of 0.1M HCl is about the same as that of 0.1M H$_2$SO$_4$, namely 1, and both of these have lower pH values than 0.1M HC$_2$H$_3$O$_2$, which has a pH of about 3. This indicates that the acid strength of hydrochloric acid is about the same as that of sulfuric acid, but both of these are stronger acids than acetic acid. When 40-mL samples of these acids are titrated with sodium hydroxide solution to a phephthalein end point, the hydrochloric acid requires about the same volume of NaOH as does the acetic acid, and the sulfuric acid requires twice as much as either of the other acids. This titration distinguishes acid equivalence from acid strength and acid concentration. All three have the same molar concentration, yet the amount of NaOH required for neutralization of one is different from that of the other two. The two acids of similar strength, HCl and H_2SO_4, as indicated by their pH values, require different amounts of NaOH for neutralization. Two acids that require the same amount of NaOH for neutralization, HCl and $HC_2H_3O_2$, have different pH values and, therefore, different strengths. The results of the demonstration can be summarized in tabular form to highlight the differences among concentration, strength, and equivalence:

Property:	concentration	strength	equivalence
Expressed through:	molarity	pH	mL of titrant
HCl	0.1	1	40
H_2SO_4	0.1	1	80
$HC_2H_3O_2$	0.1	3	40

The neutralization of sulfuric acid requires twice as much sodium hydroxide as the other two acids, because it is a diprotic acid and the other two are monoprotic. This means that each mole of sulfuric acid contains 2 moles of replaceable hydrogens, whereas each mole of hydrochloric acid and acetic acid has only 1 mole of replaceable hydrogens. Even though sulfuric acid is a diprotic acid, the pH of the sulfuric acid is nearly identical with that of the hydrochloric acid, because the first ionization of sulfuric acid, which is complete, suppresses the second ionization, which has an ionization constant of $K_a = 2 \times 10^{-2}$ [3]. The pH of acetic acid is higher than that of the other monoprotic acid, hydrochloric acid, because it is a weak acid.

$$HC_2H_3O_2(aq) \leftrightarrows H^+(aq) + C_2H_3O_2^-(aq)$$

$$K_a = \frac{[H^+][C_2H_3O_2^-]}{[HC_2H_3O_2]} = 1.8 \times 10^{-5}$$

Acetic acid is only partly ionized in solution, while hydrochloric acid is completely ionized.

For bases, a similar pattern is demonstrated in Procedure B, comparing the amounts of hydrochloric acid required to neutralize volumes of 0.1M NaOH, 0.1M NH_3, and 0.1M $Ba(OH)_2$. In this case, the pH of $Ba(OH)_2$, which requires twice as much acid for neutralization as the NaOH or NH_3, is about 0.3 of a unit higher than that of the NaOH. However, this pH difference is much smaller than that between NaOH and NH_3, which require the same amount of acid for neutralization. Although the demonstration of the differences among concentration, strength, and equivalence with these bases is not as clear-cut as with the acids, it is still effective.

REFERENCES

1. R. Perkins, *Chem13News* 147 (Feb.):11 (1984).
2. M. J. Webb, *J. Chem. Educ.* 58:193 (1981).
3. J. A. Dean, Ed., *Lange's Handbook of Chemistry*, 13th ed., McGraw-Hill Book Co.: New York (1985).

66. Silver Ion Solubilities: Red and White Precipitates

A few drops of a solution are added to a second solution, and a deep red precipitate forms. When a few drops of a third solution are added to the same solution, a white precipitate forms. As more solution is added, the red precipitate dissolves and leaves only the white precipitate.

Procedure

1. Place 200 mL of silver nitrate solution in a 250-mL Erlenmeyer flask or a large beaker.
2. Add 2 drops of the yellow sodium chromate solution. Note the immediate formation of a RED precipitate.
3. Add, dropwise, 1-2 mL of sodium chloride solution. Notice the formation of a WHITE precipitate.
4. After observing and discussing the two precipitates, add 50 mL of sodium chloride solution. Observe that the red precipitate dissolves and leaves only the white precipitate.

Reactions

This demonstration involves the formation of two precipitates from silver nitrate solution.

1. $2Ag^+(aq) + CrO_4^{2-}(aq) \rightleftharpoons Ag_2CrO_4(s)$
   red

2. $Ag^+(aq) + Cl^-(aq) \rightleftharpoons AgCl(s)$
   white

Because the solubility of red silver chromate is greater than that of silver chloride, more white silver chloride is formed as more chloride is added.

Solutions

1. The silver nitrate concentration is 0.05 M: Dissolve 0.85 g of $AgNO_3$ in 100 mL of solution.
2. The sodium chloride concentration is 0.2 M: Dissolve 1.2 g of NaCl in 100 mL of solution.
3. The sodium chromate concentration is 0.5 M: Dissolve 8.1 g of Na_2CrO_4 in 100 mL of solution.

Teaching Tips

NOTES

1. The solubility of silver chromate is 1.4×10^{-3} g/100 mL of water. The solubility of silver chloride is 8.9×10^{-6} g/100 mL of water.
2. This demonstration offers an effective way to introduce equilibrium. As more chloride is added, more silver ions react. This reaction shifts the equilibrium from the solid silver chromate to form more silver ions and chromate ions. This shift should produce a slight yellow (chromate) tint to the solution.

3. See Appendix 4 for disposal procedures.
4. Allow the precipitate to settle for an hour or so to make the extent of precipitation more obvious.

QUESTIONS FOR STUDENTS

1. Explain how each precipitate forms in this demonstration.
2. Why does the red precipitate dissolve?
3. Is there a slight yellow color in the final solution? If so, why?
4. Theoretically (but not practically), how would you dissolve the white precipitate?

45. The Colors of Some Chromium and Manganese Ions

Two solids are added to watch glasses containing MnO$_2$ and Cr$_2$O$_3$ and heated. When water is added, a bright yellow solution is produced on one glass and a deep green solution is produced on the other. Addition of acid causes these solutions to change to purple and orange, respectively.

Procedure

1. Place about ¼ tsp of NaOH and KNO$_3$ on each of two watch glasses and mix.
2. Add ¼ tsp of MnO$_2$ to one of the watch glasses and mix thoroughly. Place the watch glasses on a tripod and carefully heat until fusion occurs. Set the watch glasses aside to cool.
3. Add ¼ tsp of Cr$_2$O$_3$ to the other watch glass and mix thoroughly. Heat until fusion occurs and set the watch glass aside to cool.
4. When the watch glasses are cool, add water to each watch glass from a wash bottle; tilt each watch glass so that the "solution" runs into separate beakers.
5. The green solution is the manganate ion. To this solution add a few drops of dilute HNO$_3$; the purple color of the permanganate occurs because the green manganate is stable only in alkaline solution.
6. The yellow solution is the chromate ion. To this solution add a few drops of dilute HNO$_3$; the orange color of the dichromate ion occurs.

Solution

Add 2 parts of concentrated nitric acid to 3 parts of water to make a dilute solution of HNO$_3$.

Reactions

1. $MnO_2(s) + 2OH^-(aq) + NO_3^-(aq) \longrightarrow MnO_4^{2-} + NO_2^-(aq) + H_2O(\ell)$
   black mixture → green

   $3MnO_4^{2-}(aq) + 4H^+(aq) \longrightarrow 2MnO_4^-(aq) + MnO_2(s) + 2H_2O(\ell)$
   green → purple

2. $Cr_2O_3(s) + 4OH^-(aq) + 3NO_3^-(aq) \longrightarrow 2CrO_4^{2-}(aq) + NO_2^-(aq) + H_2O(\ell)$
   dark green mixture → yellow

   $2CrO_4^{2-}(aq) + 2H^+(aq) \longrightarrow Cr_2O_7^{2-}(aq) + H_2O(\ell)$
   yellow → orange

Teaching Tips

NOTES

1. Select two of your most used watch glasses. The process almost always causes the watch glass to crack upon cooling. Be sure not to add the water until they are cooled. Cooling may take 4 or 5 min. *Splattering will occur if the watch glasses are not cooled.* Do not use evaporating dishes or casseroles. The glazing will be removed and the dishes ruined.

2. The strong oxidizing effect of permanganate makes it useful for treatment of athlete's foot, for treatment of rattlesnake bite, and as an antidote for some poisons.
3. MnO_2 is the black material in dry cells. It is useful because of its strong oxidizing power and its high conductivity.
4. Chromium does not occur as the element in nature.

QUESTIONS FOR STUDENTS

1. Write the equations for the reactions.
2. Are the ions used in this demonstration only stable in basic or acidic solutions? Which in which?
3. Would another acid have worked just as well?
4. What was the role of the nitrate ion?
5. Could you have used something other than nitrate? What?

4.8

Precipitates and Complexes of Copper(II)

Different reagents are added sequentially to two beakers containing aqueous solutions of $CuSO_4$ and $Cu(NO_3)_2$. The color and solubility of a series of Cu species including $Cu(H_2O)_6^{2+}$, $Cu(NH_3)_4^{2+}$, $Cu(en)_2^{2+}$, $Cu(EDTA)^{2-}$, $Cu(CN)_3^{2-}$, $Cu(OH)_2$, CuO, $CuSO_4 \cdot 3Cu(OH)_2$, and CuS are observed.

MATERIALS

500 ml distilled water

50 ml 0.1M copper sulfate, $CuSO_4$ (To prepare 1 liter of 0.1M stock solution, dissolve 25 g $CuSO_4 \cdot 5H_2O$ in distilled water and dilute to 1 liter.)

50 ml 0.1M copper nitrate, $Cu(NO_3)_2$ (To prepare 1 liter of 0.1M stock solution, dissolve 24.2 g $Cu(NO_3)_2 \cdot 3H_2O$ in distilled water and dilute to 1 liter.)

ca. 5 g copper sulfate pentahydrate, $CuSO_4 \cdot 5H_2O$

1 ml 1M sulfuric acid, H_2SO_4

ca. 5 g copper sulfate, anhydrous, $CuSO_4$ (optional)

50 ml 5M aqueous ammonia, NH_3 (To prepare 1 liter of 5M stock solution, dilute 333 ml concentrated ammonia, 15M, to 1 liter with distilled water.)

50 ml 25% ethylenediamine, $H_2NCH_2CH_2NH_2$ (To prepare 1 liter of 25% stock solution, dilute 250 ml $C_2H_8N_2$ to 1 liter with distilled water.)

50 ml 0.1M disodiumdihydrogenethylenediaminetetraacetate, Na_2H_2EDTA. (To prepare 1 liter of 0.1M stock solution, dissolve 37.2 g $Na_2H_2EDTA \cdot 2H_2O$ in distilled water and dilute to 1 liter.)

25 ml 1.0M potassium cyanide, KCN (To prepare 1 liter of 1.0M stock solution, dissolve 65 g KCN in distilled water and dilute to 1 liter.) **(See Hazards section before handling cyanide salts.)**

25 ml 0.1M sodium sulfide, Na_2S (To prepare 1 liter of 0.1M stock solution, dissolve 24 g $Na_2S \cdot 9H_2O$ in distilled water and dilute to 1 liter.)

2 hot plates with magnetic stirrers and stirring bars

3 600-ml beakers, preferably tall-form

3 100-ml graduated cylinders

casserole, 3–4 inches in diameter

Meker burner

ring stand

iron ring to hold casserole

heat-protective glove

2 droppers

100-ml beaker

2 10-ml graduated cylinders

PROCEDURE

Set up two hot plates with magnetic stirrers and place on each a 600-ml beaker containing a stirring bar. Add to each beaker 75 ml of distilled water, turn on the stirrers and adjust them to a moderate stirring rate. To one beaker add 25 ml of 0.10M $CuSO_4$, and to the other add 25 ml of 0.1M $Cu(NO_3)_2$. Turn the hot plates to "high" to heat the solutions rapidly.

While the solutions are heating, you can demonstrate that the blue color characteristic of Cu(II) solutions depends on the presence of the water. Place about 5 g of ground $CuSO_4 \cdot 5H_2O$ crystals in a casserole and heat over a Meker burner. Occasionally show observers the contents, so they can follow the color changes. In a few minutes, the solid becomes almost white. Let it cool slightly, then add the solid to a 100-ml beaker containing 75 ml of distilled water and a couple of drops of 1M sulfuric acid. The solid turns blue on contact with the solution and then dissolves to yield a blue solution. You can conclude that the two original solutions contain the species that is represented as $Cu^{2+}(aq)$, regardless of its coordination number. Commercial anhydrous copper sulfate can be used to demonstrate the generation of the blue color on contact with water.

When the two solutions are boiling moderately, add 5M ammonia by drops to each and note the precipitation of a light blue solid. Continue the addition until the first purplish color of the copper-ammonia complex appears in the copper sulfate beaker and until the precipitate in the copper nitrate beaker turns brownish black. This takes approximately 3 medicine droppers of 5M ammonia in each case, often slightly less for the copper sulfate solution and slightly more for the copper nitrate solution. Cool the solutions, or allow them to cool.

To the cooled suspensions, add 10 ml of 5M ammonia and stir. The bluish green precipitate in the copper sulfate beaker dissolves, but the brownish black precipitate in the copper nitrate beaker does not. Let the suspension with the dark precipitate stir while you proceed with the solution in the copper sulfate beaker. In all likelihood, the color in that beaker is so intense that it cannot be perceived well. Discard about one half to three quarters of the solution and dilute with distilled water until the purple of the $Cu(NH_3)_4^{2+}$ can be seen. You may wish to add a little more ammonia.

To this purple solution, add about 2 ml of 25% ethylenediamine solution. If the solution does not become redder, add more ethylenediamine in small increments until it does. Be careful not to add a large excess of ethylenediamine. By inference, the color change indicates the displacement of ammonia from the inner coordination sphere of Cu(II) by ethylenediamine.

When that color change is complete, begin adding 2–3 ml portions of 0.10M EDTA solution, allowing time for reactions to occur after each addition. The solution changes to a light blue, reminiscent of the $Cu^{2+}(aq)$ color but distinguishable from it.

Returning to the copper nitrate beaker, note that the brownish black precipitate has not dissolved completely in the 5M ammonia. Some color may develop in the solution, but the solid is highly resistant to attack.

Since the addition of ethylenediamine to the ammoniacal solution of Cu(II) in the $CuSO_4$ beaker resulted in a reaction, add about 2 ml of the 25% ethylenediamine to the stirred suspension in the $Cu(NO_3)_2$ beaker. Given a little time, it dissolves the precipitate. From then on, the two solutions behave the same, and the ethylenediamine complex can be converted to the ethylenediaminetetraacetato complex.

This system can be confusing, partly because the relative stabilities of the ethylenediamine and EDTA complexes are nearly the same, and partly because the EDTA solution reacts more slowly with the Cu(II) oxide than does ethylenediamine. Sometimes, when the EDTA solution is added to the suspension of Cu(II) oxide, no reaction is apparent for some time, but then, upon the addition of ethylenediamine, the precipitate dissolves, only to yield the light blue of the EDTA complex. These reactions depend on the balance of reagents, since one complex can be converted to the other simply by adding a sufficient excess of the appropriate ligand.

You now have two solutions with essentially the same appearance, that of the EDTA complex of Cu(II). Stir the solutions and add 3-4 ml of 1.0M KCN solution to one of them. Watch closely because the initial darkening of the solution color is often quite transitory. The solution then becomes essentially colorless. If you want a second look at this phenomenon, add KCN solution to the second beaker of EDTA complex.

Alternatively, you can add to the second solution 3-4 ml of 0.10M sodium sulfide solution to create a black precipitate of CuS. If you choose this route, follow the addition of the sulfide solution by adding 3-4 ml of 1.0M KCN, which dissolves black copper sulfide. You can then return to the beaker previously treated with KCN and add Na_2S solution. As the preceding observation implies, no sulfide precipitate forms.

HAZARDS

Cyanide salts, their solutions, and hydrogen cyanide gas produced by the reaction of cyanides with acids are all extremely poisonous. Hydrogen cyanide is among the most toxic and rapidly acting of all poisons. The solutions and the gas can be absorbed through the skin. Solutions are irritating to the skin, nose, and eyes. Cyanide compounds and acids must not be stored or transported together. An open bottle of potassium cyanide can generate HCN in moist air.

Early symptoms of cyanide poisoning are weakness, difficult breathing, headache, dizziness, nausea, and vomiting; these may be followed by unconsciousness, cessation of breathing, and death.

Anyone exposed to hydrogen cyanide should be removed from the contaminated atmosphere immediately. Amyl nitrite should be held under the person's nose for not more than 15 seconds per minute, and oxygen should be administered in the intervals. If the person is not breathing, artificial resuscitation by the Silvester method (not mouth to mouth) should be attempted immediately.

Sodium sulfide solutions must not be acidified since toxic hydrogen sulfide gas will be produced. The solid or solution can cause severe burns of the eyes and skin.

Copper compounds are harmful if taken internally. Dust from copper compounds can irritate mucous membranes.

Concentrated aqueous ammonia solution can cause burns and is irritating to the skin, eyes, and respiratory system. Like ammonia, ethylenediamine (1,2-diamino-ethane) is caustic and has similar toxic properties.

Because sulfuric acid is a strong acid and a powerful dehydrating agent, it can cause burns. Spills should be neutralized with an appropriate agent, such as sodium bicarbonate, and then rinsed clean.

DISPOSAL

Solutions containing cyanide ions should be mixed with an excess of sodium hydroxide solution plus sodium hypochlorite solution (household bleach) and allowed to stand for a few hours. Flush the drain with water to eliminate any residual acid and then flush the sodium hydroxide–bleach mixture down the drain with excess water.

After flushing the drain with water to remove residual acids, any remaining solutions should be flushed down the drain.

DISCUSSION

Aside from the dehydration and rehydration of copper sulfate, this demonstration consists of a series of displacement reactions in which the copper(II) is bound ever more tightly, either in an insoluble substance or in a soluble complex. In the following net ionic equations for the reactions involved, en = ethylenediamine, $H_2NCH_2CH_2NH_2$, and EDTA = H_4Y = ethylenediaminetetraacetic acid, $(HOOCCH_2)_2NCH_2CH_2N(CH_2COOH)_2$:

$$CuSO_4 \cdot 5H_2O(s) \xrightarrow{\Delta} CuSO_4(s) + 5\ H_2O(g) \tag{1}$$

(Some well-defined hydrates occur between these two limits, but they are of no significance if the heating is vigorous.)

$$CuSO_4(s) + H_2O(l) \longrightarrow CuSO_4 \cdot xH_2O(s) \longrightarrow$$
$$Cu(H_2O)_6^{2+}(aq) + SO_4^{2-}(aq) \tag{2}$$

$$4\ Cu^{2+}(aq) + SO_4^{2-}(aq) + 6\ H_2O(l) + 6\ NH_3(aq) \longrightarrow$$
$$CuSO_4 \cdot 3Cu(OH)_2(s) + 6\ NH_4^+(aq) \tag{3}$$

$$Cu^{2+}(aq) + 2\ NH_3(aq) + 2\ H_2O(l) \longrightarrow Cu(OH)_2(s) + 2\ NH_4^+(aq)$$
$$K_{eq} = 1.1 \times 10^9 \tag{4}$$

$$Cu(OH)_2(s) + 4\ NH_3(aq) \longrightarrow Cu(NH_3)_4^{2+}(aq) + 2\ OH^-(aq)$$
$$K_{eq} = 3 \times 10^{-6} \tag{5}$$

$$CuSO_4 \cdot 3Cu(OH)_2(s) + 16\ NH_3(aq) \longrightarrow$$
$$4\ Cu(NH_3)_4^{2+}(aq) + 6\ OH^-(aq) + SO_4^{2-}(aq) \tag{6}$$

$$Cu(NH_3)_4^{2+}(aq) + 2\ en(aq) \longrightarrow Cu(en)_2^{2+}(aq) + 4\ NH_3(aq)$$
$$K_{eq} = 4 \times 10^6 \tag{7}$$

$$CuO(s) + H_2O(l) + 2\ en(aq) \longrightarrow Cu(en)_2^{2+}(aq) + 2\ OH^-(aq)$$
$$K_{eq} = 1.2 \tag{8}$$

$$Cu(en)_2^{2+}(aq) + Y^{4-}(aq) \longrightarrow CuY^{2-}(aq) + 2\ en(aq) \quad K_{eq} = 0.12 \tag{9}$$

$$2\ CuY^{2-}(aq) + 8\ CN^-(aq) \longrightarrow 2\ Cu(CN)_3^{2-}(aq) + (CN)_2(aq) + 2\ Y^{4-}(aq) \quad (10)$$

$$CuY^{2-}(aq) + S^{2-}(aq) \longrightarrow CuS(s) + Y^{4-}(aq) \qquad K_{eq} = 2.5 \times 10^{17} \quad (11)$$

$$CuS(s) + 8\ CN^-(aq) \longrightarrow 2\ Cu(CN)_3^{2-}(aq) + (CN)_2(aq) + 2\ S^{2-}(aq) \quad (12)$$

The table contains the relevant stability constants and solubility product constants. The series of displacement reactions (equations 1 and 2) shows that the blue species characteristic of aqueous solutions is hydrated Cu(II), presumably $Cu(H_2O)_6^{2+}$. The structure of solid $CuSO_4 \cdot 5H_2O$ shows the Cu(II) in a distorted octahedral coordination with 6 oxygens [2]. The Cu(II) ions in concentrated solutions of $CuCl_2$ in aqueous HCl are octahedrally coordinated [3].

The systems obtained when aqueous NH_3 is added to boiling solutions of $CuSO_4$ and $Cu(NO_3)_2$ are complex, and all of the species involved probably cannot be identified. However, the addition of aqueous NH_3 to a boiling solution of $CuSO_4$ is the standard method of preparing the basic sulfate: $CuSO_4 \cdot 3Cu(OH)_2$. This basic sulfate is bluish green. If washed and dissolved in dilute HCl and then treated with $BaCl_2$ solution, it produces a precipitate of $BaSO_4$.

No basic salt is formed in the solution starting with $Cu(NO_3)_2$, so the initial precipitate is $Cu(OH)_2$, as in the $CuSO_4$ solution. When the suspension of $Cu(OH)_2$ is heated in the presence of some excess OH^-, the $Cu(OH)_2$ dehydrates, gradually turning into black CuO. This dehydration has been studied carefully [4], and apparently no solid phases are identifiable by x-ray diffraction other than $Cu(OH)_2 \cdot xH_2O$ and CuO. The greater stability of the CuO is demonstrated by the spontaneous conversion of the $Cu(OH)_2$ to CuO and by the difference in the solubility products of a factor of 10.

What controls the solution or nonsolution of the various precipitates is unclear. In the normal formation of $Cu(NH_3)_4^{2+}$ by the addition of fairly concentrated aqueous ammonia to a solution containing Cu(II)(aq), $Cu(OH)_2$ precipitates and then redissolves in spite of the low equilibrium constant for the formation of $Cu(NH_3)_4^{2+}$. However, CuO, which has a solubility product only 1 power of 10 smaller than $Cu(OH)_2$, fails to dissolve under the same conditions. The controlling factor appears to be kinetic. The basic sulfate, which forms in preference to $Cu(OH)_2$, also dissolves rapidly in the less concentrated ammonia used in this demonstration. In all likelihood, the failure of the CuO to dissolve in the EDTA solution, even though it dissolves fairly readily in the ethylenediamine solution, is a kinetic phenomenon as well, since the two soluble complexes have virtually the same stability constants. Whatever the explanation or the sequence of steps employed, both solutions can be brought to the same composition

Equilibrium Constants for Cu(II) Species [1]

K_{eq}	NH_3	en	EDTA	OH^-	CN^- Cu(I)	S^{2-}
β_1	1.75×10^4	3.5×10^{10}	5.0×10^{18}	2.0×10^6	—	—
β_2	6.8×10^7	4.0×10^{19}	—	6.3×10^{12}	1.6×10^{16}	—
β_3	6.3×10^{10}	—	—	3×10^{14}	4.0×10^{21}	—
β_4	1.0×10^{13}	—	—	4.0×10^{15}	1.3×10^{23}	—
β_5	2.7×10^{12}	—	—	—	—	—
β_{22}	—	—	—	1.9×10^{17}	—	—
K_{s0} (Cu(OH)$_2$)	—	—	—	3×10^{-19}	—	—
K_{s0} (CuO)	—	—	—	3×10^{-20}	—	—
K_{s0}	—	—	—	—	—	8×10^{-37}

containing predominantly the light blue EDTA complex. Ultimately, one can form the EDTA complex regardless of the path taken.

The insolubility of copper(II) sulfide is so great that it will precipitate from almost any complex. However, it will not precipitate from a cyanide solution of copper, and it will dissolve in a solution of potassium cyanide.

When excess KCN solution is added to the light blue solution containing Cu(II)-EDTA, the mixture initially darkens and then the color fades rapidly. The transient purple is due to the formation of tetracyanocuprate(II), which is unstable at room temperature, and decomposes to give cyanogen and cyanide complexes of Cu(I) [5]. At least three of these complexes are known, and more can be isolated from the solutions by the appropriate choices of cation and conditions of isolation [6]. The dominant species under the conditions of this demonstration is probably $Cu(CN)_3^{2-}$. The oxidation-reduction reactions between Cu(II) and CN^- are discussed in Demonstration 2.5.

REFERENCES

1. Smith, R. M.; Martell, A. E., Eds. "Critical Stability Constants," Vol. IV, Inorganic Complexes; Plenum Press: New York, 1976.
2. Wyckoff, R. W. G., Ed. "Crystal Structures," 2nd ed.; Interscience Publishers, John Wiley and Sons: New York, 1965; Vol. III, p 766.
3. Wertz, D. L.; Tyvoll, J. L. *J. Inorg. Nucl. Chem.* **1974**, *36*, 3713.
4. Trotman-Dickenson, A. F., Executive Ed. "Comprehensive Inorganic Chemistry"; Pergamon Press: Oxford, 1973; p 46.
5. Trotman-Dickenson, A. F., Executive Ed. "Comprehensive Inorganic Chemistry"; Pergamon Press: Oxford, 1973; p 45.
6. Trotman-Dickenson, A. F., Executive Ed. "Comprehensive Inorganic Chemistry"; Pergamon Press: Oxford, 1973; p 28.

17. Making Hydrogen Gas from an Acid and a Base

Two flasks are fitted with balloons. One flask contains hydrochloric acid and aluminum metal. The other flask contains sodium hydroxide and aluminum. As the reactions proceed, balloons on both flasks become inflated because of the production of hydrogen gas.

Procedure

1. Place 20–30 mL of 3 M hydrochloric acid in one 250-mL flask.
2. Place the same volume of 3 M sodium hydroxide in another 250-mL flask.
3. Have two large balloons and a student assistant ready.
4. Drop a loose wad of aluminum foil into each flask and stretch a balloon over each. Use the same size piece of foil in each flask.
5. Watch for evidence of a reaction, relative rates of reactions, differences in the products, and the presence of any excess reactant.
6. When the balloons are inflated, carefully remove them, tie the ends, and attach the balloons to the end of the demonstration table.
7. Using a meter stick with a candle taped to one end, carefully light each balloon. Compare the effects.

Reactions

1. Active metals, such as aluminum, react with acids to release hydrogen gas.

$$2Al(s) + 6H^+(aq) \longrightarrow 2Al^{3+}(aq) + 3H_2(g)$$

2. Aluminum also dissolves in a strong base to form hydrogen gas and tetrahydroxoaluminate(III) ion.

$$2Al(s) + 2OH^-(aq) + 6H_2O(\ell) \longrightarrow 3H_2(g) + 2[Al(OH)_4]^-(aq)$$

3. The strong acid and strong base dissolve the oxide film on aluminum; the aluminum thus comes into direct contact with the acid or base to produce hydrogen.

Solutions

1. The hydrochloric acid concentration is 3 M: *See* Appendix 2.
2. The sodium hydroxide concentration is 3 M: Dissolve 120 g of NaOH in 1 L of solution.

Teaching Tips

NOTES

1. Aluminum, zinc, and gallium are among the few metals that will release hydrogen from a base. The reaction with gallium is similar to that of aluminum; this reaction produces sodium gallate(III). Zinc produces sodium zincate. You might repeat the demonstration using zinc.
2. You cannot use nitric acid because it forms a coating of aluminum oxide on the metal.

3. Buy the larger party balloons. The balloons in regular packages are too small.
4. You might like to collect the hydrogen produced by these reactions in the conventional manner (by water displacement) and test for the properties of hydrogen gas.

QUESTIONS FOR STUDENTS

1. Show how hydrogen gas can be produced from both a base and an acid.
2. Compare the rates of reaction in the two flasks.
3. Why is aluminum left unreacted in the flask with hydrochloric acid? (Three times the amount of acid is required than that of sodium hydroxide per mole of aluminum.)
4. What difference do you notice in the flasks as the reactions proceed?
5. What evidence existed that hydrogen gas was produced?
6. What other metals will undergo these reactions?

6.19

Combining Volume of Oxygen with Sulfur

The pressure of the gas inside a flask containing burning sulfur is monitored. When the system reaches thermal equilibrium, there is no significant change in the pressure inside the flask, demonstrating that the product of the combustion is sulfur dioxide rather than sulfur trioxide [1].

MATERIALS

   cylinder of oxygen, with valve

   10 g flowers of sulfur

   50 mL tap water

   500-mL filter flask

   stand, with clamp to hold flask

   mercury-filled, U-tube manometer, open at both ends, having arms at least 80 cm long

   rubber tubing to connect filter flask to manometer

   2 solid rubber stoppers to fit flask

   combustion spoon

   rubber tubing to fit valve of oxygen cylinder

   Bunsen burner

   meter stick

PROCEDURE

Preparation

Connect the filter flask to the manometer with rubber tubing, as shown in the figure. In one of the rubber stoppers, bore a hole small enough to hold the handle of the combustion spoon. Insert the handle of the spoon through the stopper and adjust it so the spoon is at the center of the flask when the stopper is seated in the mouth of the flask.

Connect rubber tubing to the oxygen cylinder and insert the open end of the tubing into the flask so that it rests on the bottom of the flask. Open the valve of the oxygen cylinder slightly to produce a moderate flow of gas into the flask. Allow the gas to flow for a minute or two to fill the flask with oxygen by the upward displacement of air. Remove the tubing and seal the flask with the remaining stopper.

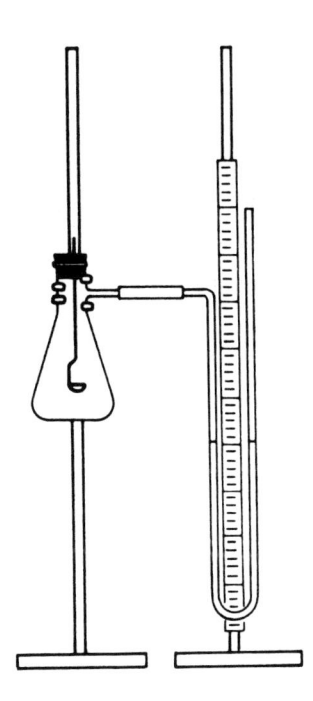

Presentation

Fill the combustion spoon with sulfur, and ignite the sulfur over a Bunsen burner. Remove the stopper from the flask and note that the levels of mercury are the same in the two arms of the manometer. Insert the spoonful of burning sulfur into the filter flask and seal the flask. The sulfur will burn for as long as several minutes, until the oxygen in the flask has been consumed.

As soon as the reaction stops, measure the difference in the levels of mercury in the arms of the manometer. Allow the flask and its contents to cool to room temperature. Again, measure the difference in the manometer's mercury levels. Open the flask, pour 50 mL of water into it, and reseal it with the solid stopper. Note what happens to the levels of the mercury in the manometer.

HAZARDS

This demonstration should be presented only in a location having adequate ventilation. The sulfur dioxide produced when the sulfur is ignited is an irritating and toxic gas. At concentrations of 3 ppm, the odor of sulfur dioxide is easily detectable. At concentrations over 8 ppm, SO_2 irritates the throat and induces coughing. Even brief exposure to concentrations over 400 ppm can be fatal.

DISCUSSION

A number of observations can be made during this demonstration including:
(a) the pressure in the flask increases during the course of the reaction;
(b) the pressure decreases as the flask cools to room temperature;
(c) when the system returns to room temperature, there is no significant difference in

the level of mercury in the two arms of the manometer, indicating that the pressure in the flask is the same after the reaction as before; and

(d) when water is added to the apparatus, the pressure inside the flask slowly decreases.

The increase in pressure during the course of the reaction may be attributed to the increase in the temperature of the gases in the flask as the exothermic combustion of sulfur takes place. Similarly, the decrease in pressure as the flask cools is due to the decrease in temperature.

The third observation may be interpreted in terms of the stoichiometry of the reaction between sulfur and oxygen. There are only two stable oxides of sulfur, SO_2 and SO_3 [2, 3]. Simple thermodynamic calculations suggest that sulfur should burn in oxygen to form sulfur trioxide rather than sulfur dioxide, because the former reaction is more highly exothermic [4].

$$2\,S(s) + 3\,O_2(g) \longrightarrow 2\,SO_3(g) \qquad \Delta H° = -395.72 \text{ kJ/mol S}$$

$$S(s) + O_2(g) \longrightarrow SO_2(g) \qquad \Delta H° = -296.83 \text{ kJ/mol S}$$

If sulfur trioxide is indeed produced in this reaction, 2 moles of SO_3 are produced for every 3 moles of O_2 consumed. This decrease in the number of moles of gas in the flask will be accompanied by a proportional decrease in the total pressure: from atmospheric pressure to two-thirds of that value, or a decrease of 250 mm Hg. If SO_2 is the product of the reaction, however, there will be no change in the number of moles of gas in the system, and, therefore, no change in the pressure. Because no change in the pressure is observed, the product of the reaction must be SO_2.

Students may misinterpret the fourth observation. They may conclude that SO_2 reacts with O_2 in the presence of water to form SO_3, or that SO_2 reacts with the water itself to form SO_3. However, when the mixture comes to equilibrium, the difference between the levels of mercury in the two arms of the manometer is greater than the 250 mm expected if SO_3 were produced. In fact, the reaction of SO_2 with O_2 is quite slow and requires high temperatures and a catalyst to occur. This reaction is carried out in the contact process for the production of sulfuric acid, in which the catalyst is finely divided platinum or vanadium(V) oxide, and the temperatures employed are about 400°C [5]. Sulfur trioxide reacts violently with water to form sulfuric acid. The fourth observation, then, must be explained through the dissolution of SO_2 in the water. The solubility of SO_2 in water is 79.8 liters of SO_2 per liter of water at 0°C and 1 atm partial pressure [6].

REFERENCES

1. H. N. Alyea and F. B. Dutton, Eds., *Tested Demonstrations in Chemistry*, 6th ed., Journal of Chemical Education: Easton, Pennsylvania (1965).
2. J. W. Mellor, *A Comprehensive Treatise on Inorganic and Theoretical Chemistry*, Vol. 10, Longmans, Green and Co.: London (1930).
3. M. Schmidt and W. Siebert, *Comprehensive Inorganic Chemistry*, Vol. 2, Pergamon Press: Oxford (1973).
4. "The NBS Tables of Chemical Thermodynamic Properties," *J. Phys. Chem. Ref. Data* 11, Supp. 2 (1982).
5. W. W. Duecker and J. R. West, Eds., *Manufacture of Sulfuric Acid*, Reinhold Book Corp.: New York (1959).
6. J. A. Dean, Ed., *Lange's Handbook of Chemistry*, 12th ed., McGraw-Hill: New York (1979).

overhead projector demonstrations

edited by
DORIS KOLB
Bradley University
Peoria, IL 61625

An Overhead Demonstration of Some Descriptive Chemistry of the Halogens and LeChatelier's Principle

Robert C. Hansen
University of Wisconsin/Platteville
Platteville, WI 53818

Chlorine, bromine, and iodine can exhibit odd number oxidation states from −1 to +7 as well as zero, but not all oxidation states have been characterized for all three of these elements. The reduction potentials of these positive oxidation states can be controlled through changes in the acidity–basicity of the solution. This control of the reduction potentials, hence of the oxidation states, particularly of iodine, and the iodine–triiodide ion equilibrium are used for a colorful demonstration to illustrate LeChatelier's principle and the solubility rule, "Like dissolves like."

Though the total chemistry involved in the demonstration, particularly that between iodide ion and bleach, is complicated[1-3] the chemistry germane to the demonstration can be summarized with only a few equations. All of the reactions/changes given involve equilibria to which LeChatelier's principle can be applied. The color changes and gradients observed show the extent of completion of the reactions as the system responds to the concentration gradients that accompany the addition and mixing of reagents. LeChatelier's principle can be used to predict the observed color changes which are then experimentally verified.

Equipment and Chemicals

An 8- × 8- × 1½-in. household Pyrex baking dish or other flat-bottom glass container of comparable surface area, 50-mL graduated cylinder, 100-mL beaker, two 400-mL beakers, stirring rod, *fresh* 5–6% commercial sodium hypochlorite solution (e.g. Hylex, Chlorox, Purex, etc.), two small polyethylene wash bottles (one filled with 6 M NaOH and the other with 6 M HCl), potassium iodide, a saturated hydrocarbon solvent of low volatility such as nonane, decane, or undecane. (**Caution:** These solvents are flammable. Exercise caution in working with them. Keep the quantity used to a minimum. It is recommended that the chlorocarbons such as carbon tetrachloride or chloroform customarily used for the color test for halogens *not* be used, even though they are nonflammable, because of the health problems associated with chlorocarbons.)

Procedure

Step 1. Fill a 400-mL beaker on the stage of the overhead projector two-thirds full with fresh bleach. For comparison, place next to it a 400-mL beaker filled to the same depth with distilled water. The light yellow-green color of the dissolved chlorine gas is readily seen on both the screen and in the beaker on the overhead.

Explanation. Sodium hypochlorite bleach is prepared by the electrolysis of cold sodium chloride solution,

$$2Cl^- + 2H_2O \xrightarrow{electrolysis} Cl_2(solution) + H_2 + 2OH^- \quad (1)$$

in a cell in which the electrolysis products, chlorine and hydroxide ion can mix.[3] Under these conditions the chlorine disproportionates to give the hypochlorite ion,

$$2OH^- + Cl_2(solution) \rightleftharpoons Cl^- + ClO^- + H_2O \quad (2)$$

Step 2. Place 5 mL of bleach solution in a 100-mL beaker on the overhead. Add two drops of 6 M HCl. (**Caution:** Do only with adequate ventilation because chlorine gas is very toxic and some people with respiratory problems are highly susceptible to it.) At the point of addition of the hydrochloric acid there is effervescence of chlorine gas. Immediately hold a moist piece of potassium iodide-starch test paper in the beaker above the effervescence. The test paper will turn dark blue–black from the formation of the starch-iodine complex.

Explanation. The extent of disproportionation of chlorine by reaction 2 is determined by the hydroxide ion concentration. At the point of addition of the hydrochloric acid, there is a high localized concentration of hydrogen ions. The hydrogen ions remove the hydroxide ions by neutralization. This causes equilibrium reaction 2 to shift to the left in accord with LeChatelier's principle to replace the hydroxide ion with a simultaneous increase in the concentration of chlorine. The effervescence results from the concentration of chlorine in solution being increased to above its equilibrium solubility. This can be denoted by a shift of reaction 3, the equation for the dissolution of chlorine, to the left.

$$Cl_2(gas) \rightleftharpoons Cl_2(solution) \quad (3)$$

The blue–black color indicates the presence of a gas (chlorine) that can dissolve in the moisture on the paper and oxidize the iodide ion present to iodine, reaction 4, which forms an intense blue complex with starch, reaction 5.

$$2I^-(aqueous) + Cl_2(gas) \rightleftharpoons 2Cl^- + I_2(aqueous) \quad (4)$$

$$I_2(aqueous) + starch \xrightarrow{iodide\ ion} I_2 \cdot starch \ (blue) \quad (5)$$

This demonstration shows why sodium hypochlorite bleach must not be mixed with an acidic cleaning agent or vinegar.

Step 3. The odor of chlorine may be observed by passing a 10-cm test tube half full of bleach around the class for the students to smell. Review the proper way to test for odors by wafting the vapors to the nose. **Caution:** Concentrated bleach solution, if spilled may quickly bleach the dyes in clothing. The slippery feel when concentrated bleach solution is rubbed between the fingers shows that it contains enough unreacted hydroxide ion to be corrosive to the skin. Have a dilute solution of sodium thiosulfate available in case of a spill.

Show a transparency copy of the caution notice on the bleach container.

Step 4. In the Pyrex baking dish on the overhead projector, dilute 5 mL of household bleach solution with 50 mL of water. Place 5 mL of a saturated alkane, such as decane, on top of the solution. Show that the two liquid phases do not mix by stirring the system. (Depending upon how level the overhead projector is, it may be necessary to place a piece of cardboard along one edge of the dish so that both phases are simultaneously visible.) Slowly add and dissolve by

---

Presented at Chemical Demonstration Conference, April 24–25, 1987, Western Illinois University, Macomb, IL.

[1] Cotton, F. A.; Wilkinson, G. *Advanced Inorganic Chemistry: A Comprehensive Test*, 4th ed.; Wiley: New York, 1980.

[2] Greenwood, N. N.; Earnshaw, A. *Chemistry of The Elements*, Pergamon: Elmsford, NY, 1984; p 999.

[3] Moeller, T. *Inorganic Chemistry: An Advanced Textbook*; Wiley: New York, 1952; p 438.

stirring 1 g of potassium iodide. The yellow-brown color of an aqueous iodine-triiodide solution forms immediately at the point of addition. If the solution does not stay a light yellow color, slowly add and dissolve by stirring more crystals of potassium iodide until it does. Stir the system until the hydrocarbon layer acquires a faint light pink tinge.

Explanation. A study of the reduction potentials[3-5] for the halogens shows that chlorine not only can oxidize iodide ion to iodine but also in a basic solution can oxidize it to hypoiodite ion and to iodate ion. In a basic solution, hypochlorite ion can also oxidize iodide ion to iodine, hypoiodite ion, and iodate ion. The hypoiodite ion, unlike the hypochlorite ion, is not stable at room temperature, but rapidly disproportionates to iodide and iodate ions.[2] In a basic solution, iodine also disproportionates to iodide and iodate ions. The yellow color is from dissolved iodine and/or triiodide ion, I_3^-. It shows that the oxidation and disproportionation reactions do not go to completion. Since the quantity of iodide ion added is in excess of what can be oxidized to iodate ion by the hypochlorite ion and dissolved chlorine in the volume of bleach solution used, the predominant ions in solution are the iodide and iodate ions. The principal equilibrium in the system is reaction 6, which is also the reaction for the disproportionation of iodine.

$$3I_2(\text{aqueous}) + 6OH^- \rightleftharpoons 5I^- + IO_3^- + 3H_2O \quad (6)$$
(yellow)

The formation of the triiodide ion is given by reaction 7.

$$I^-(\text{aqueous}) + I_2(\text{aqueous}) \rightleftharpoons I_3^-(\text{aqueous}) \quad (7)$$
(colorless)   (yellow)   (brown)

The color changes and the slow increase in the yellow color of the system as the potassium iodide is dissolved show the reversibility of reactions 6 and 7 and the validity of LeChatelier's principle.

The pink color of the hydrocarbon layer is iodine that dissolves in that phase through partition equilibrium,

$$I_2(\text{aq}) \rightleftharpoons I_2(\text{hydrocarbon}) \quad (8)$$
(yellow)   (pink)

Step 5. Add a drop or two of 6 M hydrochloric acid from the wash bottle. At the point of addition, what appears to be a black cloud forms. Mix the solution by gently tipping the dish back and forth. As the black cloud dissolves, an area of brown forms around it and increases in size. With further mixing the whole solution becomes yellower and the hydrocarbon layer becomes deeper pink. Continue the addition of hydrochloric acid while mixing the system by gently tipping the dish back and forth. The black cloud increases in size and takes longer to dissolve, and the colors of both phases of the system deepen with each addition of acid. Eventually the black cloud no longer dissolves, but coagulates into chunks. Discrete black crystals are also seen. The aqueous phase becomes lighter in color and the hydrocarbon layer deep purple. Continue the addition of hydrochloric acid until no change other than a decrease in the color of the aqueous phase due to dilution is observed.

Explanation. At the point of addition of the hydrochloric acid, the high localized concentration of hydrogen ion has neutralized the hydroxide ion and shifted reaction 6 far to the left. The black cloud, microscopic crystals of iodine, demarcates the region of shift. The iodine crystals form where the equilibrium solubility of iodine has been exceeded. This can be denoted by a shift of reaction 9, the equation for the solubility equilibrium, to the left.

$$I_2(\text{solid}) \rightleftharpoons I_2(\text{aq}) \quad (9)$$
(black)   (yellow)

At the boundary of the black cloud, as the solution is mixed, the iodine crystals contact unoxidized iodide ions with which they react to form the brown triiodide ion. The triiodide ion, being ionic, is more soluble in the polar solvent water than is the covalent nonpolar iodine molecule. The change can be explained in terms of stresses being applied to reactions 7 and 9 and their shifts to the right. The color changes and gradients correspond to the magnitude of the stresses and shifts. The formation of the chunks, coagulated crystals of iodine, occurs when the concentration of iodide ion has been lowered, by the shift of reaction 6 to the left, to the point that the tendency of reactions 7 and 9 to shift to the left predominates.

The decrease in color of the aqueous phase when the iodine precipitates out of solution shows the low solubility of the nonpolar iodine molecule in a polar solvent. The change to deep purple of the hydrocarbon phase shows that this phase, which is nonpolar, is the better solvent for the nonpolar iodine molecule. These two changes illustrate the solubility rule, "like dissolves like."

Step 6. Place 1-2 g of potassium iodide beside a chunk of coagulated iodine crystals. The chunk will dissolve, and the solution turns deep brown as dissolved iodide ions contact the surface of the solid iodine. Tip the dish gently to dissolve the potassium iodide and mix the solution. Vivid color gradients of yellow to brown form across the dish during the mixing process. Continue the addition of 1-2-g increments of potassium iodide, dissolving and mixing until all of the crystals of iodine dissolve. With each increment of potassium iodide, the aqueous phase becomes deeper brown in color and the hydrocarbon phase becomes lighter purple in color.

Explanation. The added iodide ions place stress on the left side of reaction 7, which must shift to the right to relieve the stress. This shift in turn places stresses on reactions 8 and 9. Reaction 8 must shift to the left, and reaction 9 must shift to the right in accord with LeChatelier's principle. The conversion of the iodine to the triiodide ion makes the iodine more soluble in the polar aqueous phase and less soluble in the nonpolar hydrocarbon phase in accord with the solubility rule, "like dissolves like." The color gradients again correspond to the concentration gradients that accompany the dissolving of the potassium iodide and the formation of the triiodide ion. The intensities of the colors are again indicative of the magnitude of the stresses applied to the equilibria and the extent to which they shift.

Step 7. Add a few drops of 6 M sodium hydroxide to the dish. At the point of addition, the solution in the dish turns colorless to light yellow. Tip the dish to mix the solution. Gradients of decreased color, light yellows to browns, the opposite of what was observed in step 7, form across the dish during the mixing. The color of the solution decreases. Slowly, with mixing, add sodium hydroxide solution until the solution is colorless. The area of the color gradients increases with each increment of base. The hydrocarbon phase also turns colorless.

Explanation. The added hydroxide ion places a stress on the left side of reaction 6, which must shift to the right to relieve the stress. This in turn places a stress on the right sides of reactions 7 and 8, which must shift to the left. Reaction 6 can be shifted so far to the right that there is not enough free iodine in the system to give even the faintest tinge of pink to the hydrocarbon layer because of the shift of reaction 8 to the left.

Step 8. Add a couple of drops of 6 M hydrochloric acid to the solution. The point of addition is marked by a dark brown that tapers to yellow then to colorless. Mix the solution by gently tipping the dish. The color changes observed are the reverse of those observed in step 8. Continue the addition of hydrochloric acid until no further color changes are observed. The hydrocarbon phase will take on only a faint pink tinge when the aqueous phase is deep brown.

Explanation. The hydrochloric acid neutralizes the hydroxide ion, reversing the order of stresses applied in step 8 and the shifts of the reactions. No crystals of iodine form, and the hydrocarbon phase takes on only a faint pink color because the iodine in the system is nearly quantitatively bound up as the triiodide ion due to the quantity of potassium iodide that had to be added in step 6.

Step 9. Repeat steps 8 and 9 to show the reversibility of the system.

Bromine water may be substituted for bleach solution in step 5. This change should be made with caution because of the appreciable vapor pressure of bromine over bromine water and the very corrosive properties of bromine. The substitution of potassium bromide for potassium iodide is not recommended because of the high vapor pressure of bromine encountered when the solution is acidified as in step 6. The color changes observed for the bromine system are not nearly as vivid or interesting as those for the iodine system.

---

[4] Footnote 1, p 557.
[5] Latimer, W. M.; Hildebrand, J. H. *Reference Book of Inorganic Chemistry*, 3rd ed.; Macmillan: New York, 1951.

7. Preparation of Chlorine Gas from Laundry Bleach

Chlorine gas and chlorine water are prepared by reacting laundry bleach with hydrochloric acid. The bleaching property of chlorine is shown.

Procedure

1. Prepare a gas-generating system by connecting a rubber tube to the side arm of a large filtering flask.
2. Place 30 mL of laundry bleach in the flask.
3. Add 5 mL of HCl. Stopper the flask and swirl it quickly.
4. Collect chlorine gas by upward displacement of air in several test tubes.
5. Test for the presence of chlorine by the usual methods (e.g., ability to bleach non-colorfast fabric or dyes).

Reaction

$$ClO^-(aq) + Cl^-(aq) + 2H^+(aq) \rightarrow Cl_2(g) + H_2O(l)$$

(hypochlorite)

Solution

The HCl concentration is 1.0 M (see Appendix 2). To generate chlorine faster, use more concentrated acid.

Teaching Tips

NOTES

1. Use a HOOD or well-ventilated area for this demonstration. AVOID DIRECTLY BREATHING THE CHLORINE.
2. This method is an easy way to prepare chlorine water.
3. Because bleach is prepared commercially by bubbling chlorine gas through sodium hydroxide, this demonstration is essentially the reverse of this reaction:

$$Cl_2(g) + 2OH^-(aq) \rightarrow ClO^-(aq) + Cl^-(aq) + H_2O(l)$$

4. Place a small piece of cloth soaked in turpentine in one of the tubes. CAUTION!

QUESTIONS FOR STUDENTS

1. Why was this gas not collected by the displacement of water?
2. What property of the gas allows us to collect it by the upward displacement of air?
3. Write the reaction for the preparation of chlorine gas.
4. How does bleach do what it does? (Nascent oxygen!)
5. What happened with the turpentine? Was this a *spontaneous* reaction?

28. Plastic Sulfur

Yellow sulfur is heated and forms an orange-red liquid. The liquid is poured into a beaker of water, and a flexible brown polymer of "plastic sulfur" is formed.

Procedure

1. Place a small amount of sulfur (to a height of about 1 in.) in a large test tube. Notice the properties of the sulfur.
2. Gently heat the test tube until the sulfur melts. Observe the formation of an orange liquid.
3. Continue to strongly heat the test tube. Notice the formation of a thick, dark red liquid.
4. Quickly pour the molten sulfur into a beaker of cold water. Observe the formation of a brown material, plastic sulfur.
5. Remove the rubbery material and note its properties including elasticity.

Reactions

1. Ordinary yellow sulfur, called *orthorhombic* sulfur, consists of S_8 molecules in a ring structure.
2. As the S_8 molecules are heated, the ring structure breaks and forms small 16-atom chains. This form of molten sulfur has an orange color.
3. Upon further heating (about 160 °C), the 16-atom chains break and form long chains of sulfur atoms. This structure gives the molten sulfur a thick (viscous) appearance.
4. When the long chains of sulfur are suddenly cooled, the material keeps its long-chain structure, without the ordered arrangement of sulfur atoms found in the crystalline solid. This structure gives the plastic sulfur a "rubbery" character.

Teaching Tips

NOTES

1. Another form of sulfur, monoclinic, forms when liquid sulfur gradually cools. This form of sulfur has long, needlelike crystals.
2. The different types of sulfur are *allotropes*.
3. Try stretching the plastic sulfur. Test it for other rubbery characteristics.
4. Upon standing, the plastic sulfur will revert to the original orthorhombic form.
5. The melting point of yellow sulfur is 113 °C.
6. Heating sulfur may produce some sulfur dioxide gas, which may be irritating. Do this demonstration in a hood or in a well-ventilated area.

QUESTIONS FOR STUDENTS

1. What happens, on a molecular level, as ordinary sulfur is heated?
2. Why does the plastic sulfur have properties of a rubber?
3. Is plastic sulfur a *polymer*?
4. Why does the sulfur become thick (viscous) when it is heated?

3.1

Nylon 6-10

A film of nylon is formed at the interface between two immiscible liquids. When the film is lifted from the container, it is continually replaced forming a hollow thread of polymer. The continuous thread or "rope" of nylon can be wound on a windlass until one or the other of the two reactants is exhausted.

MATERIALS

- 50 ml 0.5M hexamethylenediamine (1,6-diaminohexane), $H_2N(CH_2)_6NH_2$, in 0.5M sodium hydroxide, NaOH (To prepare, dissolve 3.0 g of $H_2N(CH_2)_6NH_2$ plus 1.0 g NaOH in 50 ml distilled water. Hexamethylenediamine can be dispensed by placing the reagent bottle in hot water until sufficient solid has melted and can be decanted. The melting point is 39–40°C.)
- 50 ml 0.2M sebacoyl chloride, $ClCO(CH_2)_8COCl$, in hexane (To prepare, dissolve 1.5 ml to 2.0 ml sebacoyl chloride in 50 ml hexane.)
- gloves, plastic or rubber
- 250-ml beaker or crystallizing dish
- forceps
- 2 stirring rods or a small windlass
- food-coloring dye (optional)
- phenolphthalein (optional)

PROCEDURE

Wearing gloves, place the hexamethylenediamine solution in a 250-ml beaker or crystallizing dish. *Slowly* pour the sebacoyl chloride solution as a second layer on top of the diamine solution, taking care to minimize agitation at the interface. With forceps, grasp the polymer film that forms at the interface of the two solutions and pull it carefully from the center of the beaker. Wind the polymer thread on a stirring rod or a small windlass. Wash the polymer thoroughly with water or ethanol before handling.

Food coloring dyes or phenolphthalein can be added to the lower (aqueous) phase to enhance the visibility of the liquid interface. The upper phase can also be colored with dyes such as azobenzene [1], but observation of the polymer film at the interface is somewhat obscured. Some of the dye will be taken up with the polymer but can be removed by washing.

HAZARDS

Hexamethylenediamine (1,6-diaminohexane) is irritating to the skin, eyes, and respiratory system. Sodium hydroxide is extremely caustic and can cause severe burns. Contact with the skin and eyes must be prevented.

Sebacoyl chloride is corrosive and irritating to the skin, eyes, and respiratory system. Hexane is extremely flammable. Hexane vapor can irritate the respiratory tract and, in high concentrations, can be narcotic.

DISPOSAL

Any remaining reactants should be mixed thoroughly to produce nylon. The solid nylon should be washed before being discarded in a solid waste container.

Any remaining liquid should be discarded in a solvent waste container or should be neutralized with either sodium bisulfate (if basic) or sodium carbonate (if acidic) and flushed down the drain with water.

DISCUSSION

The word "nylon" is used to represent synthetic polyamides. The various nylons are described by a numbering system that indicates the number of carbon atoms in the monomer chains. Nylons from diamines and dibasic acids are designated by two numbers, the first representing the diamine and the second the dibasic acid [2]. Thus, 6-10 nylon is formed by the reaction of hexamethylenediamine and sebacic acid. In this demonstration the acid chloride, sebacoyl chloride, is used instead of sebacic acid. The equation is

$$H_2N(CH_2)_6NH_2 + ClC(O)(CH_2)_8C(O)Cl \longrightarrow \left[N(H)(CH_2)_6N(H) - C(O)(CH_2)_8C(O) \right] + 2\,HCl$$

The method of reaction used in this demonstration has been termed interfacial polycondensation. This method is useful because it is a low temperature process, it is rapid even at room temperature, and it does not depend on exact stoichiometry of reactants [3].

Many diamines and diacids or diacid chlorides can be reacted to make other condensation products that are described by the generic name "nylon." One such product is an important commercial polyamide, nylon 6-6, which can be prepared by substituting adipoyl chloride [4] for sebacoyl chloride in the procedure described here. The equation is

$$H_2N(CH_2)_6NH_2 + ClC(O)(CH_2)_4C(O)Cl \longrightarrow \left[N(H)(CH_2)_6N(H) - C(O)(CH_2)_4C(O) \right] + 2\,HCl$$

REFERENCES

1. Alyea, H. N.; Dutton, F. B., Eds. "Tested Demonstrations in Chemistry," 6th ed.; Journal of Chemical Education: Easton, Pennsylvania, 1965; p 136.
2. Ravve, A. "Organic Chemistry of Macromolecules"; Marcel Dekker: New York, 1967; Ch. 15.
3. Sorenson, W. R.; Campbell, T. W. "Preparative Methods of Polymer Chemistry"; Interscience Publishers, John Wiley and Sons: New York, 1968; pp 90–93.
4. Alyea, H. N.; and Dutton, F. B., Eds. "Tested Demonstrations in Chemistry," 6th ed.; Journal of Chemical Education: Easton, Pennsylvania, 1965; p 164.

1.32

Dehydration of Sugar by Sulfuric Acid

Concentrated sulfuric acid added to granulated sugar in a beaker produces a solid-liquid mixture that changes from white to yellow to brown to black. The mixture then expands out of the beaker accompanied by vapor and the smell of burned sugar.

MATERIALS

70 g granulated sugar, $C_{12}H_{22}O_{11}$

70 ml concentrated (18M) sulfuric acid, H_2SO_4

300-ml tall-form beaker

40-cm stirring rod

paper towels

gloves, plastic or rubber

100-ml graduated cylinder

PROCEDURE

Add the granulated sugar to the tall-form beaker and insert the stirring rod into its center. Place the beaker on a mat of paper towels. Add 70 ml of concentrated sulfuric acid to the sugar and stir briefly. As the column of black solid begins to grow, support it with the stirring rod. The column grows to twice the height of the beaker.

HAZARDS

Since concentrated sulfuric acid is a strong acid and a powerful dehydrating agent, it must be handled with great care. Spills should be neutralized with an appropriate agent, such as $NaHCO_3$, and then wiped up. The solid residue may contain unreacted acid. Prolonged contact with the steam produced in the reaction can cause burns.

DISPOSAL

Carbonaceous residue should be rinsed *thoroughly* with water and then discarded in a waste container.

DISCUSSION

Among the many important industrial properties of sulfuric acid is the ability of the concentrated acid to dehydrate materials. This property is related thermodynamically to the large energy change that occurs as the sulfuric acid becomes hydrated. The dehydration of sucrose, $C_{12}H_{22}O_{11}$, is an exothermic reaction (reaction 1). The heat of dehydration for sucrose is calculated from the standard heat of combustion of sucrose (reaction 2) and the heat of formation of carbon dioxide (reaction 3):

$$C_{12}H_{22}O_{11}(s) \longrightarrow 12\ C(graphite) + 11\ H_2O(l) \tag{1}$$

$$C_{12}H_{22}O_{11}(s) + 12\ O_2(g) \longrightarrow 12\ CO_2(g) + 11\ H_2O(l) \tag{2}$$

$$C(graphite) + O_2(g) \longrightarrow CO_2(g) \tag{3}$$

For reaction 2, the standard heat of combustion, ΔH°_{comb}, is -5640.9 kJ/mole; and for reaction 3, the standard heat of formation, ΔH°_f, of carbon dioxide is -393.5 kJ/mole [1]. To calculate ΔH°_{rxn} for reaction 1, multiply reaction 3 by 12, and subtract the product from reaction 2. Thus, ΔH°_{rxn} for reaction 1 is -918.9 kJ/mole.

In this demonstration, 70 g or 0.20 moles of sucrose are used. The heat evolved for this quantity of sugar is 180 kJ.

The water formed by the dehydration of the sugar dilutes the sulfuric acid and liberates heat:

$$H_2SO_4 \cdot nH_2O + m\ H_2O \longrightarrow H_2SO_4 \cdot n_1H_2O$$

where $n + m = n_1$.

The value of n, calculated from the density of 98% H_2SO_4, is 0.11 [2]. The reaction forms 11 times 0.20, or 2.2, moles of water; therefore $n_1 = 0.11 + 2.2 = 2.3$ moles of water.

Interpolation from a table of values of ΔH°_f for $H_2SO_4 \cdot nH_2O$ yields [3]

$$\Delta H^\circ_f(H_2SO_4 \cdot nH_2O) = -814.78\ \text{kJ/mole} \quad (n = 0.11)$$

$$\Delta H^\circ_f(H_2SO_4 \cdot nH_2O) = -855.36\ \text{kJ/mole} \quad (n = 2.3)$$

The heat of dilution for this reaction is

$$(-855.36\ \text{kJ/mole}) - (-814.78\ \text{kJ/mole}) = -40.58\ \text{kJ/mole}$$

Since 1.28 moles of $H_2SO_4 \cdot nH_2O$ are used in this demonstration, the dilution of the acid contributes $(-40.58\ \text{kJ/mole})(1.28\ \text{mole}) = -51.9$ kJ. The total heat evolved is $180\ \text{kJ} + 51.9\ \text{kJ} = 232\ \text{kJ}$.

REFERENCES

1. Weast, R. C., Ed. "CRC Handbook of Chemistry and Physics," 59th ed.; CRC Press: Cleveland, Ohio, 1978–79; pp D-69, D-325.
2. Weast; p F-7.
3. "Selected Values of Chemical Thermodynamic Properties." *Natl. Bur. Stand. (U.S.)* **1952**; Circ. 500; p 41.